Brain Networks in Neuroscience: Personalization Unveiled Via Artificial Intelligence

RIVER PUBLISHERS SERIES IN BIOTECHNOLOGY AND MEDICAL RESEARCH

Series Editors:

Aiming primarily at providing detailed snapshots of critical issues in biotechnology and medicine that are reaching a tipping point in financial investment or industrial deployment, the scope of the series encompasses various specialty areas including pharmaceutical sciences and healthcare, industrial biotechnology, and biomaterials. Areas of primary interest comprise immunology, virology, microbiology, molecular biology, stem cells, hematopoiesis, oncology, regenerative medicine, biologics, polymer science, formulation and drug delivery, renewable chemicals, manufacturing, and biorefineries.

Each volume presents comprehensive review and opinion articles covering all fundamental aspect of the focus topic. The editors/authors of each volume are experts in their respective fields and publications are peer-reviewed.

For a list of other books in this series, visit www.riverpublishers.com

Brain Networks in Neuroscience: Personalization Unveiled Via Artificial Intelligence

Editors

Md. Mehedi Hassan
Khulna University, Bangladesh

Farhana Yasmin
Nanjing University of Information Science and Technology, China

Sheikh Mohammed Shariful Islam
Deakin University, Australia

Anupam Kumar Bairagi
Khulna University, Bangladesh

Si Thu Aung
State University of New York at Buffalo, USA

River Publishers

Routledge
Taylor & Francis Group
NEW YORK AND LONDON

Published 2025 by River Publishers
River Publishers
Broagervej 10, 9260 Gistrup, Denmark
www.riverpublishers.com

Distributed exclusively by Routledge
605 Third Avenue, New York, NY 10017, USA
4 Park Square, Milton Park, Abingdon, Oxon OX14 4RN

Brain Networks in Neuroscience: Personalization Unveiled Via Artificial Intelligence / by Md. Mehedi Hassan, Farhana Yasmin, Sheikh Mohammed Shariful Islam, Anupam Kumar Bairagi, Si Thu Aung.

Routledge is an imprint of the Taylor & Francis Group, an informa business

ISBN 978-87-7004-736-4 (hardback)
ISBN 978-87-4380-077-4 (paperback)
ISBN 978-87-7004-738-8 (online)
ISBN 978-87-7004-737-1 (master ebook)

Contents

Preface

The human brain, a marvel of complexity and adaptability, serves as the bedrock of cognition, emotion, and behavior. It orchestrates our thoughts, decisions, and actions, while maintaining a delicate balance between structure and function. The study of brain networks, which underpins these intricate processes, has grown into a pivotal domain in neuroscience, bridging the gap between biology, computational sciences, and artificial intelligence (AI). In this book, *Brain Networks in Neuroscience: Personalization Unveiled Via Artificial Intelligence*, we embark on a journey to explore the fascinating world of brain connectomics and its transformation through AI.

The past decade has witnessed remarkable progress in mapping and understanding the human brain's structural and functional networks. However, one critical challenge remains—how do we personalize these insights to cater to the uniqueness of each individual? Personalization holds immense promise for advancing diagnostics, treatment strategies, and rehabilitation techniques in neuroscience. AI, with its unparalleled ability to analyze and interpret vast and complex datasets, has emerged as a transformative tool in addressing this challenge. By leveraging machine learning, deep learning, and graph-based algorithms, we are now beginning to unravel the nuances of individual variability within the brain's connectome.

This book is a product of my passion for computational neuroscience and a vision to integrate AI into the personalization of brain networks. It aims to provide a comprehensive overview of the theoretical and practical advancements in the field, highlighting how AI-driven approaches are redefining our understanding of the human brain. From foundational concepts in brain connectomics to cutting-edge applications in neurodegenerative diseases and mental health, this book delves into diverse topics with a focus on their relevance to real-world challenges.

The interdisciplinary nature of this work reflects the collaborative spirit that drives modern science. Researchers from neuroscience, computer science, engineering, and clinical fields will find this book a valuable resource. It is designed not only to inform but also to inspire—encouraging readers to think beyond conventional boundaries and explore new horizons in brain network research.

I am deeply grateful to the mentors, colleagues, and students who have influenced my journey. Their insights and encouragement have been instrumental in shaping the ideas presented in this book. I also extend my heartfelt thanks to the River Press and IEEE teams for their support and dedication throughout the publication process.

As you delve into the chapters that follow, I hope you share my excitement for the potential of AI to personalize and revolutionize neuroscience. May this book serve as a guide and a source of inspiration for your own explorations in this dynamic and ever-evolving field.

Md. Mehedi Hassan
December 2024
Bangladesh

List of Figures

List of Tables

List of Contributors

Kumar, Bairagi Anupam, *Computer Science and Engineering Discipline, Khulna University, Bangladesh*

Sharma, Chetna, *Chitkara University Institute of Engineering and Technology, Chitkara University, India*

Ayontika, Das, *Department of Computer Science and Engineering, Adamas University, India*

Khin, Eaindray Htun, *School of Information Technology, King's Mongkut University of Technology Thonburi, Thailand*

Rezuana, Haque, *Computer Science and Engineering, BGC Trust University, Bangladesh*

Tamanna, Haque Ritu, *Pharmacy Discipline, Khulna University, Bangladesh*

Mahade, Hasan, *School of Software, Nanjing University of Information Science and Technology, China*

Htun, Htun, *Department of Solution, Partner Associates Int'l Co. Ltd., Myanmar*

Asif, Karim, *Faculty of Science and Technology, Charles Darwin University, Australia*

Oo, Kyaw Kyaw, *Department of Data & Insights, Capgemini Singapore Pte. Ltd., Singapore*

Kari, Lippert, *Department of Systems Engineering, University of South Alabama, USA*

Avula, Mahathi, *Department of Computer Science and Engineering, Stanley College of Engineering and Technology for Women, India*

Reddy D., Manoj Kumar, *Department of Electrical & Electronics Engineering, Vardhaman College of Engineering, India*

Sayeda, Mayesha Yousuf, *Department of Computer Science and Engineering, Green University of Bangladesh*

Md., Mahedi Hassan, *Department of Computer Science and Engineering, Bangladesh University of Business and Technology, Bangladesh*

Md., Mehedi Hassan, *Computer Science and Engineering Discipline, Khulna University, Khulna, 9208, Bangladesh*

Hirak., Mondal, *Department of Computer Science and Engineering, North Western University, Bangladesh*

Niamat., Munsi, *Bachelor of Medicine and Bachelor of Surgery (MBBS) Sir Salimullah medical College, Bangladesh*

Rahman Md., Mushfiqur, *Department of Statistics, University of Dhaka, Bangladesh*

Anindya, Nag, *Computer Science and Engineering Discipline, Khulna University, Bangladesh*

Chander, Prabha, *Chitkara University Institute of Engineering and Technology, Chitkara University, India*

Amrina, Rahman, *Department of Management Information Systems (MIS), University of Dhaka*

Arafat, Rahman, *Computer Science and Engineering Discipline, Khulna University, Bangladesh*

Kumar, Reddy C. Kishor, *Department of Computer Science and Engineering, Stanley College of Engineering and Technology for Women, India*

Riya, Sil, *Computer Science and Engineering, Presidency University, India*

Thandiwe, Sithole Nastassia, *Department of Chemical Engineering Technology, University of Johannesburg, South Africa*

Doss, Srinath, *Faculty of Engineering and Technology, Botho University, Botswana*

Islam Md, Tanvir, *Research Fellow, Department of Software, SungkyunKwan University, South Korea*

Akshitha, Thota, *Department of Computer Science & Engineering, Stanley College of Engineering and Technology for Women, India*

Aung Si, Thu, *Department of Mathematics, The State University of New York, USA*

May, Thu Kyaw, *Department of Electronic Engineering, West Yangon Technological University, Myanmar*

Farhana, Yasmin, *School of Computer and Software, Nanjing University of Information Science and Technology, China*

Xue, Yu, *School of Software, Nanjing University of Information Science and Technology, China*

Tamanna, Zubairi Sana, *Department of Computer Science and Engineering, East Delta University, Bangladesh*

List of Abbreviations

3D-CF	3D convolutional features
3DMTM	3D motion trail model
AC	Attenuation correction
AD	Alzheimer's disease
ADHD	Attention deficit disorder
AHN	Artificial hormone network
AI	Artificial intelligence
ANN	Artificial neural network
ASD	Autism spectrum disorders
ASL	Arterial spin labelling
BC	Brain connectivity
BCI	Brain–computer interface
BCIs	Brain–computer interfaces
BD	Brain disorders
BNN	Biological neuron network
BOLD	Blood oxygen level dependent
C2	Diatomic carbon
Ca2+	Calcium ions
CD	Conduct disorder
CEN	Central executive network
Cl	Chloride ion
CNN	Convolutional neural network
CNS	Central nervous system
DBS	Deep brain stimulation
DE	Differential entropy
DL	Deep learning
DMN	Default mode network
dMRI	Diffusion magnetic resonance imaging
DNA	Deoxyribonucleic acid
DNN	Deep neural network
DS	Diffusion spectrum imaging
DSI	Diffusion spectral imaging
DT	Decision tree

DTI	Diffusion tensor imaging
EC	Effective connectivity
ECG	signals Electrocardiography signals
EEG	Electroencephalography
ELM	Extreme learning machine
EMD	Empirical mode decomposition
FA	Fractional anisotropy
FC	Functional connectivity
fMRI	Functional magnetic resonance imaging
FOG	Freezing of gait
FT	Fiber tractography
GABA	Gamma-aminobutyric acid
GAN	Generative adversarial network
GEO	Gene expression omnibus
GNN	Graph neural network
GPR	Gaussian process regression
HAR	Human activity recognition
HCI	Human–computer interaction
HCP	Human Connectome Project
HHT	Hilbert-huang transform
HP-DMI	hierarchical pyramid depth motion images
ICA	Independent component analysis
iEEG	Intracranial electroencephalograph
IMFs	Intrinsic mode functions
IoT	Internet of Things
JMOCAP	Joint motion capture (JMOCAP)
K+	Potassium ion
KNN	K-nearest neighbors
LBP	Local binary patterns
LID	Levodopa induced dyskinesia
LIME	Local interpretable model-agnostic explanations
LR	Logistic regression
LSTM	Long short-term memory
M1	Primary motor cortex
MD	Mean diffusivity
MEG	Magnetoencephalography
ML	Machine learning
MLP	Multilayer perceptron
MPFC	Medial prefrontal cortex

MRI	Magnetic resonance imaging
Na+	Sodium ion
NB	Naive Bayes
NFC	Neuro-fuzzy classifier
NITRC	Neuroimaging informatics tools and resources clearinghouse
NLP	Natural language processing
ODD	Oppositional defiant disorder
OPM	Optically driven magnetometer
OSA	Obstructive sleep apnea
PCA	Principal component analysis
PCC	Posterior cingulate cortex
PD	Parkinson's disease
PET	Positron emission tomography
PFC	Prefrontal cortex
PMC	Premotor cortex
PNS	Peripheral nervous system
PPN	Pedunculopontine nucleus
PSD	Power spectral density
RF	Random forest
RL	Reinforcement learning
RNN	Recurrent neural network
RSN	Resting-state network
R-SNARE	Resident soluble N-ethylmaleimide-sensitive factor attachment protein receptor
SC	Structural connectivity
SLFN	Single layer feedforward network
SMA	Supplementary motor area
SNARE	Soluble N-ethylmaleimide-sensitive factor attachment protein receptor
SOZ	Seizure onset zone
SQUIDs	superconducting quantum interference devices
STN	Subthalamic nucleus
STORM	Stochastic optical reconstruction microscopy
SVM	Support vector machine
SVR	Support vector regression
tDCS	Transcranial direct current stimulation
TEM	Transmission electron microscopy
TMS	Transcranial magnetic stimulation

TOM	Theory of mind
V-ATPase	Vacuolar-type H+-ATPase
VMAT	Vesicular monoamine transporter
WPD	Wavelet packet decomposition
XAI	Explainable AI
XGB	Extreme gradient boosting

1

Exploring Brain Connectivity: Structural, Functional, and Effective Networks in Neuroscience

Rezuana Haque[1], Md. Niamat Munsi[2], and Chander Prabha[3]

[1]Computer Science and Engineering, BGC Trust University, Bangladesh
[2]Bachelor of Medicine and Bachelor of Surgery (MBBS) Sir Salimullah Medical College, Bangladesh
[3]Chitkara University Institute of Engineering and Technology, Chitkara University, India
E-mail: rezuanahaque27@gmail.com; munshiniamat46@gmail.com

Abstract

This chapter explores the concept of brain connectivity (BC) and its critical role in neuroscience. BC refers to the intricate network of neural pathways that facilitate communication between different brain regions, essential for cognition, behavior, and perception. The chapter discusses three main types of brain connectivity: structural, functional, and effective. Structural connectivity examines the physical wiring of the brain through white matter tracts, studied using techniques like diffusion tensor imaging (DTI) within MRI. Functional connectivity focuses on temporal correlations between neural activities in different regions, commonly investigated with functional magnetic resonance imaging (fMRI), electroencephalography (EEG), and magnetoencephalography (MEG). Effective connectivity delves into causal interactions and directional influences between neural elements, employing methods such as dynamic causal modeling (DCM) and Granger causality analysis. Key methodologies are explained, highlighting their principles, applications, and limitations. fMRI captures blood-oxygen-level-dependent (BOLD) signals, while EEG and MEG provide high temporal resolution for recording electrical and magnetic activity. The chapter also addresses integrating these modalities to gain a comprehensive understanding of brain connectivity. By elucidating the different aspects of brain connectivity and the tools used to study them, this chapter aims to enhance our understanding of the brain's complex network architecture, with implications for diagnosing and treating neurological disorders and advancing cognitive neuroscience.

Keywords: Brain connectivity, neuroscience, neural pathways, fMRI, EEG, MEG.

1.1 Introduction

In neuroscience, BC is a fundamental concept. It refers to the network of connections and interactions among various brain areas. Neural pathways are these connections that allow various brain regions to communicate with one another. This communication helps integrate sensory information, coordinate movement, and manage cognitive processes like attention, memory, and decision-making [1]. Structural connectivity (SC), functional connectivity (FC), and effective connectivity (EC) are the three primary types of BC [2]. SC refers to the physical connections between brain regions [3]. It is important to understand how different brain regions interact and communicate with each other. SC helps identify which brain regions are involved in disorders and is often measured using diffusion tensor imaging (DTI) [3]. FC examines how different brain regions interact over time [3]. It helps identify the source of disorders by highlighting abnormal brain activity patterns associated with specific conditions. This is measured using electroencephalography (EEG) and magnetic resonance imaging (fMRI) [3], [4], [5]. EC focuses on the cause-and-effect relationships between brain regions [3], [6]. EC shows how information flows in the brain and helps pinpoint the origins of disorders. Magnetoencephalography (MEG) and positron emission tomography (PET) are frequently used to measure effective connectivity. Studying brain connectivity (BC) helps identify the sources of brain disorders (BD) and guides brain surgery. For example, the cerebral cortex processes sensory information, controls movement, and enables thought and behavior [7]. Damage to part can cause epilepsy, stroke, and brain injury [8]. The cerebellum helps coordinate movement and maintains balance [9]. Cerebellum damage can lead to movement disorders like brain tumor, Parkinson's disease, Huntington's disease, stroke, etc. [9]. The brain and spinal cord are connected by the brainstem [10]. It manages automatic functions, such as heart rate and respiration [10].

Damage to the brainstem can cause severe conditions such as coma and death [11]. BC is very important for cognitive functions. It helps different brain areas work together for tasks like attention, memory, and decision-making. A neural network is a set of nerve pathways connecting various brain regions [12]. Neural pathways allow information exchange and coordination. These networks transmit signals quickly, organizing and analyzing information in just milliseconds [12]. Network analysis helps map and study these connections. FC examines how brain regions interact over time using fMRI and EEG while SC looks at physical connections using DTI. Disruptions in these connections can cause many disorders. By studying how brain connectivity and cognitive functions are related, researchers can learn how different brain processes work. This knowledge helps them develop better treatments for cognitive and neurological disorders.

This chapter discusses the following points:

1. What is BC and why is it important for integrating sensory information, coordinating motor functions, and regulating cognitive processes?
2. What are structural, functional, and effective connectivity?
3. How do methods like fMRI, EEG, MEG, and MRI help study brain connectivity?
4. How are disruptions in brain connectivity linked to different brain disorders?

1.2 Brain Connectivity

Brain connectivity explores how different parts of the brain are connected and communicate with each other. On the other hand, brain dynamics involves understanding how brain activity evolves over time, how different brain regions synchronize their activity, and how these patterns support cognitive functions and behavior. Brain connectivity can be modeled using graph theory [13]. With this method, the brain is represented as a network of nodes and edges [14]. The nodes represent distinct brain regions or groups of neurons, and the edges represent the connections between these nodes. These connections can be either structural, like physical connections such as axons, or functional, like synchronized activity between regions. For instance, each node in a network diagram can represent a brain region, while lines between nodes denote their connections. Researchers can analyze brain connectivity by applying graph theory metrics like degree centrality, clustering coefficient, and path length. Degree centrality measures the number of connections a node (brain region) has. Nodes with a high degree are considered central or important within the network. The clustering coefficient measures how connected a node's neighbors are to each other. A high clustering coefficient indicates that the neighbors of a node tend to form a tightly knit group. By calculating the clustering coefficient, researchers can locate areas where local connectivity is high, indicating regions that handle specific tasks like visual processing or language. Path length is the shortest distance between two nodes in the network. By analyzing path length, researchers can determine how quickly and efficiently different parts of the brain can communicate.

Understanding brain connectivity is important for gaining insights into how the brain functions normally. It helps us see how the brain changes as we grow and age. This knowledge is also important for identifying issues related to neurological disorders. By studying brain connectivity, we can develop better diagnostic tools and treatments for specific conditions.

1.3 Techniques for Brain Connectivity Measure

Several methods are used in BC analysis for studying the connections between different brain areas. Important techniques for researching BC include MEG, fMRI, MRI, and EEG. Researchers can understand brain connections and interactions using these methods.

1.3.1 Functional magnetic resonance imaging (fMRI)

fMRI is a method that uses differences in oxygen and blood flow to generate images of the brain. It helps study brain connectivity by showing how different brain regions work together. fMRI provides detailed images because it has high spatial resolution.

1.3.2 Electroencephalography (EEG)

A non-invasive technique to measure brain electrical activity is electroencephalography (EEG). It studies brain connectivity by observing synchronized brain waves and captures rapid neural activity quickly. EEG has high temporal resolution. Using an EEG, seizures and epilepsy can be diagnosed [15].

1.3.3 Magnetoencephalography (MEG)

MEG measures magnetic fields from brain activity non-invasively. It tracks brain connectivity by observing synchronized activity between regions and also captures rapid neural activity with high temporal resolution. Most epilepsy and brain tumor patients have MEG scans [16].

1.3.4 Magnetic resonance imaging (MRI)

MRI generates highly precise pictures of the body's internal architecture using strong magnetic fields and radio waves. It is commonly used to identify and track illnesses, including tumors, neurological diseases, and injuries [17].

1.4 Neurological Disorders

Neurological disorders are a major health issue worldwide. It often involves disruptions in brain connectivity. Here, we will discuss Alzheimer's disease (AD), Parkinson's disease, and schizophrenia, three brain disorders that are related to brain connectivity issues.

1.4.1 Alzheimer's disease (AD)

AD is a brain disorder that worsens over time [13]. It causes dementia by forming beta-amyloid plaques and neurofibrillary tangles [18]. The symptoms of AD change with the progression of the disease and is divided into stages based on how much cognitive function and ability are lost. The stages include:

- Preclinical stage: No symptoms are present yet.
- Mild cognitive impairment stage: There are some memory problems.
- Dementia stage: Significant memory loss occurs [19].

Brain connectivity is affected by AD, especially in the default mode network (DMN). The DMN is a network of brain regions that is more active at rest than during task performance [20]. The DMN is responsible for activities like remembering past events, thinking about oneself, daydreaming, and using imagination. Research shows that AD reduces FC between brain regions, including the DMN. This condition leads to cognitive decline and memory loss [21].

1.4.2 Parkinson's disease (PD)

PD is a neurological condition that slowly kills dopamine-producing neurons in the brain [22]. This causes symptoms like shaking, stiffness, and slow movement [23]. Research shows that PD alters the connectivity patterns in different brain areas, like the DMN, sensorimotor cortex, and basal ganglia. For example, one study found reduced FC in PD patients [24], while another discovered increased resting-state connectivity in cortical areas early in the disease [25]. This means that even in the early stages, the brain regions of PD patients show stronger connections when at rest. These changes are linked to motor symptoms and cognitive decline.

1.4.3 Schizophrenia

Schizophrenia is a serious mental disorder. The symptoms of this disease are hallucinations, delusions, disorganized speech and behavior, and a lack of interest [26]. It disrupts brain connectivity, especially in the prefrontal cortex (PFC) and temporal lobe, which are important for attention, memory, and language [27]. Studies show that schizophrenia causes both increased and decreased connectivity in brain networks. This disruption leads to cognitive problems and negative symptoms in patients. Schizophrenia patients often show higher connectivity within the default mode network (DMN) but lower connectivity between the DMN and other networks [28]. Changes in sensory networks can cause hallucinations. Schizophrenia is also linked to white matter issues, especially in the frontal, temporal, and parietal lobes [29]. These connectivity changes start early and get worse as the disease progresses.

1.5 Conclusion

Brain connectivity is the key to processing sensory information, coordinating movement, and managing cognitive functions. This chapter explains how brain connectivity is important for understanding how the brain works and how it affects various disorders. It explains the different types of brain connectivity including structural, functional, and effective. The chapter describes methods like fMRI, EEG, MEG, and MRI used to measure these connections. It also explains how conditions like AD, PD, and schizophrenia affect brain connectivity. These changes lead to issues with thinking, memory, and movement. Understanding these patterns helps in diagnosing and treating these conditions better. This knowledge can lead to new and more effective treatments. Studying brain connectivity provides valuable insights and holds promise for improving diagnosis and treatment.

Acknowledgments

We would like to thank the editors for their invaluable help in writing this chapter. Their guidance and expertise were essential in shaping the content and ensuring the accuracy and clarity of the information presented.

References

[1] E. L. Parks and D. J. Madden, "Brain Connectivity and Visual Attention," Brain Connect., vol. 3, no. 4, pp. 317–338, Aug. 2013, doi: 10.1089/brain.2012.0139.

[2] "Structural and Functional Brain Networks: From Connections to Cognition | Science." Accessed: Jun. 13, 2024. [Online]. Available: https://www.science.org/doi/10.1126/science.1238411

[3] P. Babaeeghazvini, L. M. Rueda-Delgado, J. Gooijers, S. P. Swinnen, and A. Daffertshofer, "Brain Structural and Functional Connectivity: A Review of Combined Works of Diffusion Magnetic Resonance Imaging and Electro-Encephalography," Front. Hum. Neurosci., vol. 15, Oct. 2021, doi:10.3389/fnhum.2021.721206.

[4] R. Mohanty, W. A. Sethares, V. A. Nair, and V. Prabhakaran, "Rethinking Measures of Functional Connectivity via Feature Extraction," Sci. Rep., vol. 10, no. 1, p. 1298, Jan. 2020, doi:10.1038/s41598-020-57915-w.

[5] E. W. Lang, A. M. Tomé, I. R. Keck, J. M. Górriz-Sáez, and C. G. Puntonet, "Brain Connectivity Analysis: A Short Survey," Comput. Intell. Neurosci., vol. 2012, p. 412512, 2012, doi: 10.1155/2012/412512.

[6] A. K. Abbas, G. Azemi, S. Amiri, S. Ravanshadi, and A. Omidvarnia, "Effective connectivity in brain networks estimated using EEG signals is altered in children with ADHD," Comput. Biol. Med., vol. 134, p. 104515, Jul. 2021, doi: 10.1016/j.compbiomed.2021.104515.

[7] "Cerebral cortex | Description, Anatomy, Function, & Disease | Britannica." Accessed: Jun. 14, 2024. [Online]. Available: https://www.britannica.com/science/cerebral-cortex

[8] "Neurological Disorders." Accessed: Jun. 14, 2024. [Online]. Available: https://dphhs.mt.gov/schoolhealth/chronichealth/neurologicaldisorders

[9] "The Location and Function of the Cerebellum in the Brain," Verywell Mind. Accessed: Jun. 14, 2024. [Online]. Available: https://www.verywellmind.com/what-is-the-cerebellum-2794964

[10] H. Basinger and J. P. Hogg, "Neuroanatomy, Brainstem," in StatPearls, Treasure Island (FL): StatPearls Publishing, 2024. Accessed: Jun. 14, 2024. [Online]. Available: http://www.ncbi.nlm.nih.gov/books/NBK544297/

[11] R. Mathur, C. Balucani, A. Elmashala, and R. Geocadin, "Coma, Disorders of Consciousness, and Brain Death," in Emergency Neurology, K. L. Roos, Ed., Cham: Springer International Publishing, 2021, pp. 345–372. doi: 10.1007/978-3-030-75778-6_17.

[12] D. A. Kelly, "Communicating Across Neural Networks." Accessed: Jun. 15, 2024. [Online]. Available: https://www.brainfacts.org:443/brain-anatomy-and-function/cells-and-circuits/2022/neural-networks-110822

[13] L. E. Ismail and W. Karwowski, "A Graph Theory-Based Modeling of Functional Brain Connectivity Based on EEG: A Systematic Review in the Context of Neuroergonomics," IEEE Access, vol. 8, pp. 155103–155135, 2020, doi: 10.1109/ACCESS.2020.3018995.

[14] M. Mijalkov, E. Kakaei, J. B. Pereira, E. Westman, G. Volpe, and Alzheimer's Disease Neuroimaging Initiative, "BRAPH: A graph theory software for the analysis of brain connectivity," PloS One, vol. 12, no. 8, p. e0178798, 2017, doi: 10.1371/journal.pone.0178798.

[15] "NeuroWave-Net: Enhancing epileptic seizure detection from EEG brain signals via advanced convolutional and long short-term memory networks." Accessed: Jun. 15, 2024. [Online]. Available: https://www.aimspress.com/article/doi/10.3934/bioeng.2024006

[16] M. M. School, "Magnetoencephalography (MEG) | UTHealth Neurosciences," Neurology, Neurosurgery, Spine Care | UTHealth Neurosciences. Accessed: Jun. 15, 2024. [Online]. Available: https://med.uth.edu/neurosciences/conditions-and-treatments/magnetoencephalography-meg/

[17] "NeuroNet19: an explainable deep neural network model for the classification of brain tumors using magnetic resonance imaging data | Scientific Reports." Accessed: Jun. 15, 2024. [Online]. Available: https://www.nature.com/articles/s41598-024-51867-1

[18] S. Kocahan and Z. Doğan, "Mechanisms of Alzheimer's Disease Pathogenesis and Prevention: The Brain, Neural Pathology, N-methyl-D-aspartate Receptors, Tau Protein and Other Risk Factors," Clin. Psychopharmacol. Neurosci., vol. 15, no. 1, pp. 1–8, Feb. 2017, doi: 10.9758/cpn.2017.15.1.1.

[19] A. Kumar, J. Sidhu, F. Lui, and J. W. Tsao, "Alzheimer Disease," in StatPearls, Treasure Island (FL): StatPearls Publishing, 2024. Accessed:

Jun. 15, 2024. [Online]. Available: http://www.ncbi.nlm.nih.gov/books/NBK499922/

[20] A. Mohan et al., "The Significance of the Default Mode Network (DMN) in Neurological and Neuropsychiatric Disorders: A Review," Yale J. Biol. Med., vol. 89, no. 1, pp. 49–57, Mar. 2016.

[21] M. Yu, O. Sporns, and A. J. Saykin, "The human connectome in Alzheimer disease — relationship to biomarkers and genetics," Nat. Rev. Neurol., vol. 17, no. 9, pp. 545–563, Sep. 2021, doi:10.1038/s41582-021-00529-1.

[22] Z. D. Zhou, L. X. Yi, D. Q. Wang, T. M. Lim, and E. K. Tan, "Role of dopamine in the pathophysiology of Parkinson's disease," Transl. Neurodegener., vol. 12, p. 44, Sep. 2023, doi:10.1186/s40035-023-00378-6.

[23] S. Zafar and S. S. Yaddanapudi, "Parkinson Disease," in StatPearls, Treasure Island (FL): StatPearls Publishing, 2024. Accessed: Jun. 15, 2024. [Online]. Available: http://www.ncbi.nlm.nih.gov/books/NBK470193/

[24] L. Albano et al., "Functional connectivity in Parkinson's disease candidates for deep brain stimulation," NPJ Park. Dis., vol. 8, p. 4, Jan. 2022, doi: 10.1038/s41531-021-00268-6.

[25] A. Cerasa, F. Novellino, and A. Quattrone, "Connectivity Changes in Parkinson's Disease," Curr. Neurol. Neurosci. Rep., vol. 16, no. 10, p. 91, Aug. 2016, doi: 10.1007/s11910-016-0687-9.

[26] "What is Schizophrenia?" Accessed: Jun. 15, 2024. [Online]. Available: https://www.psychiatry.org:443/patients-families/schizophrenia/what-is-schizophrenia

[27] Y. He et al., "Organized Resting-state Functional Dysconnectivity of the Prefrontal Cortex in Patients with Schizophrenia," Neuroscience, vol. 446, pp. 14–27, Oct. 2020, doi: 10.1016/j.neuroscience.2020.08.021.

[28] P. Skudlarski et al., "Brain connectivity is not only lower but different in schizophrenia: a combined anatomical and functional approach," Biol. Psychiatry, vol. 68, no. 1, pp. 61–69, Jul. 2010, doi: 10.1016/j.biopsych.2010.03.035.

[29] E. Luvsannyam et al., "Neurobiology of Schizophrenia: A Comprehensive Review," Cureus, vol. 14, no. 4, p. e23959, doi:10.7759/cureus.23959.

2

Advances in Brain Imaging Technologies: A Comprehensive Overview

Mahade Hasan[1], Farhana Yasmin[2], Xue Yu[1], and Asif Karim[3]

[1]School of Software, Nanjing University of Information Science and Technology, China
[2]School of Computer and Software, Nanjing University of Information Science and Technology, China
[3]Faculty of Science and Technology, Charles Darwin University, Australia
E-mail: farabi.eu.bd@gmail.com; farhana005.cse@gmail.com; xueyu@nuist.edu.cn; asif.karim@cdu.edu.au

Abstract

Advances in brain imaging technology have significantly enhanced the diagnosis and understanding of neurological disorders. This study explores the clinical roles and advancements of five key brain imaging modalities: functional magnetic resonance imaging (fMRI), electroencephalography (EEG), magnetoencephalography (MEG), positron emission tomography (PET), and diffusion tensor imaging (DTI). These technologies play pivotal roles in neuroscience, from mapping brain activity and functional connectivity to visualizing metabolic processes in the brain. The integration of deep learning and machine learning algorithms with these imaging methods has developed their clinical applications. These advancements have improved diagnostic accuracy, enabled early disease detection, and facilitated the development of personalized treatment strategies. By implementing artificial intelligence, these imaging techniques can now provide more precise disease characterization and better insights into neurological conditions such as Alzheimer's disease, epilepsy, multiple sclerosis, and traumatic brain injury. This study highlights the transformative impact of combining advanced imaging technologies with artificial intelligence, emphasizing their importance in modern neuroscience and clinical practice.

Keywords: Advanced technology, brain imaging, DTI, EEG, fMRI, MEG, PET.

2.1 Introduction

2.1.1 Overview of brain imaging technologies

Brain imaging technologies include a diverse array of techniques that allow for the observation and investigation of the human body function, and connections of the brain. Structural imaging modalities, such as MRI and CT, as well as functional techniques like fMRI and PET, offer researchers and clinicians invaluable tools to study the brain at different levels of detail and complexity [1]. This section will explore the fundamental concepts that underlie various brain imaging techniques and provide a concise summary of their uses and constraints.

2.1.2 Importance of brain imaging in neuroscience and clinical practice

Brain imaging is essential to improving our understanding of brain function and dysfunction. These tools in neuroscience aid in the investigation of neural systems that underlie cognition, emotion, and behavior. They help researchers in understanding the complexity of the human brain [2, 3]. Brain imaging plays a crucial role in clinical practice by assisting in the diagnosis, treatment planning, and monitoring of neurological and psychiatric illnesses. It allows clinicians to detect any abnormalities in the structure or function of the brain and customize treatment properly.

2.1.3 Objectives

The primary objectives of this chapter are to provide an in-depth overview of modern brain imaging technologies, to emphasize the significance of these technologies in both neuroscience research and clinical practice, and to examine the future potential and problems in the area. The chapter is structured in a way that encourages a comprehensive understanding of various types of imaging and their uses.

An overview of structural brain imaging methods, including MRI and CT, is given first as they are essential for displaying the anatomical characteristics of the brain. Subsequently, an investigation of functional brain imaging methods is conducted, including fMRI and PET, which allow the analysis of brain activity and metabolic processes. Next, we explore molecular imaging and electrophysiological approaches, which provide information on the chemical and electrical processes occurring in the brain.

The chapter also covers the use of artificial intelligence and machine learning in brain imaging, highlighting the novel capacity of these technologies to automate picture interpretation and enhance diagnostic precision. We will explore challenges and future pathways in brain imaging, with a specific emphasis on technological progress, ethical discussions, and the use of research discoveries in medical settings.

Finally, this study ends by providing a concise overview of significant progress, discussing the potential impact on the fields of neuroscience and medicine, and suggesting areas for further investigation. This systematic approach guarantees a comprehensive comprehension of the present state and future potential of brain imaging technology.

2.2 Integration of AI and ML in Brain Imaging

The development of ML and AI has become crucial to the advancement of brain imaging technology. This section covers deep learning techniques for brain image processing, the role of AI and ML in neuroimaging, issues facing AI-assisted neuroimaging, and potential future paths in this field.

2.2.1 Role of AI and ML in neuroimaging

The automated processing and interpretation of intricate brain imaging data made possible by AI and ML has completely changed the field of neuroimaging [4], [5].

Figure 2.1 highlights the significant roles that artificial intelligence (AI) and machine learning (ML) play in neuroimaging. AI and ML techniques are

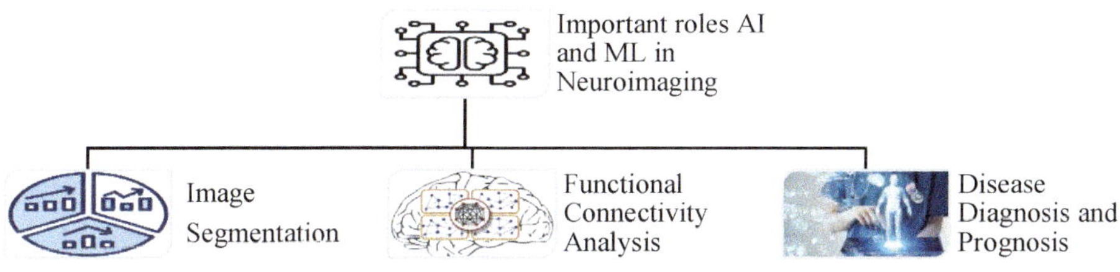

Figure 2.1 Important roles consist of AI and ML in neuroimaging.

employed in image segmentation. This process involves the precise delineation of anatomical structures and pathological regions within the brain. Accurate image segmentation is crucial for proper diagnosis, treatment planning, and monitoring of neurological conditions [6]. AI and ML algorithms are used to analyze the brain's functional connectivity. This analysis examines the interactions and communication between different brain regions, helping researchers and clinicians understand the brain's functional organization and identify abnormalities associated with neurological and psychiatric disorders. AI and ML models play an essential role in disease diagnosis and prognosis. By processing large volumes of neuroimaging data, these models can detect patterns and biomarkers indicative of specific conditions. This capability aids in early diagnosis, the development of personalized treatment plans, and the prediction of disease outcomes, ultimately improving patient care [7, 8].

2.2.2 Deep learning approaches for brain image analysis

Deep learning, a subset of AI, has shown remarkable success in brain image analysis due to its ability to learn complex representations from large datasets [9]. Some prominent deep learning approaches are discussed in this section.

Figure 2.2 illustrates the use of various neural networks for brain health care monitoring. The process begins with inputting raw neuroimaging data and other relevant patient information into the neural network system. This data is then processed using deep learning techniques, which involve multiple layers of neural networks to extract and analyze complex features.

Among the neural networks used, CNNs are extensively used for tasks such as image classification, segmentation, and detection in neuroimaging. They can automatically learn spatial hierarchies and extract relevant features from brain images [10]. Recurrent neural networks (RNNs) on the other hand, including long short-term memory (LSTM) networks, are applied to analyze

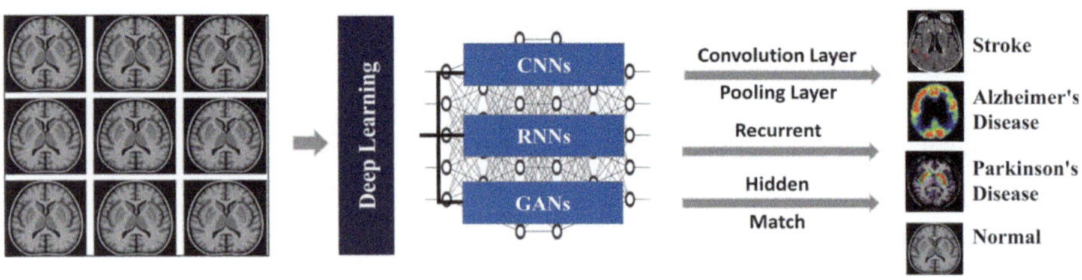

Figure 2.2 Uses of various neuron networks for brain health care monitoring.

temporal sequences in functional imaging data (for example fMRI), capturing the dynamic aspects of brain activity [11]. Generative adversarial networks (GANs) are another type of neural network utilized in this context. GANs are used for data augmentation, generating synthetic brain images to overcome the challenge of limited training data. They are also employed for image reconstruction and enhancement [12].

The processed data, now enhanced through various layers such as convolution, pooling, recurrent, hidden, and match layers, is generated as output. This output is then used for accurate diagnosis, treatment planning, and monitoring of brain health, ultimately contributing to improved patient care.

2.3 Functional Magnetic Resonance Imaging

The non-invasive method of measuring and mapping brain activity is known as fMRI [13]. Its accuracy is dependent on its ability to identify brain activity-induced changes in blood oxygenation and flow [14]. fMRI is a powerful tool in neuroscience and clinical research, and this section will explore its basics, current technological advances, and many applications.

Figure 2.3 illustrates the structure of a functional magnetic resonance imaging (fMRI) machine. Central to the device is a powerful superconducting

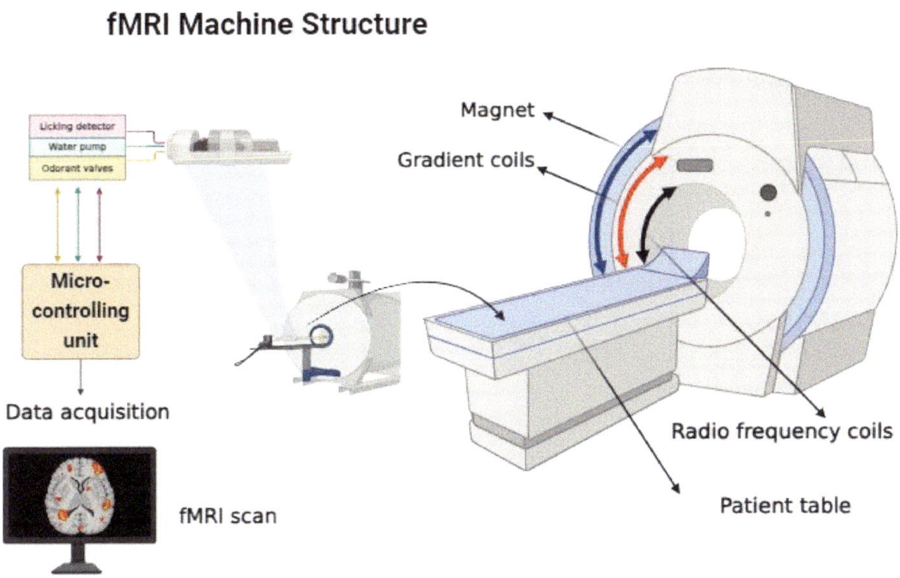

Figure 2.3 Functional magnetic resonance imaging machine structure.

magnet that aligns protons in the body. Gradient coils create varying magnetic fields for spatial encoding, while radio frequency coils transmit pulses to excite protons and receive the emitted signals. A computer system controls the imaging process, reconstructing signals into images using advanced algorithms. The patient lies on a motorized bed that slides into the magnet's bore, with the control panel and display allowing operators to manage and monitor the process in real time. Effective shielding prevents external interference and contains the magnetic field, ensuring safety. This structure enables the fMRI machine to capture detailed images of brain activity non-invasively, aiding in the diagnosis and treatment of neurological conditions.

2.3.1 Principles of fMRI

fMRI works by detecting contrast that is blood oxygen levels dependent (BOLD) [15, 16]. An area of the brain uses more oxygen while it is working harder. An increase in blood flow meets this elevated oxygen demand, which in turn causes alterations in the ratio of oxygenated to deoxygenated hemoglobin [17]. fMRI keeps track of these shifts, and the resulting pictures show patterns of brain activity linked to different kinds of motor and cognitive work.

The primary objective of an fMRI study is to create an environment or activity that is intended to cause a certain brain response. A quick sequence of high-resolution photos is captured throughout this task. Subsequently, the research concentrates on pinpointing regions of brain activity within these images that exhibit notable changes in BOLD values.

2.3.2 Advancements in fMRI technology

Recent years have seen remarkable advancements in fMRI technology, leading to significant improvements in spatial and temporal resolution, sensitivity, and overall utility for studying the brain. Noteworthy progressions include the use of ultra-high field strength magnets, with magnetic fields of 7 Tesla or higher, to enhance signal-to-noise ratio and spatial resolution, enabling detailed imaging of complex brain areas. Multiband imaging has emerged as a method to accelerate data collection by simultaneously capturing multiple brain slices, thus enhancing temporal resolution and facilitating the study of rapid neurological processes. Additionally, the application of advanced analytical methods such as machine learning and sophisticated statistical approaches has enhanced the analysis of fMRI data by enabling the identification and interpretation of subtle patterns of brain activity. Furthermore, the adoption of Resting-State fMRI, which explores brain activity during rest, has unveiled intrinsic connectivity networks essential for understanding brain function in both healthy and diseased states. These advancements in fMRI technology have significantly deepened our understanding of brain function and pathophysiology, offering valuable insights for precise diagnostics and targeted treatments in the field of neuroscience.

2.3.3 Real-time fMRI

Real-time fMRI is a cutting-edge technique that allows for the immediate monitoring and analysis of brain activity as it occurs. By employing rapid image reconstruction and analysis pipelines, real-time fMRI enables researchers and clinicians to provide immediate feedback to participants or patients. This approach has diverse applications, including neuroscience research, brain-computer interfaces, and clinical interventions for neurological and psychiatric disorders. Challenges in real-time fMRI include the need for efficient data processing pipelines and validation of clinical efficacy. Despite these challenges, ongoing advancements in technology and methodology hold promise for the continued development and utilization of real-time fMRI in understanding and modulating brain function in real-time contexts.

2.3.4 Applications of fMRI in neuroscience and clinical research

fMRI has a wide range of applications in both neuroscience research and clinical practice. In neuroscience, fMRI is used to investigate the neural basis of cognitive functions such as memory, language, and emotion. It allows researchers to explore how different brain regions interact and how these interactions change with learning, development, and aging. fMRI can identify abnormal brain activity patterns associated with disorders such as epilepsy, multiple sclerosis, and brain tumors. fMRI helps map critical brain areas, such as those involved in language and motor functions, to avoid damage during neurosurgery. By comparing pre- and post-treatment fMRI scans, clinicians can assess the effectiveness of interventions for conditions like depression, anxiety, and chronic pain. fMRI biomarkers are used to predict the onset and progression of neurodegenerative diseases, such as Alzheimer's and Parkinson's.

Agarwal et al. [18] evaluated multiple brain atlases and connection matrices while analyzing fMRI data for ADHD diagnosis using image-based

Table 2.1 Machine and deep learning publications related to fMRI.

Ref.	**Subjects/features**	**Objective**	**Model and accuracy (%)**
Agarwal et al. [9]	fMRI data of ADHD patients and healthy controls	Enhance ADHD diagnosis using fMRI and deep learning	Image-based and network-based models; accuracy: 74.48%–90.90%
Wu et al. [10]	Task-based fMRI data, dynamic brain networks	Improve understanding of cognitive functions through dynamic brain networks	New task-based fMRI decoder; improved accuracy and efficiency
Vedaei et al. [11]	PET and rs-fMRI data of mTBI patients vs. healthy controls	Classify mTBI using multimodal imaging and deep learning	Multimodal approach with autoencoders; accuracy: 95.83%
Elakkiya and Dejey [12]	fMRI data for autism screening using CNN	Address fMRI dataset complexity in autism screening	AutiNet dataset: 77.78%, MinAutiNet dataset: 88.89%
Dai et al. [13]	Large-scale rs-fMRI data of MDD patients	Improve MDD diagnosis and understanding through functional connectivity	Transformer-encoder; accuracy: 68.61% (general), 78.11% (recurrent MDD)
Lin et al. [14]	rs-fMRI data of 83 LLD patients	Predict suicide risk in older adults with LLD	CSE and 3D CNNs; accuracy: Over 75% in key networks
Kamboj et al. [15]	Rs-fMRI data and expert-guided connectomics of 52 children with RE	Improve SOZ localization for RE treatment	DL and expert-guided model; accuracy: 84.8%, F1 score: 91.7%

and network-based methods. This study used sophisticated deep-learning techniques. Showing up to 64% variation in diagnostic precision, diagnostic accuracy between 74.48% and 90.90% was achieved, exceeding conventional deep learning techniques. Wu et al. [19] developed a task-based fMRI data decoder that analyses dynamic brain networks by means of a thorough deep learning methodology. The decoder gave new knowledge about human cognition by producing results that were more accurate and efficient than those of conventional machine learning. Using a multimodal approach and PET and resting-state fMRI data, mTBI patients and healthy controls were separated from one another [20]. Classification accuracy increased by up to 95.83% with the multimodal approach compared to 79%–91.67% with single-modality models. Elakkiya and Dejey [21], using dynamic activation functions and the simplified LeNet-5 architecture, developed AutiNet and MinAutiNet deep learning models for autism screening. Present benchmarks were achieved with an accuracy of 77.78% with AutiNet and 88.89% with MinAutiNet. Applying large-scale rs-fMRI datasets, this study developed a transformer-encoder method for MDD classification with an emphasis on unsupervised pre-training and simplified design [22]. For a better knowledge of MDD, identified impacted brain regions and achieved 68.61% accuracy, with 78.11% for recurrent MDD. Using resting-state fMRI data, suicidal preferences in LLD patients were classified using CSE and 3D CNNs. Identifying significant brain networks and areas predictive of suicide risk, identified suicidal tendencies with over 75% accuracy [23]. To identify the seizure onset zone (SOZ) for the treatment of refractory epilepsy, deep learning was combined with expert-guided rs-fMRI connectomics [24]. Examining to deep learning alone, achieved 84.8% accuracy and 91.7% F1 score, much improved SOZ localization. Table 2.1 shows fMRI advanced technology-related work.

Therefore, fMRI is a versatile and continually evolving tool that has significantly advanced our understanding of brain function and has become indispensable in both research and clinical diagnostics.

2.4 Diffusion Tensor Imaging (DTI)

In biological tissues, particularly the brain, diffusion tensor imaging (DTI) is a type of magnetic resonance imaging (MRI) that measures the diffusion of water molecules [25].

2.4.1 Principles of DTI

DTI evaluates the path and rate of water molecule diffusion in brain tissue. Understanding white matter tracts can benefit from an understanding of the way axon fiber orientation blocks water diffusion. The diffusion tensor,

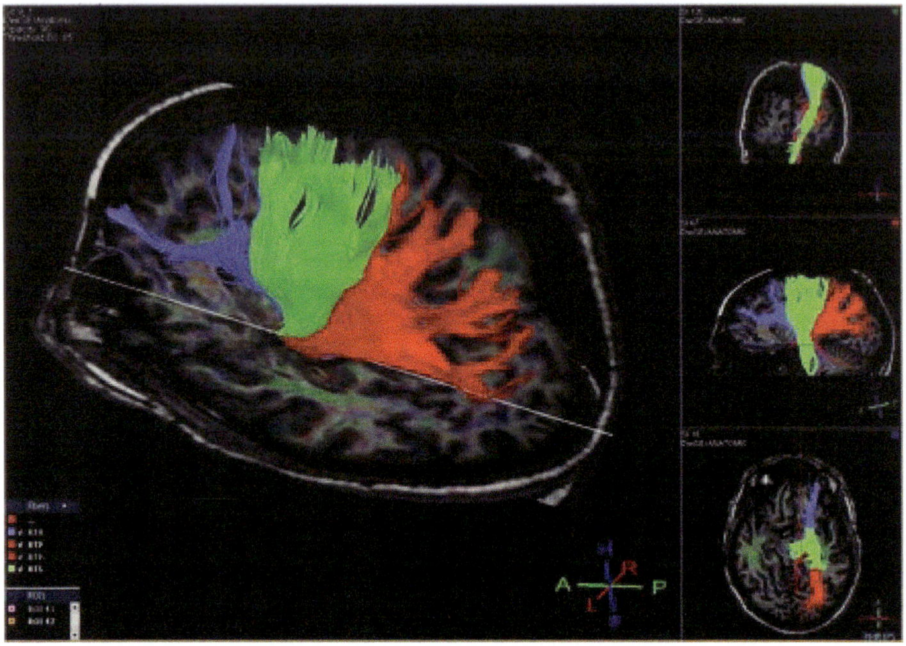

Figure 2.4 Diffusion tensor imaging [29].

which measures diffusion in three dimensions, is the main output of DTI. As illustrated in Figure 2.4, a diffusion tensor imaging is shown that uses anisotropic diffusion to estimate the axonal (white matter) organization of the brain and fiber tractography (FT) is a 3D reconstruction technique to assess neural tracts using data collected by diffusion tensor imaging.

The values of axial and radial diffusivity, mean diffusivity (MD), and fractional anisotropy (FA), which show tissue structure and integrity, can be computed using the diffusion tensor [17, 18, 28].

2.4.2 Recent developments in DTI techniques

DTI measures the diffusion of water molecules in brain tissue and provides information on the direction and intensity of water transport [30]. Since the amount of water that may permeate is limited by axonal fiber orientation, this is especially useful for studying white matter pathways. The diffusion tensor can be utilized for calculating a number of metrics that provide details about the microstructure and integrity of tissue [31].

- **Diffusion imaging with high angular resolution (HARDI):** Crossing fibers may be accurately imaged using q-ball and diffusion spectrum imaging by reconstructing different fiber orientations inside a voxel.
- **Advanced diffusion models:** Diffusion kurtosis imaging (DKI) and neurite orientation distribution and density imaging (NODDI) are two advanced diffusion models that provide more information about tissue structure and neurite architecture than the tensor model.
- **Machine learning techniques:** Deep learning has improved the automation of illness classification, tractography, and segmentation in DTI data analysis.

2.4.3 Clinical applications of DTI in psychiatry and neurology

DTI is beneficial for monitoring treatment and offering helpful data for diagnosis, prognosis, and treatment monitoring in neurology and psychiatry.

- **Neurological disorders**: Multiple sclerosis, stroke, traumatic brain injury, and neurological disorders like Parkinson's and Alzheimer's are among the conditions where DTI is used in neurology to assess the integrity of the white matter.
- **Psychiatric disorders**: Research using DTI have shown that conditions such as autism spectrum disorder, bipolar disorder, schizophrenia, and major depressive disorder affect white matter connectivity. These results provide novel insights into the neurology of these illnesses and point to putative biomarkers for diagnosis and therapy response forecasting.
- **Surgical planning**: DTI-based tractography helps neurologists plan ahead and reduces the risk of postoperative neurological disorders by finding white matter tracts nearby lesion sites.

Using attention processes, transformer-based deep learning algorithms were used for analyzing DTI data [32]. As a result, the scanning time was shortened and FA, AxD, and MD values could be measured. Better results were achieved than conventional techniques for the ADNI dataset's early Alzheimer's disease detection. Imaging models based on DKI and DTI characteristics were employed to segment the renal parenchyma using the Swin UNETR model in order to predict DKD in individuals with type 2 diabetes [33]. The integrated radiomics model performed better than the separate parameter maps in terms of DKD prediction. The Graphcut Hidden Markov Model (Graph_HMM) is a hybrid method that uses DTI and HMM to predict AD incidence [34]. It displayed outstanding specificity, sensitivity, and accuracy on the ADNI database. Graph_HMM displays

Table 2.2 Machine and deep learning publications related to DTI.

Ref.	**Subjects/features**	**Objective**	**Model and accuracy (%)**
Tiwari et al. [32]	Transformer-based deep learning techniques applied to DTI data	Enhance early detection of Alzheimer's disease	Superior performance over traditional methods on ADNI dataset
Yang et al. [33]	Radiomics models from DKI and DTI parameters for predicting DKD in T2DM patients	Improve early detection and management of DKD	Combined radiomics model outperforms individual parameter maps (AUC: 0.918)
Sikkandar et al. [34]	Combined approach using HMM and DTI to predict AD occurrence	Evaluate DTI-based approaches as biomarkers for AD diagnosis	High accuracy, sensitivity, and specificity on ADNI database
Gou et al. [35]	RATC-MTL for predicting AD progression	Achieve superior predictive performance compared to single-task and multi-task algorithms	Not specified
Taoka et al. [36]	Examination of ALPS-index as an indicator of conditions related to the glymphatic system	Highlight the complexity of evaluating the glymphatic system	Not specified

positive results, showing the potential use of DTI-based methods as biomarkers for AD diagnosis. Relation-aware tensor completion multitask learning (RATC-MTL) was developed to predict the progression of AD [35]. RTC-MTL calculates correlations between and within modes in cognitive scores. RATC-MTL performed better in prediction than both single-task and state-of-the-art multi-task algorithms. The study of the ALPS-index as a tool for glymphatic system conditions, emphasized that it should be used cautiously when interpreting glymphatic function [36]. Draw attention to the difficulties in assessing the lymphatic system and the requirement for further evaluation techniques. Table 2.2 shows DTI advanced technology related work in machine learning and deep learning.

A comprehensive overview of current developments in diffusion tensor imaging (DTI) technologies is given in Table 2.2, with an emphasis on the combination of deep learning and machine learning techniques. The research included in the table uses DTI data for a range of neurological and medical uses. This table summarizes the significant developments in the field and illustrates the potential of AI with enhanced imaging for clinical diagnosis.

2.5 Positron Emission Tomography (PET)

Positron emission tomography (PET) is a nuclear imaging method that offers important metabolic and functional data about tissues, with a special focus on brain research and clinical diagnostics. This section contains a summary of the fundamental concepts of PET imaging, current progress in the development of novel radio signals and imaging techniques, and the various ways in which PET imaging is used in the fields of neurological diseases and cancers.

2.5.1 Principles of PET imaging

PET imaging uses radiotracers, which decay to release positrons. Radiotracers accumulate in tissues based on metabolic activity after injection. Positrons are emitted and annihilated with electrons as the radiotracer decays, creating opposite-direction gamma rays. PET scanners use gamma rays to provide 3D pictures of tracer distribution in the body [37].

As illustrated in Figure 2.5, the working principle of a PET scan begins with the injection of a radioactive tracer into the subject. This tracer

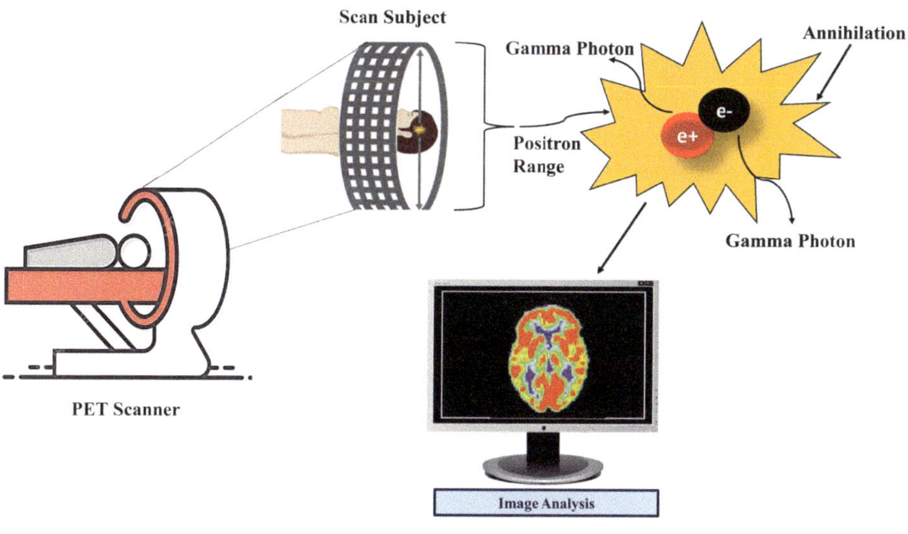

Figure 2.5 Working principle of positron emission tomography.

accumulates in metabolically active tissues. As the tracer decays, it emits positrons that travel a short distance (positron range) before annihilating with electrons, resulting in the production of gamma photon pairs. The PET scanner, equipped with a ring of detectors, captures these coincident gamma photons to pinpoint the tracer's location. This data is processed to reconstruct images displaying the tracer distribution. Subsequent image analysis assesses metabolic activity, identifying high uptake areas associated with diseases such as cancer, neurological disorders, and cardiovascular conditions. PET scans thus provide detailed metabolic images, aiding in diagnosis, monitoring, and research.

2.5.2 Novel radiotracers and imaging modalities in PET

Novel radiotracers and imaging modalities have improved diagnostic accuracy and practical uses in PET technology:

- **Novel radiotracers**: Novel radiotracers target molecular pathways and biomarkers. Amyloid and beta tracers are used to image Alzheimer's disease, while PSMA tracers are utilized to image prostate cancer.
- **Hybrid imaging modalities**: PET combined with CT or MRI improves anatomical localization and offers structural information. PET/CT and PET/MRI are now clinical standards, improving diagnostics.
- **Quantitative PET imaging**: More precise radiotracer uptake measures improve disease progression and therapy response assessment.

2.5.3 PET applications in neurodegenerative diseases and oncology

In neurodegeneration and oncology, PET imaging can reveal disease causes, diagnosis, and treatment monitoring:

- PET is widely used to diagnose and investigate neurodegenerative illnesses like Alzheimer's, Parkinson's, and Huntington's. PET imaging of amyloid and tau is useful for diagnosing and monitoring Alzheimer's disease pathology. PET measures neuroinflammation and synaptic density, revealing disease progression and treatment effects.
- **Oncology:** PET is essential for detecting, staging, and monitoring many malignancies. FDG-PET is frequently used to detect malignancies, assess metastasis, and evaluate therapy response. New tracers targeting cancer biomarkers improve tumor visualization and quantification.

This study compared machine learning models to statistical approaches for myocardial perfusion abnormalities using rubidium-82 PET data [38]. Machine learning models, particularly random forest, demonstrated higher receiver-operator characteristic area under the curve (AUC) values (0.92–0.95) for per-patient illness diagnosis than traditional approaches 0.87, although they performed comparably for per-vessel ischemia or scar location The results suggest that random forest machine learning improves cardiac perfusion imaging diagnosis. This paper proposes utilizing a deep neural network (DNN) model trained with PET and MRI data to improve PET image signal-to-noise ratio (SNR) [39]. The DNN model, which uses three modified U-Nets (3U-net), reduces noise from PET pictures reconstructed using FBP and MLEM. Using digital brain phantoms, SNR improved by 31.3% and MSE decreased by 34.0% for 1U-net and 3U-net, respectively, demonstrating the potential of MRI information to improve PET picture quality. This study compares MRI, CT, and PET data fusion approaches for generating attenuation correction (AC) pictures in cranial imaging [40]. The accuracy of indirect, direct, and direct with high-resolution convolution (HRC) was examined using normalized mean squared error (NMSE) and structural similarity. While visual inspection showed no difference, the direct+HRC approach had the highest SSIM of 0.975 and the lowest NMSE of 0.0138, suggesting it may be better at obtaining exact AC images for clinical PET imaging. This work examines how successively deploying DL-Enhancement (DLE) and DL-based time-of-flight (ToF) (DLT) deep learning algorithms improve FDG PET-CT image quality [41]. Reduced image noise and increased lesion detectability, diagnostic confidence, and SUVmax were observed in ToF-OSEM + DLE + DLT reconstructions, with an average increase of 28% $\pm$ 14% and 11% $\pm$ 5% for D710 and DMI data, respectively. This study compares six partial volume correction (PVC) approaches for PET picture quality [42]. Structural similarity index (SSIM) values of 0.63 for LD and 0.29 for FD in geometric transfer matrix (GTM) and 0.63 for LD and 0.67 for FD in iterative Yang (IY) suggest lower error levels. Multi-target correction (MTC) and Richardson-Lucy (RL) have bigger quantitative errors, with MTC having 2.71 RMSE for LD and 2.45 for FD and RL having 5 for LD and 3.27 for FD PVC methods. Table 2.3 presents an overview of publications on machine and deep learning applied to PET, detailing publication history, subjects/features, objectives, and model accuracy (%).

In summary, PET imaging is a dynamic and growing technology that gives unique metabolic and functional cancer and brain illness findings.

Table 2.3 Machine and deep learning publications related to PET.

Ref.	Subjects/features	Objective	Model accuracy (%)
Berman et al. [28]	Myocardial perfusion abnormalities using rubidium-82 PET data	Compare ML models to statistical approaches	Random forest: AUC 0.92–0.95
Chih-Chieh Liu and Jinyi Qi [29]	PET and MRI data	Enhance PET image SNR with DNN	SNR 31.3%, MSE 34.0%
Ueda et al. [30]	MRI, CT, and PET data	Compare fusion approaches for AC pictures	Direct+HRC: SSIM 0.975, NMSE 0.0138
Dedja et al. [31]	FDG PET-CT images	Evaluate DLE and DLT DL algorithms	Lesion detectability, diagnostic confidence, SUVmax ↑
Azimi et al. [32]	PET image quality	Compare six PVC approaches	GTM, IY: SSIM LD 0.63, FD 0.29–0.67; MTC, RL: RMSE LD 2.71–5, FD 2.45–3.27

Advances in radiotracer development and hybrid imaging should improve its clinical utility and diagnostic accuracy.

2.6 Magnetoencephalography (MEG)

Magnetoencephalography (MEG) measures brain magnetic fields non-invasively [43]. MEG fundamentals, technological developments, and brain dynamics and connectivity applications are covered in this section.

2.6.1 MEG principles

MEG detects magnetic fields from neuronal currents outside the scalp. The technology uses sensitive superconducting quantum interference devices (SQUIDs) to monitor small magnetic fields with high temporal resolution. MEG fundamentals include:

The working principle of MEG technology is shown in Figure 2.6. It begins with MEG sensors that detect magnetic fields generated by neural activity. Neurons produce these magnetic fields perpendicular to the electric currents they carry. The MEG sensors capture this data, free from distortion

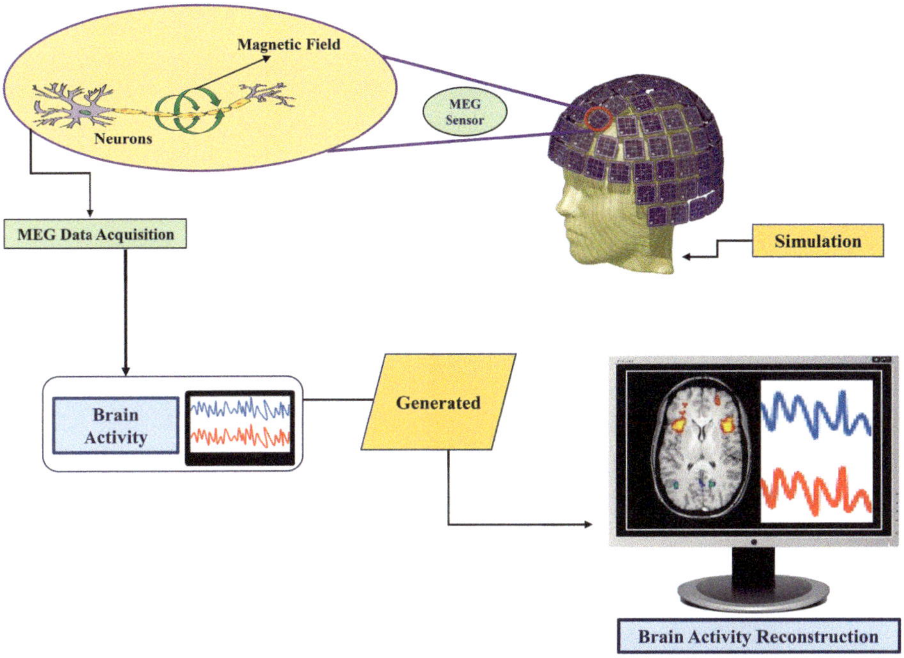

Figure 2.6 Working principle of MEG technology.

by the skull or scalp, directly measuring neural activity. Next, the captured MEG data is processed to identify the sources of these magnetic signals within the brain. Advanced techniques such as beamforming and minimum-norm estimation are used to reconstruct brain activity from the raw MEG data. This process generates detailed geographical maps of brain activity, providing valuable insights into neural function and aiding in both clinical and research applications.

2.6.2 Recent advances in MEG technology

Magnetoencephalography (MEG) has been greatly improved in terms of sensitivity, accuracy in space, and clinical and research settings by recent developments. The use of optically delivered Magnetometers (OPMs), innovative sensors with high sensitivity and room temperature operation, is one notable advancement. The wide range and accessibility of MEG systems could be enhanced by these sensors. Furthermore, the development of wearable MEG systems, which are comfortable and portable, has enormous potential for researching pediatric populations and naturalistic behaviors.

Additionally, the interpretation of MEG data and source localization have been greatly enhanced by developments in computational technologies, such as machine learning, which has increased the overall efficacy and usefulness of MEG in a variety of applications.

2.6.3 MEG in studying brain dynamics and connectivity

MEG is uniquely suited for studying the temporal dynamics and connectivity of brain networks due to its high temporal resolution. Key applications include:

- **Cognitive and sensory processes:** MEG evaluates brain activation timing and sequence during cognitive tasks like language processing, memory, and sensory perception.
- **Brain connectivity:** MEG examines functional and effective connectivity between brain areas. Coherence analysis and Granger causality study brain network directed interactions.
- **Clinical applications:** MEG reduces surgical risks by mapping eloquent cortical areas in epilepsy and brain tumor patients before surgery. Autism, schizophrenia, and traumatic brain injury brain abnormalities are studied with it.

MEG-based brain-computer interfaces have improved because of deep learning [44]. Novel preprocessing pipelines and convolutional neural networks helped MEG-RPSnet classify Rock-Paper-Scissors gestures with 85.56% accuracy. Even with regional sensor subsets, MEG-RPSnet outperformed EEG-based BCI architectures and classical machine learning techniques for non-invasive BCI. These findings suggest MEG-based BCI hand-gesture decoding systems are feasible. This machine learning study classified face and scrambled face visual stimuli using magnetoencephalography (MEG) signals from 16 subjects [45]. Riemannian feature extraction gave the LSTM and GRU models 92.99% and 91.66% classification accuracy. The accuracy exceeds linear discriminant analysis. Positive results show deep learning improves non-invasive MEG-based brain decoding and BCI. MEG waves from 16 participants were utilized to classify face and scrambled face visual stimuli using machine learning. Using Riemannian feature extraction, LSTM and GRU models outperformed LDA with classification accuracies of 92.99% and 91.66%, respectively [46]. These encouraging results demonstrate that deep learning can improve non-invasive MEG-based brain decoding and BCI. This study introduces magnetoencephalography

Table 2.4 Machine and deep learning publications related to MEG.

Ref.	Subjects/features	Objective	Model and its accuracy (%)
Yifeng Bu et al. [34]	12 subjects; MEG signals	Classify Rock-Paper-Scissors gestures using MEG signals	MEG-RPSnet (CNN-based): 85.56% accuracy. Better than EEG-based and traditional methods
Zeynep et al. [35]	16 subjects; MEG signals	Classify face and scrambled face visual stimuli using MEG signals	LSTM: 92.99%, GRU: 91.66%. Exceeds LDA. Deep learning shows superior performance in MEG-based decoding
Hamdan et al. [36]	16 subjects; MEG signals	Classify face and scrambled face visual stimuli using MEG signals	LSTM: 92.99%, GRU: 91.66%. Demonstrates deep learning's effectiveness over traditional methods for MEG decoding
Fan et al. [37]	MEG signals	Interpret MEG predictions with feature attribution	AUDC: 0.005, high attribution accuracy. Enables data compression with minimal loss (0.19%) in classification accuracy

(MEG) feature attribution as a novel interpretation paradigm for individual predictions[47]. The method translates MEG data into feature sets and assigns contribution weights using optimal Shapley values, achieving an area under the deletion test curve (AUDC) of 0.005, exceeding current computer vision algorithms in attribution accuracy. Visualization illustrates that crucial elements fit neurophysiological theories and that data may be compacted without sacrificing classification accuracy. This model-agnostic method enhances BCI system and decoding model interpretation and application. Table 2.4 shows MEG-related activities in machine learning and deep learning.

In summary, MEG provides unparalleled temporal resolution in the study of brain function, making it a powerful tool for both research and clinical applications. Ongoing advancements in sensor technology and analytical methods continue to expand its capabilities and potential impact.

2.7 Electroencephalography (EEG)

A brain electroencephalogram (EEG) measures electrical activity. This test is an EEG. Small metal disks called electrodes attach to the scalp during the test. The waves on an EEG represent brain cell communication via electrical impulses [48, 49, 50].

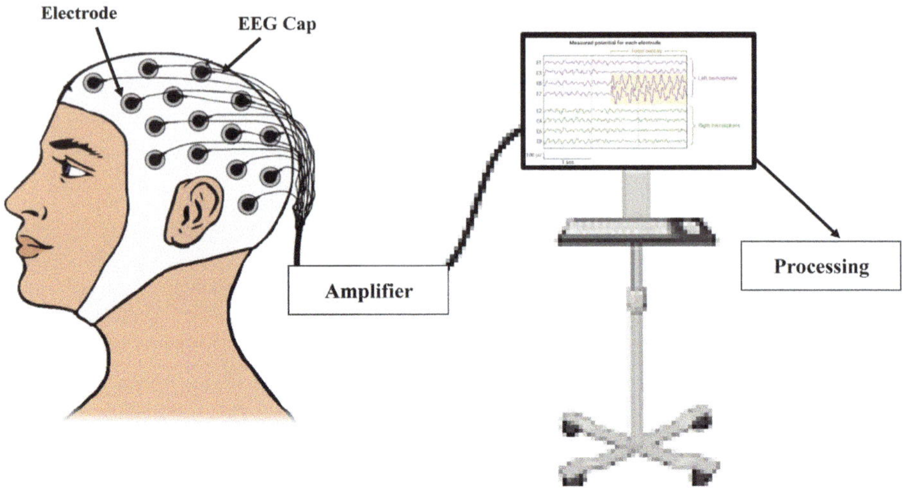

Figure 2.7 Working principle of EEG technology.

2.7.1 Principles of EEG

Brain activity is measured via scalp electrodes in EEG. Brain cell firing electrical potentials can be recorded using this method. Thousands to millions of coordinated neurons generate EEG signals. Alpha, beta, theta, and delta frequency bands affect waking, sleep, and cognitive states. EEG is used in neuroscience and clinical research to evaluate brain function, connectivity, and dynamics.

As depicted in Figure 2.7, EEG detects electrical activity in the brain using scalp-placed electrodes. These electrodes measure and amplify electrical signals, which are then recorded and processed to remove noise. EEG analyzes brain waves across frequency bands like delta, theta, alpha, beta, and gamma, each linked to different brain activities. Advanced techniques such as inverse modeling and dipole source analysis localize the origins of these signals, providing insights into neural processes. EEG is crucial for diagnosing neurological disorders, conducting sleep studies, and advancing cognitive research.

2.7.2 Advancements of EEG in technology

Technological advances in EEG have improved signal quality, resolution of space, and data analysis approaches. High-density EEG systems with more electrodes provide better brain activity localization due to their enhanced coverage of space and resolution. Signal-to-noise ratios have improved due to amplifier design and noise reduction techniques, allowing the detection of small brain activity changes. Time-frequency analysis, source localization, and connection metrics have helped characterize dynamic brain networks and interactions.

2.7.3 Applications of EEG in clinical research

EEG has neuroscience and scientific uses. EEG is needed to identify, monitor, and evaluate epilepsy seizure activity, origin areas, and treatment. Long-term active EEG monitoring records brain activity to identify epilepsy. EEG detects hypoxia and paralysis by monitoring sleep stages, awakenings, and abnormalities. Neuroscience diagnoses depression, anxiety, mental illness, cognitive performance, and therapeutic effects with EEG. EEG-based brain-computer interface (BCI) systems enable motor-disabled patients to use external equipment for neurorehabilitation and assistive technologies.

Wang et al. [51] study cognitive fatigue state classifications matched self-reported exhaustion states with 88.85% accuracy. This high accuracy shows that the wireless gadget and analysis approach can objectively diagnose construction worker mental exhaustion in real time. The exact detection of fatigue has major implications for safety management and risk reduction in dynamic work contexts like construction sites. This study shows that time sequence and time-frequency-image modifications of EEG signals can detect epileptic seizures [52]. The model uses public datasets including CHB-MIT, Bern-Barcelona, and Bonn EEG records to obtain high accuracy. In the Bonn dataset, scalogram and spectrogram images have average binary classification accuracy of 99.07% and 99.28% and multiple classification accuracy of 97.60% and 98.56%. The Bern-Barcelona and CHB-MIT datasets have 95.46% and 96.23% accuracy ratings. The model has good accuracy rates (99.21%($\pm$ 0.56), 99.50%($\pm$ 0.45), and 98.84%($\pm$ 1.58) for detecting epileptic seizure activity, localizing epileptic regions, and classifying EEG data after 8-fold cross-validation. This study detects emotions using EEG data and a novel spatial and temporal feature extraction method [53]. The proposed model classifies well using differential entropy, convolutional encoding, band attention, and long short-term memory networks. Benchmark emotion databases DEAP and SEED have stellar accuracy ratings of 85.86%, 84.27%, and 92.47%. This deep learning study used several features derived from strategically positioned EEG sensors to recognize speech-based brain wave patterns. Wavelet scattering technique reduces dimensionality and complexity

Table 2.5 Machine and deep learning publications related to EEG.

Ref.	Subjects/features	Objective	Model and accuracy (%)
Wang et al. [51]	Cognitive fatigue states, wireless gadget data	Objective diagnosis of construction worker mental exhaustion	88.85% accuracy in cognitive fatigue state classification
Varlı and Yılmaz [52]	EEG signals from CHB-MIT, Bern-Barcelona, and Bonn datasets	Detection of epileptic seizures using EEG signals	Bonn dataset: 99.07% (scalogram), 99.28% (spectrogram) accuracy
Zhang et al. [53]	DEAP and SEED benchmark emotion databases	Emotion detection using EEG data	DEAP: 85.86% accuracy, SEED: 84.27% accuracy
Abdulghani et al. [54]	EEG signals with strategically positioned sensors	Recognition of speech-based brain wave patterns	92.50% classification accuracy
Zeynali et al. [55]	EEG signals	Categorization of EEG signals using Transformer models	Ensemble model: 96.1% accuracy

and prevents deep learning algorithm overfitting [54]. By converting EEG signals into auditory commands like "up," "down," "left," and "right," the suggested approach achieves 92.50% classification accuracy. A transformer-based EEG signal categorization model using temporal and spectral data is shown in this study [55]. Use power spectral density (PSD) to transform EEG signals into the frequency domain before using transformer models to improve categorization. Deep ensemble learning techniques improve model generalization, resulting in 96.1%, 94.20%, and 93.60% ensemble model, temporal transformer, and spectral transformer accuracies. These results show that the proposed model can classify EEG signals in BCI applications reliably. Several current studies related to deep learning and deep learning technologies are shown in Table 2.5.

A comprehensive overview of current developments in EEG technology is given in Table 2.5, with a focus on the use of deep learning and machine learning techniques. The research included in the table uses EEG data for a range of neurological and medical diagnoses.

2.8 Challenges and Opportunities

The objective of current brain imaging research is to develop novel approaches for improving the accuracy and practicality of different imaging modalities. The goal of developing optically driven magnetometers (OPMs) for MEG is to provide room-temperature operation with increased sensitivity and flexibility. Another important area of development is wearable MEG systems, which are intended to make research of naturalistic behaviors easier, particularly in pediatric populations. The way that relation-aware tensor completion multitask learning (RATC-MTL) and Transformer-based models integrate machine learning and deep learning is changing the way that modalities like diffusion tensor imaging (DTI) handle data and forecast neurological disorders like Alzheimer's.

2.9 Future Directions

Using the advantages of each modality, multimodal imaging – which combines methods becoming a more thorough method of evaluating the structure and function of the brain. Utilizing imaging data to personalize treatments based on individual patient profiles, forecast disease development, and track beneficial responses, personalized medicine is growing in popularity. It is expected that future research will focus on developing hybrid models that enhance diagnostic precision and predictive skills by combining different machine learning approaches with conventional imaging analysis. Surgical and operational decision-making are about going through an innovation due to real-time imaging technologies that allow for immediate study of brain activity. Important topics for future development consist of combining and standardizing imaging data, finding new biomarkers for neurological diseases, and developing wearable brain imaging devices for long-term monitoring outside of clinical settings. These methods have the potential to significantly enhance neuroscience's capacity for diagnosis, prediction, and treatment.

2.10 Conclusion

The application of machine learning (ML) and artificial intelligence (AI) to brain imaging has created new opportunities for personalized therapy, automated image processing, and illness diagnosis. With these challenges, it is expected that future studies and technology advancements will strengthen and make AI-assisted neuroimaging even more accessible. There are still challenges, namely with regard to data standardization, legal compliance, ethical issues, and technical constraints. It will take interdisciplinary cooperation as well as ongoing improvements in imaging technology and analytical techniques to address these issues. Looking in advance, brain imaging has a bright future since cutting-edge technologies have the potential to completely transform the industry. Big data techniques, wearable imaging devices, multimodal integration, and high-field MRI are probably going to be important tools for furthering our understanding of the brain and enhancing clinical outcomes. The field of brain imaging technology has advanced significantly, offering vital resources for investigating the intricacies of the human brain. It will take sustained innovation and interdisciplinary cooperation to overcome current obstacles and open up new avenues for neuroscience and clinical practice.

Acknowledgments

We would like to express our heartfelt appreciation to the editors for their invaluable contributions to this chapter. Their expertise and meticulous feedback were instrumental in refining our ideas and ensuring the highest standards in our work. Their guidance throughout the editorial process was invaluable, providing clarity and coherence to our research findings. We are deeply thankful for their unwavering support and commitment to excellence. This chapter would not have been possible without their dedication and expertise.

References

[1] C. Yen, C.-L. Lin, and M.-C. Chiang, "Exploring the Frontiers of Neuroimaging: A Review of Recent Advances in Understanding Brain Functioning and Disorders," Life (Basel), vol. 13, no. 7, p. 1472, Jun. 2023, doi: 10.3390/life13071472.

[2] "Simultaneous electroencephalography-functional magnetic resonance imaging for assessment of human brain function - PMC." Accessed: May 18, 2024. [Online]. Available: https://www.ncbi.nlm.nih.gov/pmc/articles/PMC9371554/

[3] T. Morita, M. Asada, and E. Naito, "Contribution of Neuroimaging Studies to Understanding Development of Human Cognitive

Brain Functions," Front Hum Neurosci, vol. 10, p. 464, 2016, doi: 10.3389/fnhum.2016.00464.

[4] S. Lemm, B. Blankertz, T. Dickhaus, and K.-R. Müller, "Introduction to machine learning for brain imaging," NeuroImage, vol. 56, no. 2, pp. 387–399, May 2011, doi: 10.1016/j.neuroimage.2010.11.004.

[5] R. Monsour, M. Dutta, A.-Z. Mohamed, A. Borkowski, and N. A. Viswanadhan, "Neuroimaging in the Era of Artificial Intelligence: Current Applications," Fed Pract, vol. 39, no. Suppl 1, pp. S14–S20, Apr. 2022, doi: 10.12788/fp.0231.

[6] F. Yasmin et al., "PoxNet22: A Fine-Tuned Model for the Classification of Monkeypox Disease Using Transfer Learning," IEEE Access, vol. 11, pp. 24053–24076, 2023, doi: 10.1109/ACCESS.2023.3253868.

[7] F. Yasmin et al., "AQIPred: A Hybrid Model for High Precision Time Specific Forecasting of Air Quality Index with Cluster Analysis," Hum-Cent Intell Syst, vol. 3, no. 3, pp. 275–295, Sep. 2023, doi: 10.1007/s44230-023-00039-x.

[8] M. Hasan, F. Yasmin, and L. Deng, "Rule Mining of Early Diabetes Symptom and Applied Supervised Machine Learning and Cross Validation Approaches based on the Most Important Features to Predict Early-stage Diabetes," IJIRMPS - International Journal of Innovative Research in Engineering & Multidisciplinary Physical Sciences, vol. 11, no. 3, Jun. 2023, doi: 10.37082/IJIRMPS.v11.i3.230225.

[9] F. Yasmin et al., "GastroNet: Gastrointestinal Polyp and Abnormal Feature Detection and Classification With Deep Learning Approach," IEEE Access, vol. 11, pp. 97605–97624, 2023, doi: 10.1109/ACCESS.2023.3312729.

[10] Z. Cui, J. Yang, and Y. Qiao, "Brain MRI segmentation with patch-based CNN approach," in 2016 35th Chinese Control Conference (CCC), Jul. 2016, pp. 7026–7031. doi: 10.1109/ChiCC.2016.7554465.

[11] C. Aarthi, V. J. Ramya, P. Falkowski-Gilski, and P. B. Divakarachari, "Balanced Spider Monkey Optimization with Bi-LSTM for Sustainable Air Quality Prediction," Sustainability, vol. 15, no. 2, Art. no. 2, Jan. 2023, doi: 10.3390/su15021637.

[12] S. Bengesi, H. El-Sayed, M. K. Sarker, Y. Houkpati, J. Irungu, and T. Oladunni, "Advancements in Generative AI: A Comprehensive Review of GANs, GPT, Autoencoders, Diffusion Model, and Transformers," IEEE Access, vol. 12, pp. 69812–69837, 2024, doi: 10.1109/ACCESS.2024.3397775.

[13] M. A. Salehi et al., "Brain-based sex differences in schizophrenia: A systematic review of fMRI studies," Human Brain Mapping, vol. 45, no. 5, p. e26664, 2024, doi: 10.1002/hbm.26664.

[14] S. Amemiya, H. Takao, and O. Abe, "Resting-State fMRI: Emerging Concepts for Future Clinical Application," Journal of Magnetic Resonance Imaging, vol. 59, no. 4, pp. 1135–1148, 2024, doi: 10.1002/jmri.28894.

[15] "Increased Risk of Recurrent Stroke in Symptomatic Large Vessel Disease With Impaired BOLD Cerebrovascular Reactivity | Stroke." Accessed: May 18, 2024. [Online]. Available: https://www.ahajournals.org/doi/full/10.1161/STROKEAHA.123.044259

[16] X. Yu et al., "Blood Oxygen Level-Dependent MR Imaging of Lower Extremities in Peripheral Artery Disease and Its Correlation With Walking Performance," Journal of Magnetic Resonance Imaging, vol. 59, no. 1, pp. 192–200, Jan. 2024, doi: 10.1002/jmri.28757.

[17] M. De Simone, G. Iaconetta, G. Palermo, A. Fiorindi, K. Schaller, and L. De Maria, "Clustering Functional Magnetic Resonance Imaging Time Series in Glioblastoma Characterization: A Review of the Evolution, Applications, and Potentials," Brain Sciences, vol. 14, no. 3, Art. no. 3, Mar. 2024, doi: 10.3390/brainsci14030296.

[18] S. Agarwal, A. Raj, A. Chowdhury, G. Aich, R. Chatterjee, and K. Ghosh, "Investigating the impact of standard brain atlases and connectivity measures on the accuracy of ADHD detection from fMRI data using deep learning," Multimed Tools Appl, Jan. 2024, doi: 10.1007/s11042-023-17962-7.

[19] Y. Wu, Y. Wang, B. Qiu, and X. Wang, "Volume-Wise Decoding of Human Brain's Task fMRI via Deep Learning," in 2024 4th International Conference on Neural Networks, Information and Communication (NNICE), Guangzhou, China: IEEE, Jan. 2024, pp. 259–263. doi: 10.1109/NNICE61279.2024.10498777.

[20] F. Vedaei et al., "Deep learning-based multimodality classification of chronic mild traumatic brain injury using resting-state functional MRI and PET imaging," Front. Neurosci., vol. 17, Jan. 2024, doi: 10.3389/fnins.2023.1333725.

[21] M. K. Elakkiya and Dejey, "Novel deep learning models with novel integrated activation functions for autism screening: AutiNet and MinAutiNet," Expert Systems with Applications, vol. 238, p. 122102, Mar. 2024, doi: 10.1016/j.eswa.2023.122102.

[22] P. Dai et al., "Classification of MDD using a Transformer classifier with large-scale multisite resting-state fMRI data," Human Brain Mapping, vol. 45, no. 1, p. e26542, 2024, doi: 10.1002/hbm.26542.

[23] C. Lin et al., "Predicting suicidality in late-life depression by 3D convolutional neural network and cross-sample entropy analysis of resting-state fMRI," Brain and Behavior, vol. 14, no. 1, p. e3348, 2024, doi: 10.1002/brb3.3348.

[24] P. Kamboj, A. Banerjee, V. L. Boerwinkle, and S. K. S. Gupta, "The expert's knowledge combined with AI outperforms AI alone in seizure onset zone localization using resting state fMRI," Front. Neurol., vol. 14, Jan. 2024, doi: 10.3389/fneur.2023.1324461.

[25] J. Huang et al., "Deep learning-based diffusion tensor cardiac magnetic resonance reconstruction: a comparison study," Sci Rep, vol. 14, no. 1, p. 5658, Mar. 2024, doi: 10.1038/s41598-024-55880-2.

[26] "Frontiers | Structural volumetric and Periodic Table DTI patterns in Complex Normal Pressure Hydrocephalus—Toward the principles of a translational taxonomy." Accessed: Jun. 06, 2024. [Online]. Available: https://www.frontiersin.org/articles/10.3389/fnhum.2024.1188533/full

[27] "From images to insights: a neuroradiologist's practical guide on white matter fiber tract anatomy and DTI patterns for pre-surgical planning | Neuroradiology." Accessed: Jun. 06, 2024. [Online]. Available: https://link.springer.com/article/10.1007/s00234-024-03362-7

[28] "Evaluation of Diffusion Tensor Imaging Analysis Along the Perivascular Space as a Marker of the Glymphatic System | Radiology." Accessed: Jun. 06, 2024. [Online]. Available: https://pubs.rsna.org/doi/full/10.1148/radiol.232899

[29] "Diffusion tensor imaging and fibre tractography | Radiology Reference Article | Radiopaedia.org." Accessed: Jun. 07, 2024. [Online]. Available: https://radiopaedia.org/articles/diffusion-tensor-imaging-and-fibre-tractography-1?lang=gb

[30] "Application study of DTI combined with ASL in the crossed cerebellar diaschisis after subacute cerebral hemorrhage | Neurological Sciences." Accessed: Jun. 06, 2024. [Online]. Available: https://link.springer.com/article/10.1007/s10072-023-06908-4

[31] "Frontiers | Machine learning research based on diffusion tensor images to distinguish between anorexia nervosa and bulimia nervosa." Accessed: Jun. 06, 2024. [Online]. Available: https://www.frontiersin.org/journals/psychiatry/articles/10.3389/fpsyt.2023.1326271/full

[32] A. Tiwari, A. Singhal, S. J. Shigwan, and R. K. Singh, "Early Diagnosis of Alzheimer through Swin-Transformer-Based Deep Learning Framework using Sparse Diffusion Measures," in Proceedings of the 15th Asian Conference on Machine Learning, PMLR, Feb. 2024, pp. 1369–1384. Accessed: Jun. 06, 2024. [Online]. Available: https://proceedings.mlr.press/v222/tiwari24a.html

[33] D. Yang et al., "Diffusion Tensor and Kurtosis MRI-Based Radiomics Analysis of Kidney Injury in Type 2 Diabetes," Journal of Magnetic Resonance Imaging, vol. n/a, no. n/a, doi: 10.1002/jmri.29263.

[34] M. Y. Sikkandar et al., "Hidden Markov Model based Predicting of Alzheimer's Disease with graph cut segmentation using MR Diffusion Tensor Imaging (DTI)," Journal of Intelligent & Fuzzy Systems, vol. 46, no. 2, pp. 4277–4289, Jan. 2024, doi: 10.3233/JIFS-234613.

[35] Y. Gou, Y. Liu, F. He, B. Hunyadi, and C. Zhu, "Tensor Completion for Alzheimer's Disease Prediction From Diffusion Tensor Imaging," IEEE Transactions on Biomedical Engineering, pp. 1–12, 2024, doi: 10.1109/TBME.2024.3365131.

[36] T. Taoka, R. Ito, R. Nakamichi, T. Nakane, H. Kawai, and S. Naganawa, "Diffusion Tensor Image Analysis ALong the Perivascular Space (DTI-ALPS): Revisiting the Meaning and Significance of the Method," Magnetic Resonance in Medical Sciences, vol. advpub, 2024, doi: 10.2463/mrms.rev.2023-0175.

[37] F. Hashimoto, Y. Onishi, K. Ote, H. Tashima, A. J. Reader, and T. Yamaya, "Deep learning-based PET image denoising and reconstruction: a review," Radiol Phys Technol, vol. 17, no. 1, pp. 24–46, Mar. 2024, doi: 10.1007/s12194-024-00780-3.

[38] D. Berman et al., "Machine and deep learning models for accurate detection of ischemia and scar with myocardial blood flow positron emission tomography imaging," Journal of Nuclear Cardiology, vol. 32, p. 101797, Feb. 2024, doi: 10.1016/j.nuclcard.2024.101797.

[39] "Higher SNR PET image prediction using a deep learning model and MRI image - IOPscience." Accessed: Jun. 06, 2024. [Online]. Available: https://iopscience.iop.org/article/10.1088/1361-6560/ab0dc0/meta

[40] T. Ueda et al., "Feasibility of direct brain 18F-fluorodeoxyglucose-positron emission tomography attenuation and high-resolution correction methods using deep learning," Asia Oceania Journal of Nuclear Medicine and Biology, Jan. 2024, doi: 10.22038/aojnmb.2024.74875.1522.

[41] M. Dedja et al., "Sequential deep learning image enhancement models improve diagnostic confidence, lesion detectability, and image reconstruction time in PET," EJNMMI Phys, vol. 11, no. 1, p. 28, Mar. 2024, doi: 10.1186/s40658-024-00632-4.

[42] M.-S. Azimi et al., "Deep learning-based partial volume correction in standard and low-dose positron emission tomography-computed tomography imaging," Quant Imaging Med Surg, vol. 14, no. 3, pp. 2146–2164, Mar. 2024, doi: 10.21037/qims-23-871.

[43] "Fast and accurate MEG source localization using deep learning." Accessed: Jun. 06, 2024. [Online]. Available: https://www.spiedigitallibrary.org/conference-proceedings-of-spie/12925/129254E/Fast-and-accurate-MEG-source-localization-using-deep-learning/10.1117/12.3006914.short#_=_

[44] "Magnetoencephalogram-based brain–computer interface for hand-gesture decoding using deep learning | Cerebral Cortex | Oxford Academic." Accessed: Jun. 06, 2024. [Online]. Available: https://academic.oup.com/cercor/article/33/14/8942/7161766

[45] "Balkan Journal of Electrical and Computer EngineeringÂă≫ SubmissionÂă≫ Brain Decoding over the MEG Signals Using Riemannian Approach and Machine Learning." Accessed: Jun. 06, 2024. [Online]. Available: https://dergipark.org.tr/en/pub/bajece/issue/79097/1144279

[46] S. Hamdan, K. DuBray, J. Treutel, R. Paudyal, and K. Poudel, "Reducing MEG interference using machine learning," Machine Learning with Applications, vol. 12, p. 100462, Jun. 2023, doi: 10.1016/j.mlwa.2023.100462.

[47] Y. Fan, H. Mao, and Q. Li, "A Model-Agnostic Feature Attribution Approach to Magnetoencephalography Predictions Based on Shapley Value," IEEE Journal of Biomedical and Health Informatics, vol. 27, no. 5, pp. 2524–2535, May 2023, doi: 10.1109/JBHI.2023.3248139.

[48] "Deep learning techniques for classification of electroencephalogram (EEG) motor imagery (MI) signals: a review | Neural Computing and Applications." Accessed: Jun. 06, 2024. [Online]. Available: https://link.springer.com/article/10.1007/s00521-021-06352-5

[49] M. Jafari et al., "Emotion recognition in EEG signals using deep learning methods: A review," Computers in Biology and Medicine, vol. 165, p. 107450, Oct. 2023, doi: 10.1016/j.compbiomed.2023.107450.

[50] "Frontiers | Deep learning-based EEG emotion recognition: Current trends and future perspectives." Accessed: Jun. 06, 2024. [Online].

Available: https://www.frontiersin.org/journals/psychology/articles/10.3389/fpsyg.2023.1126994/full

[51] Y. Wang, Y. Huang, B. Gu, S. Cao, and D. Fang, “Identifying mental fatigue of construction workers using EEG and deep learning,” Automation in Construction, vol. 151, p. 104887, Jul. 2023, doi: 10.1016/j.autcon.2023.104887.

[52] M. Varlı and H. Yılmaz, “Multiple classification of EEG signals and epileptic seizure diagnosis with combined deep learning,” Journal of Computational Science, vol. 67, p. 101943, Mar. 2023, doi: 10.1016/j.jocs.2023.101943.

[53] Y. Zhang, Y. Zhang, and S. Wang, “An attention-based hybrid deep learning model for EEG emotion recognition,” SIViP, vol. 17, no. 5, pp. 2305–2313, Jul. 2023, doi: 10.1007/s11760-022-02447-1.

[54] M. M. Abdulghani, W. L. Walters, and K. H. Abed, “Imagined Speech Classification Using EEG and Deep Learning,” Bioengineering, vol. 10, no. 6, Art. no. 6, Jun. 2023, doi: 10.3390/bioengineering10060649.

[55] M. Zeynali, H. Seyedarabi, and R. Afrouzian, “Classification of EEG signals using Transformer based deep learning and ensemble models,” Biomedical Signal Processing and Control, vol. 86, p. 105130, Sep. 2023, doi: 10.1016/j.bspc.2023.105130.

3

Understanding Brain Connectivity: From Synapses to Networks

Sayeda Mayesha Yousuf[1] and Amrina Rahman[2]

[1]Department of Computer Science and Engineering,
Green University of Bangladesh
[2]Department of Management Information Systems (MIS),
University of Dhaka
E-mail: mayeshayousuf16@gmail.com; rahmanamrina187@gmail.com

Abstract

In a world where 57.8 million adults have some form of mental illness, understanding the architecture and functionalities of the human brain has become more than necessary in this age. Neurons make the fundamental units of the human brain and execute the transmission of brain signals throughout the entire body. As there is no physical connection between neurons, a junction between these neurons called a synapse makes it possible. This chapter has shed light both on the microscopic structure of neurons and synapses and the macroscopic organization of the brain network and connectivity. The discussion of the anatomy and the types of neurons enhances the understanding of the units of the nervous system. The chapter has also depicted the building blocks of communication synapses, their structure, synaptic cleft, dynamics of resting and action potential, and the roles of neurotransmitters. All these descriptions of this study have built the theoretical foundation of the concept of synaptic transmission for a reader and through this transmission brain signals are carried throughout the entire body. An intricate detail of the synaptic vesicle cycle and the key functioning vesicle proteins hold an important part of this study which further describes the functionalities of synapses in the human body. Applications of artificial intelligence (AI) in brain networks have enriched the theoretical concepts gained traditionally by experts over the years, and our study contains a dedicated section on AI implications in brain networks. Besides presenting a snapshot of the previous relevant works, the chapter has shown what lies ahead in this field.

Keywords: Brain network, neurons, synapses, connectivity, synaptic cleft, action potential, synaptic transmission.

3.1 Introduction

According to the Queensland Brain Institute, neurons function as the fundamental units of the brain and nervous system that receive sensory information from the external environment, send motor commands to the muscles, and transform and relay the electrical signals in the process. The human brain employs neurons to establish connections and perform functions and these neurons are generally classified as glia [1]. Dendrites, axons, and a cell body or soma indicate the three main parts of a neuron. Here, axons pass the signals to other neurons which are received by another neural component – dendrite. Neurons use electrical and chemical signals to transfer information in the brain and also in the whole body. The electrical signals are known as action potentials which are rapid changes in membrane voltage, dependent on the relative ratio and permeability of ions. On the other hand, these chemical messengers of the nervous system are called neurotransmitters which are the basis of communication among neurons throughout the body. They make the brain able to perform various functions with the help of the process of chemical synaptic transmission. There is a gap junction between two neurons called synaptic cleft or synaptic gap which is described later in this article. In this synaptic gap, the release of neurotransmitters converts the electrical signals that pass the axons to chemical ones [2]. Some of the neurotransmitters include acetylcholine, norepinephrine, dopamine, gamma-aminobutyric acid (GABA), glutamate, serotonin, histamine etc. Neurotransmitters transfer signals from one neuron to the next whereas action potentials carry them within a neuron. In terms of carrying signals, neurons are not connected. Synapses are the places where neurons connect and communicate with each other to transfer neural signals. There are approximately hundreds of thousands of synaptic connections in each neuron [3]. The electrical signals in the presynaptic terminal are converted into chemical signals and further converted into electrical ones in the postsynaptic terminal [4]. In this way, messages are carried across different parts of the body in the form of signals.

Another important area of the biological neural network is the structural-functional connectivity in the brain. Cognitive functions stem from the activities taking place in different regions of the brain [5]. According to Bressler, cognition is the effect of interconnected actions in distributed brain networks that function in large-scale networks, for example, networks of vision, motion, memory, and attention [6]. All these indicate that an understanding of the underlying formation of the structural connection is crucial to knowing about functional interactions in various brain areas. As per the study by Wang et al. [7], structural connectivity (SC) involves white matter anatomic connections that connect different brain areas. To understand these connections, techniques including noninvasive diffusion imaging techniques like, diffusion tensor imaging (DTI) and diffusion spectral imaging (DSI) [8], computational tractography approaches [9], and probabilistic diffusion tractography [10] are utilized. On the other hand, according to the study by Friston et al. [11], functional connectivity (FC) indicates the temporal correlations or statistical dependencies between detected neurophysiologic phenomena in spatially remote brain areas. Several linear and nonlinear procedures are used to estimate FC. However, both structural and functional networks form the brain network. Here, two important concepts in the network connectivity of the brain are nodes and edges. Nodes are groups of brain areas, while edges indicate the measurement of functional connectivity between a pair of nodes. Nodes and edges provide a useful framework to define and measure SC and FC.

This study has emphasized the microscopic neuronal and synaptic architecture while there has also been a focus on the transmissions within and between neurons. There is an attempt to shed light on some key research areas:

- The chapter has discussed the functionalities of different types of neurons in the human body.
- It has also depicted the influence of the structure of a synapse in synaptic transmission of brain signals.
- The implications of AI in recent years in the field of brain networks and the future scopes lies in brain network connectivity also hold a significant part of this study.

3.2 Related Works

Synaptic connectivity and brain networks have been an area of study by experts over the years. This section will discuss the key works that have improved this field and researchers' understanding of the related concepts. According to Shephard et al. [12], Santiago Ramón y Cajal conceptualized neurons as the building blocks of the nervous system in his work "Neuron Doctrine." His meticulous drawing laid the foundation for the understanding of synaptic connectivity. After that, there has been a significant improvement in the field of connectivity over the years with the help of advances in technology. Sigal et al. [13] depicted synaptic proteins on a nanometer scale using STORM (stochastic optical reconstruction microscopy) that unleashed the molecular structure of synapses. Besides, Tao et al. [14] conducted another study significant for the understanding of synaptic vesicle release. This study developed an illustration of synaptic vesicles and the presynaptic release machinery which were in native state at that time. There has also been work on cell-type-specific connectivity or microcircuit connectivity. The connectivity of various interneuron types located in the visual cortex was displayed by Jiang et al. [15] utilizing optogenetics, electrophysiology, and calcium imaging. Researchers have also used AI and large-scale electron microscopy combination that reconstructed a cubic millimeter of a fly brain [16]. Apart from that, Glasser et al. [17] worked on multi-modal MRI data which was sourced from the Human Connectome Project. This study built a foundation for the study of structural and functional connectivity from the insights collected from 180 parted areas of the human cortex.

J. B. Furness et al. [18] studied types of neurons for understanding the enteric nervous system. He tried to describe 14 functionally defined neuron types. In addition, he evidenced the morphologies, projections, major neurotransmitters, and physiological identification of all the neurons in the guinea pig small intestine enteric nervous system that has now been determined. Sergio E. Galindo et al. [19] represented a model SynCoPa which is the visualizing connectivity paths and synapses. This chapter introduces SynCoPa, a tool intended to close representational gaps by offering methods that enable merging precise morphological neuron representations with the synapse-level observation of neuronal connections. Another author Luiz Pessoa et al. [20] reviewed the mapping between brain structure and function. She aimed to discuss broad issues surrounding the link between structure and function in the brain that will motivate a network perspective. M. S. Bassi et al. [21] reviewed the complex processes by which synaptic plasticity modifies the neural network wiring of the brain. They studied the complexity of brain functions – including cognition, behavior, and pathology – and needed an understanding of how synaptic plasticity and network structure interact.

3.3 Anatomy of Neurons

Nerve cells called neurons carry messages throughout our body, enabling us to perform a wide range of functions including breathing, speaking, eating, walking, and thinking. The human brain contains 100 billion neurons. Approximately 10,000 additional neurons can connect to one neuron. As shown in Figure 3.1. If every single neuron in our brain were to connect with every single other neuron, the diameter of our brain would be 12.5 miles [22]. This measures the distance from one end to the other of Manhattan Island. Consequently, neurons often establish communication primarily with nearby neurons, whereas long-distance connections are exceptional rather than the norm.

All neurons have substantially the same structure. They comprise three components: a cell body (or soma), dendrites, and an axon. Although neurons share the same fundamental structure and function, it is crucial to highlight that there are some substantial distinctions between various types of neurons in terms of the spatial arrangements of the dendrites and axons.

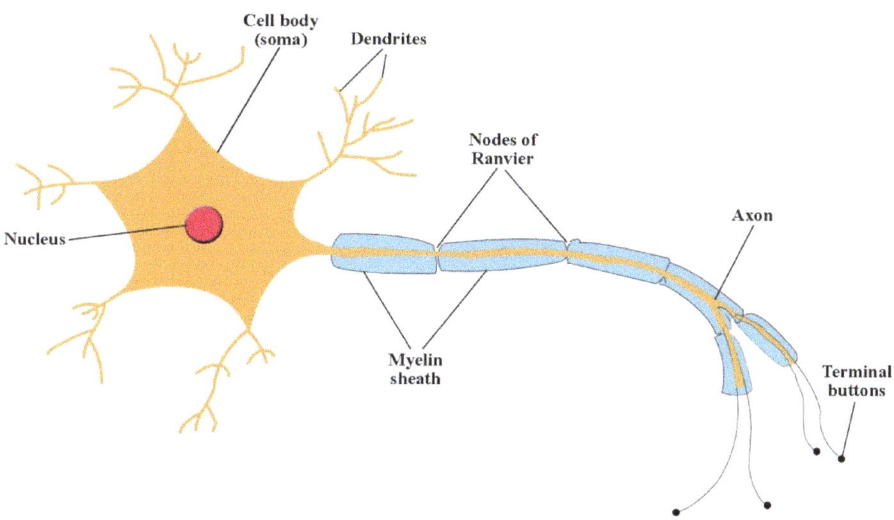

Figure 3.1 Anatomy of a neuron.

The cell body carries the nucleus and other organelles. The nucleus holds the genetic code, which is responsible for protein production. The axon is described as the lengthy neural process that ensures the transfer of information from the cell body to the nerve terminal [23]. Various branching extensions called dendrites extend from the cell body to facilitate communication between neurons.

Neurons receive information from other neurons and decide how to process it before transmitting it to other neurons. Dendrites receive signals from neurons nearby. The quantity and signature of the dendritic branch differ greatly depending on the kind of neuron. This is why the discussion of different types of neurons is the next subsection of this chapter. Besides shedding light on individual structural characteristics, it also describes the different functionalities offered by them.

3.4 Types of Neurons

The human nervous system comprises thousands of distinct types of neurons, each with unique structures and functions. Despite this diversity, many neurons can be categorized based on the direction in which they transmit nerve impulses, providing a functional classification. Therefore, they can be divided into three fundamental classes according to function.

3.4.1 Sensory neurons

There are five fundamental senses – taste, touch, hearing, sight, and smell [24]. From their surroundings, these neurons receive sensory data. The structural composition of sensory neurons comprises a cell body housing essential organelles such as the nucleus, dendrites, and axons. Additionally, these neurons may exhibit specialized terminations, such as free nerve endings or encapsulated receptors, which facilitate the detection of diverse stimuli encompassing touch, pressure, temperature, pain, and chemical signals [25].

3.4.2 Motor neurons

Motor neurons are primarily found in the spinal cord and brainstem, where they originate, but they also extend into various regions of the brain, including the motor cortex. The motor cortex, located in the outer layer of the brain (cerebral cortex), plays a crucial role in the planning, control, and execution of voluntary movements. While the majority of motor neurons reside in the spinal cord and brainstem, their connections extend to the motor cortex, forming the neural circuitry responsible for motor control and coordination. These neurons facilitate the transmission of information from the central nervous system to various tissues and organs, thereby enabling motor function.

3.4.3 Interneurons

Interneurons are a type of nerve cell, predominantly situated within the integrative centers of the central nervous system. Their axons and dendrites are typically localized within specific regions of the brain [26]. Interneurons are functional before vertically moving pyramidal neurons, and they create the first functional connections in the developing hippocampus [27]. In the body, these neurons comprise the vast majority of neurons. They act as intermediaries, passing information from the motor to sensory neurons. Learning, memory, and planning are all significantly impacted by them.

The network of neurons builds the foundation of the entire nervous system but there is no specific physical connection between two neurons. To ensure the transmission of signals from the brain, neurons make the utilization of synapses which ensure seamless transmission. The next section scrutinizes the structure of a synapse and how synaptic connectivity ensures the transmission of brain signals.

3.5 What Are Synapses?

Synapses are the connection between neurons and the brain. Synapses are the points at which neurons can communicate with one another. In the context of information flow, the two neurons involved in a synapse are called the presynaptic and postsynaptic neurons. The presynaptic neuron is the stage before it's going to initiate the process.

On the other hand, the postsynaptic neuron will receive the information. A gap known as the synaptic gap or synaptic cleft must be crossed whenever an action potential or nerve signal travels along an axon and reaches the end or distal component of that axon called the axon terminal. To accomplish it, we must go through a process known as synaptic transmission. Although the small connection that passes information from one neuron to the next may appear trivial, many intricate processes can occur between those cells. There are thousands of synapses in a typical neuron [28]. Therefore, from both the computational and the visualization perspectives, it is challenging to represent and interactively visualize the intricate morphologies and connectome features of nontrivial neural circuits [29].

3.5.1 Synaptic cleft

Now, (detail about synaptic gap). Neurons are not physically connected. A gap junction between two neurons is called a synaptic gap or synaptic cleft. The dimensions of subcellular components, such as the length of the synaptic cleft, are expressed in nanometers (1 nm = 10^{-9} m), whereas the units of measurement for cellular components are centimeters (1 cm = 10^{-2} m) and micrometers (1 μm = 10^{-6} m) [30]. The signal passes through the synaptic gap from the presynaptic neuron's axon terminals to the postsynaptic neuron's dendrite. Figure 3.2 illustrates the Synaptic Cleft.

3.5.2 Electrical signaling and action potential

The main concern is how to control the neuron's action, or how to stimulate it. Acetylcholine is a good topic to discuss here. A neurotransmitter that

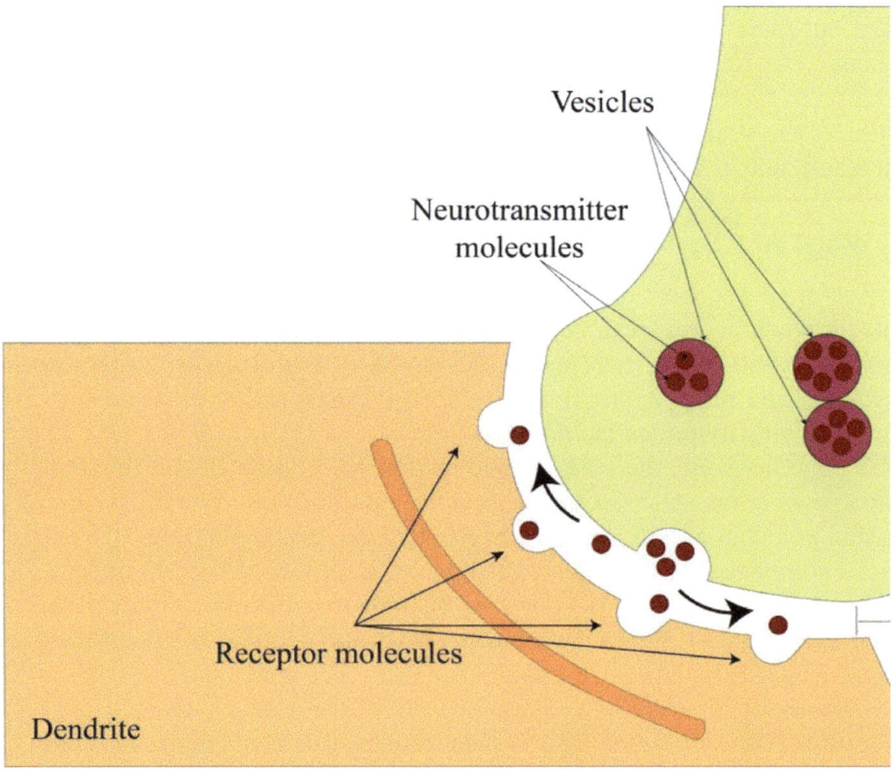

Figure 3.2 Synaptic cleft.

regulates our movement is acetylcholine. It tells our muscles to contract when we want to reach for an object or to lean over.

However, the neurons are shielded by the membrane. The cell membrane that envelops each neuron acts as a selective barrier, regulating the entry and exit of various substances as shown in Figure 3.3. Specific protein molecules embedded within the membrane function as gatekeepers, permitting certain substances to pass through under particular conditions. Among these substances are charged ions, such as sodium (Na^+) and potassium (K^+) ions, which play crucial roles in neuronal function and signal transmission.

There is usually a resting potential of −70 mV across the membrane because of the balance of ions inside and outside the cell. The inside of the cell is negative compared to the outside. The generation of an action potential relies heavily on voltage-gated ion channels. Because they belong only to axons, only axons can generate action potentials. There must be some sequence of events in action potential.

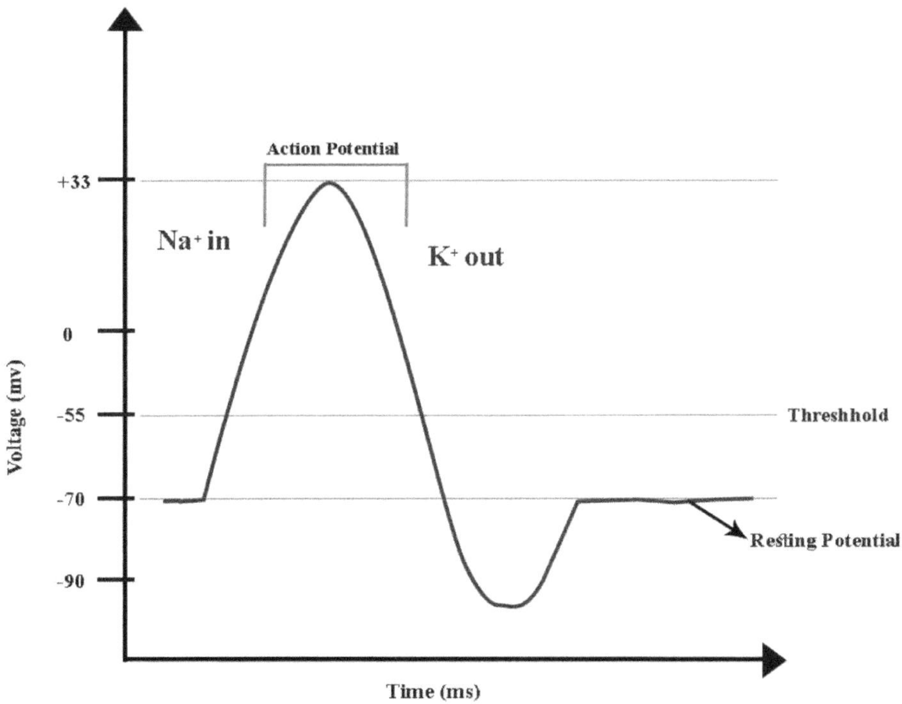

Figure 3.3 Na^+ in & K^+ out curves of action potential, resting potential, threshold.

1. The voltage-gated Na^+ channels initiate to open when a passive current of sufficient amplitude flows through the axon membrane.
2. Upon opening the channel, Na+ ions can enter the cell, resulting in a reduction of the negative potential typically present on the inside. This phenomenon is referred to as depolarization of the cell, which is the next phase of action potential.
3. The cell's negative potential is restored by K+ moving out through voltage-gated K+ channels and Na+ moving out through voltage-gated Na+ channels.
4. Hyperpolarization – the interior is more negative than at rest – occurs for a short duration. This increases the difficulty of the axon to depolarize right away, therefore, stopping the action potential from going backward.

An action potential in one section of the axon triggers the activation of neighboring voltage-sensitive Na^+ channels, causing the action potential to propagate throughout the whole length of the axon, originating at the cell body and terminating at the axon terminal. When an axon is myelinated, the conduction of the action potential along it can be accelerated. Myelin, a multilamellar membrane, is synthesized by oligodendrocytes within the central nervous system (CNS) and Schwann cells in the peripheral nervous system (PNS). Its primary function is to enwrap axons in segments, which are demarcated by the nodes of Ranvier [31].

3.5.3 Resting membrane potential

The electrical potential difference across the plasma membrane of a non-excited cell is known as its resting membrane potential. A cell's intracellular potential as compared to its extracellular potential has long been used as a measure of the electrical potential gradient across its membrane [32–33]. When a neuron is at rest, its membrane displays concentration gradients for Na^+ and K^+. The resting potential is the result of the separation of charges caused by ions moving down their gradients via channels. Every cell in the body has a gradient of electrical energy across its plasma membrane, which affects the flow of a wide range of nutrients into and out of cells. It also plays a crucial role in the movement of salt (and therefore water) across cell membranes and between organ-based compartments, is a crucial component of the signaling processes linked to the coordinated movements of cells and organisms, and is ultimately the foundation of all cognitive processes [34].

3.5.4 Chemical signaling and the postsynaptic neurons

When the action potential reaches the axon terminal, an electrical signal triggers a series of events that result in the release of neurotransmitters into the synaptic cleft. Neurotransmitters are bound to protein receptors in the membrane of postsynaptic neurons. Several receptors function as ion channels that are activated by neurotransmitters, which should not be mistaken for the voltage-gated ion channels found in the axon. This establishes a localized flow of sodium ions (Na^+), potassium ions (K^+), or chloride ions (Cl^-), which generates the synaptic potential. Within excitatory synapses in the brain, there are specialized receptors located in the postsynaptic membrane that are prepared to react to the release of the neurotransmitter glutamate from the presynaptic terminal [35]. When stimulated, these glutamate receptors initiate various metabolic processes that transmit messages to the postsynaptic neuron.

3.5.5 Synaptic vesicles

Synaptic vesicles serve a crucial function in synaptic transmission. They are considered essential organelles participating in synaptic activities such as absorption, storage, and stimulus-dependent neurotransmitter release [36]. Moreover, some proteins connected to the synaptic vesicle membrane regulate axonal transit and recycling. A variety of membrane integral proteins are found in synaptic vesicles. The membrane topography is primarily dependent on one or several trans-membrane domains ascribed to these proteins by molecular research. Some of the key proteins found in synaptic vesicles are synaptotagmins, synaptobrevins, synaptophysin, synapsins, vesicular monoamine transporters (VMATs), and acidification transporters. Figure 3.4 illustrates the synaptic vesicle cycle.

Synaptotagmin is a transmembrane protein found in synaptic vesicles, which are small membrane-bound organelles that store and release neurotransmitters at synapses. Its discovery marked a significant advancement in understanding the molecular mechanisms underlying neurotransmitter release and synaptic transmission [37]. Through its two C2 domains, the synaptic vesicle Ca^{2+} sensor synaptotagmin binds Ca^{2+} to initiate membrane contacts. Certain conditions for synaptotagmin activity remain to be determined, aside from membrane insertion by the C2 domains [38]. The highly conserved protein machinery known as the SNARE complex regulates the fusion of synaptic vesicles with the plasma membrane [39]. Neurons have developed additional mechanisms beyond the SNARE complex to regulate the precise timing and location of membrane fusion events. Among these regulatory mechanisms, the synaptotagmin (Syt) family of calcium sensors stands out as a crucial player [40].

Synaptobrevin is a tiny integral membrane protein that possesses a SNARE motif, known as a vesicular SNARE (vesicular SNARE, sometimes known as R-SNARE due to a conserved arginine in the middle of the SNARE motif) [41]. The two rapid synapse-specific membrane trafficking events that synaptobrevin is required for are the rapid exocytosis that releases neurotransmitters and the rapid endocytosis that allows for the reuse of synaptic vesicles [42]. This process is fundamental for neuronal communication and the proper functioning of the nervous system.

Synaptophysin is a glycoprotein that is found in the presynaptic vesicles of neurons and analogous vesicles of the adrenal medulla. It is an integral membrane glycoprotein with a molecular weight of 38,000 [43]. The presence of synaptophysin in analogous vesicles of the adrenal medulla suggests its broader involvement in secretory processes beyond the central nervous system. This dual role highlights the versatility and importance of synaptophysin

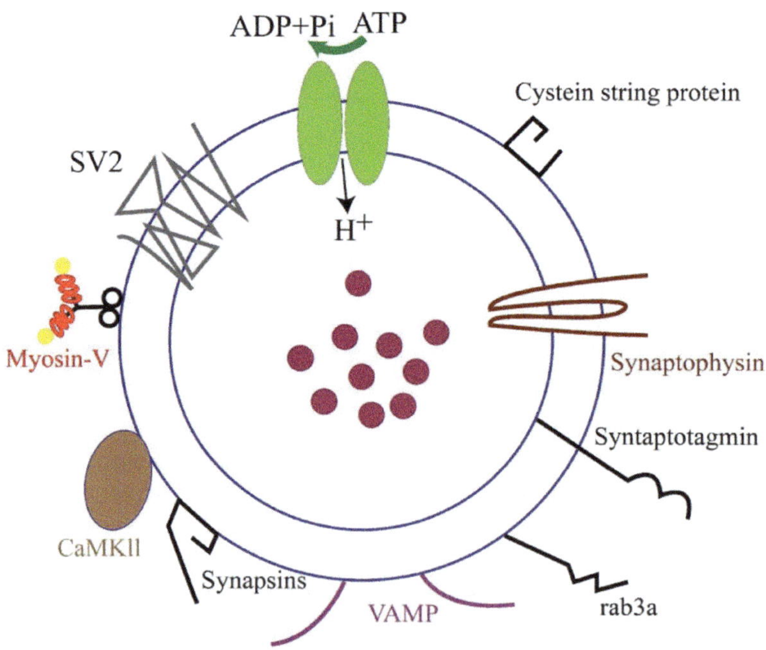

Figure 3.4 The synaptic vesicle cycles.

across various physiological contexts, from neuronal communication to endocrine regulation.

Synapsins are prominent presynaptic proteins that are highly abundant on synaptic vesicles and play a significant role in facilitating synaptic communication [44]. The synapsins are a group of proteins that have been consistently linked to the regulation of neurotransmitter release at synapses. At the maximum degree of expression, synapsins I and II are found as well. They are specifically restricted to presynaptic terminals, whereas synapsin III is expressed in inadequate quantities and is also detected in cell bodies and growth cones [45].

Vesicular monoamine transporters (VMATs) play a critical role in the regulation of monoamine neurotransmitters within the central nervous system. These transporters are primarily responsible for the packaging of cytosolic monoamines, such as dopamine, serotonin, and norepinephrine, into synaptic vesicles within monoaminergic neurons [46].

Acidification of synaptic vesicles is essential for neurotransmitter loading through the synaptic vesicle cycle. Synaptic vesicles are tiny, membrane-bound spaces found in nerve terminals. They contain neurotransmitters, which are chemical messengers that send signals across synapses from one neuron to another or from a neuron to a target cell [47]. The loading of neurotransmitters into synaptic vesicles is a multistep process that includes the acidity of the vesicle interior. This acidification is principally mediated by a proton pump called the vacuolar-type H+-ATPase (V-ATPase), which actively pumps protons (H^+) into the vesicle lumen, increasing its acidity.

At this point, it is also important to know about anatomical and functional synaptic connectivity as it helps to understand the roles of synapses in performing brain functions. Disruptions in this connectivity can result in different neurological disorders, for example, autism, schizophrenia, and Alzheimer's disease.

3.5.6 Anatomical and functional synaptic connectivity

Neural connectivity in the human brain helps experts to better understand the brain architecture, its information processing, and other functionalities. The entire neuronal network is connected by special junctions called "Synapses" and they can be both electrical and chemical, passing the signals throughout the body. According to Jirsa et al. [48], synaptic connectivity can be defined as the ensemble of direct chemical and electrical connections between neurons. Synaptic connectivity can be viewed from different perspectives –

anatomical, structural, and functional synaptic connectivity to have a better understanding.

3.5.6.1 Anatomical synaptic connectivity

Anatomical synaptic connectivity involves physical connections between neurons in the brain that are formed by synapses. It ensures the proper signals and responses from the brain necessary for the body. He et al. [49] first showed a "small world" architecture in the anatomical network of the human brain. The study involved an assumption that speculated that covariation in cortical thickness would mirror the existence of anatomical connections between brain regions. At least 104 pairs of thickness correlations were found significant in terms of anatomical connectivity. According to the measurement employing human diffusion imaging, short- and long-range connections were found present in intra- and inter-hemispheric areas. Moreover, maximum significant connections seemed to have shorter anatomical distances where there were also some long-range connections. However, methods including tract-tracing techniques, transmission electron microscopy (TEM), diffusion tensor imaging (DTI) and tractography, viral tracing circuit mapping, etc. have been utilized to analyze anatomical synaptic connectivity.

3.5.6.2 Functional synaptic connectivity

Functional connectivity provides the measurement of functional communication that occurs between specific brain areas. It indicates the statistical dependencies between different brain activity patterns [50]. This study demonstrated that functional networks had features of high clustering and short path lengths similar to the anatomical networks, based on data regarding epileptiform activity in cortical areas of the macaque cortex. Employing different analytical methods, the cortical network of functional interactions was not found to be homogeneous. Rather it demonstrated a distinct classification into functional assemblies of mutually interacting regions. The assemblies recommended that visual, somatomotor, and orbito-temporo-insular systems in the cortex could be considered as individual divisions but unlike that, motor and somatosensory regions were found related to each other [51].

3.5.7 Synaptic transmission

Nerve cells communicate with one another in two ways, known as rapid and slow synaptic transmission [52]. Imagine the bustling streets of a city, where information flows in different lanes, each serving a distinct purpose in keeping the city alive. In the realm of the brain, nerve cells, or neurons, communicate through two distinct pathways akin to these lanes: speedy and slow synaptic transmission.

3.5.7.1 Rapid synaptic transmission

Rapid synaptic transmission is analogous to the city's express lane, providing for quick and efficient communication across neurons. Neurotransmitters travel across synapses at breakneck speed, much like cars on a highway. This form of transmission is distinguished by its high speed and precision, making it perfect for quick responses and reactive activities. Consider it the fast lane for urgent messages, with quick coordination of motions and sensory impressions.

3.5.7.2 Slow synaptic transmission

As shown in Figure 3.5, slow synaptic transmission, on the other hand, takes a scenic route, allowing for a more leisurely exchange of information. Unlike its faster predecessor, this pathway employs more complex signaling systems, simulating a winding road through scenic scenery. While slower,

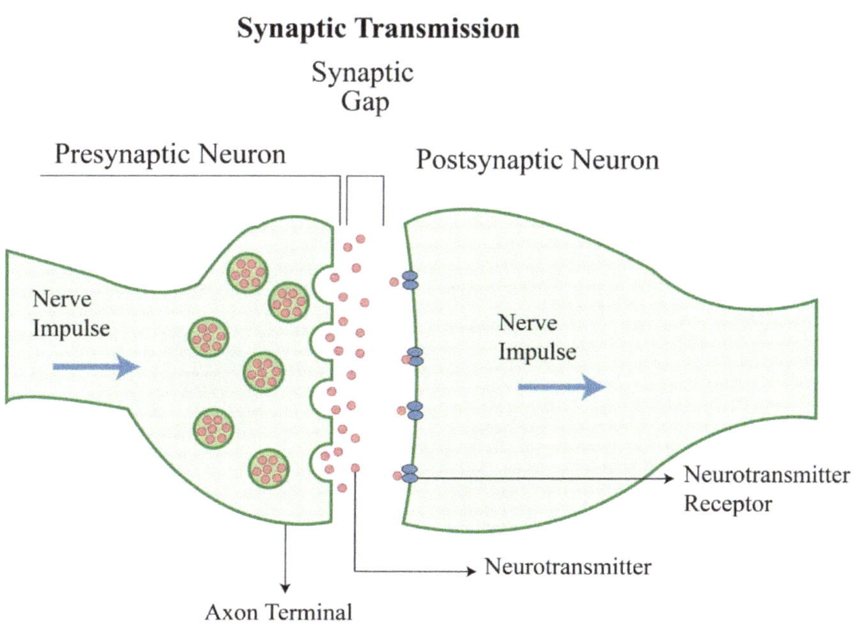

Figure 3.5 Receptor molecules, synapse vesicles, and synapse transmission.

this transmission style allows for more subtle communication, which aids in activities such as learning, memory formation, and emotional reactions. It is the path of deep contemplation, where thoughts wander and connections emerge over time.

In the central nervous system, the postsynaptic reaction to an action potential is varied. Neurotransmitter release is a probabilistic approach and the postsynaptic response to neurotransmitter release has diverse timing and magnitude. Synaptic transmission is the consequence of a series of responsive and diffusive molecular processes (such as conformational changes, binding events, and diffusion) that exhibit stochastic features on the molecular scale [53].

Experts are deploying modern technologies like artificial intelligence (AI) to gain more insights about synaptic transmission and functionalities. Using AI models, brain networks can be simulated and classified, which also opens doors for the personification of these networks.

3.6 AI in Brain Network

AI techniques can analyze mapping data of connectivity which can generate new information employing machine learning and deep learning. AI algorithms are utilized to extract significant features from the data of multi-modal imaging which unleashes information about brain network connectivity.

These models and techniques can also create notable impacts to gain a deeper understanding of the functional and structural connectivity of the

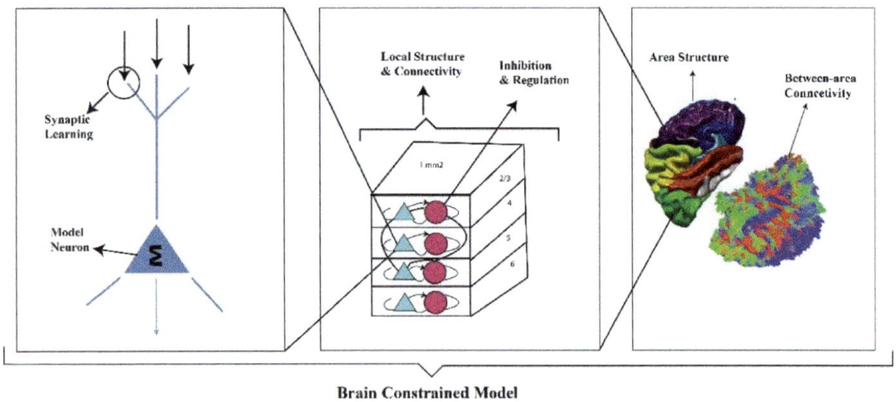

Figure 3.6 Brain activity and behavior during perceptual and cognitive activities can be simulated, explained, and predicted.

brain. Based on the previous research works, the following areas can be explored as they hold immense possibilities.

3.6.1 Brain network simulation using generative adversarial networks (GANs)

To simulate structural and functional connectivity data in brain networks, different generative AI models are quite helpful. For example, as a deep learning technique, experts use generative adversarial networks (GANs) to address computer vision challenges in the field of neuroscience. According to Laino et al. [54], GANs have several applications in the field of neuroradiology, for example, image reconstruction, image synthesis, image-to-image and cross-modality synthesis, etc. Two neural networks – a generator G and a discriminator D are employed in this method while a generator network produces new data instances and a discriminator network tries to differentiate between actual data from the training set and artificial or generated data by the generator. Upon simultaneous training, both of the networks get improved. The generator network proceeds towards the goal of producing indistinguishable data or images with each feedback and the discriminator one gets better at identifying real and fake data.

Among many GANs applications, one notable implication is image-to-image translation and cross-modality synthesis. As per the study by Armanious et al. [55], GAN can be used in producing synthetic CT images. The deep learning model employed in this study, utilized the training data of non-attenuation corrected PET data and the related CT data. These CT data were further used for PET AC. Color-coded different maps and visual scrutinies were used for the qualitative assessment of data generated by CT and PET. The entire technique resulted in acceptable accuracy in terms of PET measurement.

3.6.2 Applications related to personalized brain network models comparing SVR, GPR, and DNN models

Personalized models can consider the individuality in network connectivity. Personalized disease risk prediction, cognitive performance, etc. can be better understood using AI models. A study conducted by Liem et al. [56] showed the use of single and multimodal brain-imaging data (MRI, DTI, rs-fMRI) to improve age prediction. Four ML models including ridge regression, support vector regression (SVR), Gaussian process regression (GPR), and deep neural

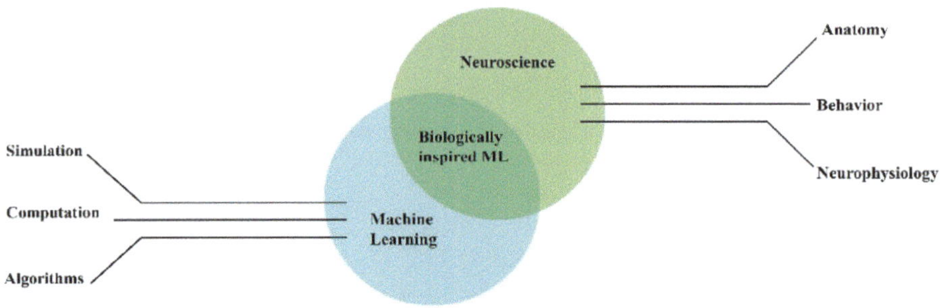

Figure 3.7 Biologically inspired systems.

networks (DNN) were compared considering chronological age as the output variable. The prediction of age led to the considered brain age. The sum of squared error was reduced by ridge regression. Besides, with the help of the radial basis function kernel, a nonlinear relationship between image variables and age was developed. Additionally, GPR measured a likelihood function and determined the covariance of an earlier distribution. Finally, DNN was employed to design single and multimodal features. The study discovered that the difference between predicted age and chronological age leads to cognitive impairment. In another study, Payan et al. [57] conducted a study that employed sparse autoencoders and 3D convolutional neural networks that predicted the Alzheimer's disease condition of a patient. In this case, the models used MRI data of the brain.

3.6.3 Biomarker discovery

Biomarkers help to define specific physiological reactions to certain conditions or diseases using measurements related to specific happenings in a cell. They can serve in early disease prediction or prevent the health from further damage by sending warnings. For example, high cholesterol levels are often a common biomarker indicating risks associated with heart diseases whole serum LDL indicates cholesterol and blood pressure. P53 genes and MMPs work as tumor indicators for cancer. According to the National Institute of Environmental Health Sciences, biomarkers function in measuring environmental or external chemicals and indicate negative effects on health and susceptibility to health dangers due to environmental exposure.

Several studies have been conducted on brain signatures which are important biomarkers. Woo et al. [58] have conducted a study that developed such brain signatures or biomarkers using multivariate pattern-recognition approaches. In developing brain signatures, broad exploration, demonstration of diagnostic accuracy, characterization, and surrogacy are important stages. The exploration stage tries different ML models to identify the best one for the problem. The most potential models require different tests to show diagnostic accuracy. These biomarkers are then required to generate their findings in terms of different samples and settings in the characterization stage. This ultimately leads to the implications of these biomarkers as surrogate measures. The chapter also evaluated and increased the value of these biomarkers. This can ultimately lead to early diagnosis of neurological disorders, assessment of the cognitive states, and personalized analysis. Moreover, these studies have utilized features related to network connectivity to classify persons with Alzheimer's disease, autism spectrum disorder, or schizophrenia.

Integration of knowledge derived from neuroscience and the application of AI models can create wonders in brain connectivity. This can ultimately lead to solutions for neurological disorders. There also lies a scope for better utilization of the existing neurological to gain revolutionary insights.

3.7 Future Scope

The future opportunities of this study lie in an advanced understanding of the brain structure and functions, cognitive processes, and neurological disorders. There can be an increased opportunity for better integration of connectivity data, ranging from the synaptic level to the entire brain network level. These studies [59–60] have portrayed implications of different methods including electron microscopy, super-resolution imaging, and diffusion MRI in this case. In the future, there are scopes to develop computational models to gain more insights about neural connectivity. Also, there should be more work on the dynamic connectivity of the brain as, historically, there has been an inclined focus towards the static one. On top of that, different neurological and psychiatric disorders, for example, autism spectrum disorder, Alzheimer's disease, etc. can stem from unusual brain connections [61–62]. To ensure early detection of these diseases, determining the connectivity biomarkers is very crucial. This can also contribute to designing the proper treatment plans for such neurological disorders. The correlation between brain connectivity and aging is another potential area to work on as it can unleash valuable information about neurological disorders. All these point to future research scopes related to our study.

3.8 Conclusion

In this article, we have tried to shed light on the anatomical, structural, and functional connectivity of synapses which play important roles in our brain activities. Synapses are the basic units of communication in the nervous system, allowing neurons to connect and support all cognitive processes and activities. The complicated network of synapses serves as the foundation of brain connectivity, allowing information to be transmitted via both rapid and slow synaptic transmission channels. Nerve cells communicate with one another via synapses, establishing complex neuronal circuits that support a variety of activities, including sensory perception, motor control, memory creation, and emotional regulation. The dynamic nature of synaptic connections enables plasticity, or the brain's ability to adapt and rearrange in response to experiences and learning. Understanding synapse function and brain connectivity is critical for unlocking the secrets of the brain and treating neurological illnesses caused by synaptic malfunction. Advances in neuroscience continue to shed information on the complexity of synapse biology and its consequences for brain function and malfunction, paving the door for novel therapies and treatments. Synapses, then, are more than simply places where neurons come into touch; they are the links that link the enormous terrain of the mind, arranging the symphony of neuronal activity that characterizes human thought and behavior. However, AI connection in brain networking is like orchestrating a symphony where artificial intelligence conducts the intricate dance of neurons. It tunes into the brain's rhythms, decoding its mysteries to compose a melody of understanding. Through this harmonious collaboration, neural interfaces are fine-tuned to perfection, facilitating seamless communication between mind and machine. This symphony doesn't just resonate within the realms of healthcare and neuroscience; it crescendos into new frontiers, orchestrating innovations that redefine what's possible in technology and beyond.

Acknowledgments

We would like to express our gratitude to the editors for their tremendous help in writing this chapter. Their guidance was crucial in determining the precision and transparency of this chapter.

References

[1] "What is a neuron?," Queensland Brain Institute - University of Queensland https://qbi.uq.edu.au/brain/brain-anatomy/what-neuron (accessed Jun. 13, 2024).

[2] "What are neurotransmitters?," Queensland Brain Institute - University of Queensland, https://qbi.uq.edu.au/brain/brain-functions/what-are-neurotransmitters (accessed Jun. 13, 2024).

[3] M. J. Caire, "Physiology, synapse," StatPearls [Internet]., https://www.ncbi.nlm.nih.gov/books/NBK526047/ (accessed Jun. 13, 2024).

[4] G. M. Shepherd, "The synaptic organization of the brain," Google Books, https://books.google.com/books/about/The_Synaptic_Organization_of_the_Brain.html?id=xlCqS0EODUoC (accessed Jun. 13, 2024).

[5] S. Zeki and S. Shipp, "The functional logic of cortical connections," *Nature*, vol. 335, no. 6188, pp. 311–317, Sep. 1988.

[6] S. L. Bressler and V. Menon, "Large-scale brain networks in cognition: Emerging methods and principles," *Trends in Cognitive Sciences*, vol. 14, no. 6, pp. 277–290, Jun. 2010.

[7] Z. Wang, Z. Dai, G. Gong, C. Zhou, and Y. He, "Understanding structural-functional relationships in the human brain," *The Neuroscientist*, vol. 21, no. 3, pp. 290–305, June. 2014.

[8] D. Le Bihan, "Looking into the functional architecture of the brain with diffusion MRI," *Nature Reviews Neuroscience*, vol. 4, no. 6, pp. 469–480, Jun. 2003.

[9] T. E. Conturo, N. F. Lori, T. S. Cull, E. Akbudak, A. Z. Snyder, J. S. Shimony, R. C. McKinstry, H. Burton, and M. E. Raichle, "Tracking neuronal fiber pathways in the living human brain," *Proceedings of the National Academy of Sciences*, vol. 96, no. 18, pp. 10422–10427, Aug. 1999.

[10] G. J. M. Parker and D. C. Alexander, "Probabilistic anatomical connectivity derived from the microscopic persistent angular structure of cerebral tissue," *Philosophical Transactions of the Royal Society B: Biological Sciences*, vol. 360, no. 1457, pp. 893–902, May 2005. doi: 10.1098/rstb.2005.1639

[11] K. J. Friston, "Functional and effective connectivity in neuroimaging: A synthesis," *Human Brain Mapping*, vol. 2, no. 1–2, pp. 56–78, Jan. 1994.

[12] G. M. Shepherd, *Foundations of the Neuron Doctrine*. New York: Oxford University Press, 2023.

[13] Y. M. Sigal, C. M. Speer, H. P. Babcock, and X. Zhuang, "Mapping synaptic input fields of neurons with super-resolution imaging," *Cell*, vol. 163, no. 2, pp. 493–505, Oct. 2015.

[14] T. E. Conturo, N. F. Lori, T. S. Cull, E. Akbudak, A. Z. Snyder, J. S. Shimony, R. C. McKinstry, H. Burton, and M. E. Raichle, "Tracking neuronal fiber pathways in the living human brain," in Proceedings of the National Academy of Sciences, vol. 96, no. 18, pp. 10422-10427, Aug. 31, 1999.

[15] X. Jiang, H. Shen, Y. S. M. Li, C. Xu, L. N. Jiang, Y. Yang, A. M. A. Mancarci, M. Li, Z. Wang, S. Lv, C. Li, H. Zhou, W. Gu, D. Zhang, H. Shen, L. Hong, G. A. G. Ascoli, and Y. J. Weng, "Principles of connectivity among morphologically defined cell types in adult neocortex," *Science*, vol. 350, no. 6264, Nov. 2015.

[16] S. Dorkenwald, M. Cossell, A. Iacaruso, T. Chartrand, A. C. S. Brown, M. Chung, ... M. Hafting, "Binary and analog variation of synapses between cortical pyramidal neurons," Dec. 2019.

[17] M. F. Glasser, T. S. Coalson, E. C. Robinson, C. D. Hacker, J. Harwell, E. Yacoub, K. Ugurbil, J. Andersson, C. F. Beckmann, M. Jenkinson, S. M. Smith, and D. C. Van Essen, "A multi-modal parcellation of human cerebral cortex," *Nature*, vol. 536, no. 7615, pp. 171–178, Jul. 2016.

[18] J. B. Furness, "Types of neurons in the enteric nervous system," *Journal of the Autonomic Nervous System*, vol. 81, no. 1–3, pp. 87–96, Jul. 2000.

[19] S. E. Galindo, P. Toharia, O. D. Robles, and L. Pastor, "Syncopa: Visualizing connectivity paths and synapses over detailed morphologies," *Frontiers in Neuroinformatics*, vol. 15, Dec. 2021.

[20] L. Pessoa, "Understanding brain networks and brain organization," *Physics of Life Reviews*, vol. 11, no. 3, pp. 400–435, Sep. 2014.

[21] M. Stampanoni Bassi, E. Iezzi, L. Gilio, D. Centonze, and F. Buttari, "Synaptic plasticity shapes brain connectivity: Implications for network topology," *International Journal of Molecular Sciences*, vol. 20, no. 24, p. 6193, Dec. 2019.

[22] C. S. Wilson and A. A. Mongin, "Cell Volume Control in healthy brain and neuropathologies," *Cell Volume Regulation*, pp. 385–455, 2018.

[23] D. Debanne, E. Campanac, A. Bialowas, E. Carlier, and G. Alcaraz, "Axon physiology," *Physiological Reviews*, vol. 91, no. 2, pp. 555–602, Apr. 2011.

[24] Tess, "All about neurons," TESS Research Foundation, https://www.tessresearch.org/neurons (accessed Jun. 13, 2024).
[25] S. Deuchars and J. Deuchars, "Neuroscience-a novelty for the nervous: Neuroscience (1997). Purves D, Augustine GJ, Fitzpatrick D, Katz LC, Lamantia A-S, McNamara Jo (eds). Sunderland, MA: Sinauer Associates, inc. 562 pp.," *BioEssays*, vol. 20, no. 10, pp. 871–872, Dec. 1998.
[26] T. Freund and S. Kali, "Interneurons," Scholarpedia, http://www.scholarpedia.org/article/Hippocampal_interneurons (accessed Jun. 9, 2024).
[27] Y. Ben-Ari, I. Khalilov, A. Represa, and H. Gozlan, "Interneurons set the tune of developing networks," Trends in Neurosciences, vol. 27, no. 7, pp. 422–427, Jul. 2004.
[28] "Neurons: What are they and how do they work?," Medical News Today, https://www.medicalnewstoday.com/articles/320289 (accessed Jun. 9, 2024).
[29] S. E. Galindo, P. Toharia, O. D. Robles, and L. Pastor, "Syncopa: Visualizing connectivity paths and synapses over detailed morphologies," *Frontiers in Neuroinformatics*, vol. 15, Dec. 2021.
[30] "Synaptic cleft," Synaptic Cleft - an overview | ScienceDirect Topics, https://www.sciencedirect.com/topics/mathematics/synaptic-cleft (accessed Jun. 9, 2024).
[31] S. Poliak and E. Peles, "The local differentiation of myelinated axons at Nodes of Ranvier," *Nature Reviews Neuroscience*, vol. 4, no. 12, pp. 968–980, Dec. 2003.
[32] S. H. Wright, "Generation of resting membrane potential," *Advances in Physiology Education*, vol. 28, no. 4, pp. 139–142, Dec. 2004.
[33] P. Enyedi and G. Czirják, "Molecular background of leak K+currents: Two-pore domain potassium channels," *Physiological Reviews*, vol. 90, no. 2, pp. 559–605, Apr. 2010.
[34] S. H. Wright, "Generation of resting membrane potential," *Advances in Physiology Education*, vol. 28, no. 4, pp. 139–142, Dec. 2004.
[35] M. Sheng and M. J. Kim, "Postsynaptic signaling and plasticity mechanisms," *Science*, vol. 298, no. 5594, pp. 776–780, Oct. 2002.
[36] W. Volknandt, "The synaptic vesicle and its targets," *Neuroscience*, vol. 64, no. 2, pp. 277–300, Jan. 1995.
[37] M. Geppert, B. T. Archer, and T. C. Südhof, "Synaptotagmin II. A novel differentially distributed form of synaptotagmin," *Journal of Biological Chemistry*, vol. 266, no. 21, pp. 13548–13552, Jul. 1991.

[38] Z. Guan, M. Del Vescovo, S. V. Dravid, and T. C. Südhof, "A synaptotagmin suppressor screen indicates SNARE binding controls the timing and Ca2+ cooperativity of vesicle fusion," *eLife*, vol. 6, Sep. 2017.

[39] T. C. Südhof and J. E. Rothman, "Membrane fusion: Grappling with snare and SM proteins," *Science*, vol. 323, no. 5913, pp. 474–477, Jan. 2009.

[40] E. R. Chapman, "How does Synaptotagmin Trigger neurotransmitter release?," *Annual Review of Biochemistry*, vol. 77, no. 1, pp. 615–641, Jun. 2008.

[41] "Synaptobrevin," Synaptobrevin - an overview | ScienceDirect Topics, https://www.sciencedirect.com/topics/neuroscience/synaptobrevin (accessed Jun. 9, 2024).

[42] F. Deák, S. Schoch, X. Liu, T. C. Südhof, and E. T. Kavalali, "Synaptobrevin is essential for fast synaptic-vesicle endocytosis," *Nature Cell Biology*, vol. 6, no. 11, pp. 1102–1108, Oct. 2004.

[43] B. Wiedenmann, W. W. Franke, C. Kuhn, R. Moll, and V. E. Gould, "Synaptophysin: A marker protein for neuroendocrine cells and neoplasms," Proceedings of the National Academy of Sciences of the United States of America,(accessed Jun. 9, 2024).

[44] B. Wiedenmann, W. W. Franke, C. Kuhn, R. Moll, and V. E. Gould, "Synaptophysin: A marker protein for neuroendocrine cells and neoplasms.," *Proceedings of the National Academy of Sciences*, vol. 83, no. 10, pp. 3500–3504, May 1986.

[45] "Synapsin," Synapsin - an overview | ScienceDirect Topics, https://www.sciencedirect.com/topics/biochemistry-genetics-and-molecular-biology/synapsin (accessed Jun. 9, 2024).

[46] K. Wimalasena, "Vesicular monoamine transporters: Structure-function, pharmacology, and Medicinal Chemistry," *Medicinal Research Reviews*, vol. 31, no. 4, pp. 483–519, Jun. 2011.

[47] S. Gowrisankaran and I. Milosevic, "Regulation of synaptic vesicle acidification at the neuronal synapse," *IUBMB Life*, vol. 72, no. 4, pp. 568–576, Jan. 2020.

[48] V. Jirsa, "Synaptic connectivity in neural population models," *Encyclopedia of Computational Neuroscience*, pp. 1–3, 2014.

[49] H. Y. Z. AC;, "Small-world anatomical networks in the human brain revealed by cortical thickness from MRI," Cerebral cortex (New York, N.Y. : 1991), https://pubmed.ncbi.nlm.nih.gov/17204824/ (accessed Jun. 9, 2024).

[50] M. Ueda, K. Kuroda, H. Saeki, N. Hirai, and T. Fujii, "Correlation between brain functional connectivity and neurocognitive function in patients with left frontal glioma," *Scientific Reports*, https://www.ncbi.nlm.nih.gov/pmc/articles/PMC9643499/ (accessed Jun. 9, 2024).

[51] K. E. Stephan, J. Zilles, and K. Amunts, "Computational analysis of functional connectivity between areas of primate cerebral cortex," *Philosophical Transactions of the Royal Society of London. Series B: Biological Sciences*, vol. 355, no. 1393, pp. 111–126, Jan. 2000.

[52] P. Greengard, "The neurobiology of slow synaptic transmission," *Science*, vol. 294, no. 5544, pp. 1024–1030, Nov. 2001.

[53] C. Ribrault, K. Sekimoto, and A. Triller, "From the stochasticity of molecular processes to the variability of synaptic transmission," *Nature Reviews Neuroscience*, vol. 12, no. 7, pp. 375–387, Jun. 2011.

[54] M. E. Laino, F. C. P. Ferreira, P. J. Rosa, and M. A. T. Figueiredo, "Generative adversarial networks in Brain Imaging: A narrative review," *Journal of Imaging*, vol. 8, no. 4, p. 83, Mar. 2022.

[55] K. Armanious *et al.*, "Independent attenuation correction of whole body [18F]FDG-PET using a deep learning approach with Generative Adversarial Networks," *EJNMMI Research*, vol. 10, no. 1, May 2020,

[56] F. Liem, C. Varoquaux, A. Kynast, P. Beyer, G. S. Masouleh, A. Huntenburg, S. Lampe, J. Rahim, N. Abraham, D. Craddock, M. A. T. Figueiredo, and R. E. A. E. E. Sørensen, "Predicting brain-age from Multimodal Imaging Data captures cognitive impairment," *NeuroImage*, vol. 148, pp. 179–188, Mar. 2017.

[57] C.-Y. Shen, "Learning to augment graphs: Machine-learning-based social network intervention with self-supervision," supp1-3340230.PDF.

[58] C.-W. Woo, L. J. Chang, M. A. Lindquist, and T. D. Wager, "Building better biomarkers: Brain models in translational neuroimaging," Nature neuroscience, https://www.ncbi.nlm.nih.gov/pmc/articles/PMC5988350/ (accessed Jun. 13, 2024).

[59] N. Kasthuri, K. J. Hayworth, D. R. Berger, R. L. Schalek, J.-B. Conchello, S. Knowles-Barley, D. Lee, A. Vázquez-Reina, V. Kaynig, T. R. Jones, H. S. Roberts Jr, J. L. Morgan, J. C. Tapia, H. S. Seung, and J. W. Lichtman, "Saturated reconstruction of a volume of neocortex," *Cell*, vol. 162, no. 3, pp. 648–661, Jul. 2015.

[60] M. E. Hale, "Mapping circuits beyond the models: Integrating Connectomics and Comparative Neuroscience," *Neuron*, vol. 83, no. 6, pp. 1256–1258, Sep. 2014.

[61] A. Fornito, A. Zalesky, and M. Breakspear, "The Connectomics of Brain Disorders," *Nature Reviews Neuroscience*, vol. 16, no. 3, pp. 159–172, Feb. 2015.

[62] Y. Zhu, X. Zhang, H. Wang, X. Chen, L. Wang, J. Zhang, and Y. Zhang, "Disrupted brain connectivity networks in acute ischemic stroke patients," *Brain Imaging and Behavior*, vol. 11, no. 2, pp. 444–453, Feb. 2016.

4

Artificial Intelligence in Neuroscience

Mahade Hasan[1], Farhana Yasmin[2], Xue Yu[1], and Md. Mehedi Hassan[3]

[1]School of Software, Nanjing University of Information Science and Technology, China
[2]School of Computer and Software, Nanjing University of Information Science and Technology, China
[3]Computer Science and Engineering Discipline, Khulna University, Khulna, 9208, Bangladesh
E-mail: farabi.eu.bd@gmail.com; farhana005.cse@gmail.com; xueyu@nuist.edu.cn; mehedihassan@ieee.org

Abstract

The use of artificial intelligence (AI) in neuroscience presents a dynamic advantage for the diagnosis and treatment of diseases connected to the brain. The study examines significant developments and potential paths for AI use in neuroscience. This study highlights the need for explainable AI and indicates how transparent and interpretable AI models are in building confidence in clinical decision-making processes. As a major trend, edge AI improves neurological and psychiatric care by improving real-time data processing and decision-making at the data-gathering site. AI-driven biomarker discovery is presented as revolutionary, providing individualized treatment plans based on genomic and neuroimaging data as well as insights into early disease diagnosis. The study emphasizes how AI and digital health technology could be used together to support specific medication and ongoing patient monitoring. AI has the potential to greatly enhance neurological patient outcomes by utilizing these breakthroughs in diagnosis, therapy, and patient outcomes overall. Going ahead, more advancement in AI research and development will be necessary to unleash fresh perspectives on brain pathology and function, providing the possibility for improved diagnostic and therapeutic approaches.

Keywords: Artificial intelligence, BCI, deep learning, machine learning, neuroscience.

4.1 Introduction

AI refers to the simulation of human intelligence in machines that are programmed to think and learn like humans. These intelligent systems are capable of performing tasks that typically require human intelligence, such as visual perception, speech recognition, decision-making, and language translation. AI is a broad field encompassing various sub-disciplines, including machine learning (ML), deep learning, natural language processing (NLP), robotics, and computer vision [1], [2].

4.1.1 Key components of AI

AI encompasses several key components that enable it to simulate human-like cognitive processes and perform complex tasks.

Figure 4.1 shows six key components of artificial intelligence. Neural networks form the backbone of AI, modeled after the human brain's interconnected neurons, allowing machines to learn from data through pattern recognition [3–4]. Deep learning, a subset of ML, utilizes multi-layered neural networks to extract complex features and make decisions, revolutionizing fields like image and speech recognition [5]. Machine learning algorithms power AI systems by enabling them to improve their performance over time without explicit programming. These algorithms underpin predictive analytics, recommendation systems, and autonomous decision-making in diverse applications from healthcare diagnostics to financial forecasting [6–8] . Computer vision equips machines with the ability to interpret and process visual information, crucial for tasks such as facial recognition, autonomous driving, and medical imaging analysis [9], [10]. Cognitive computing aims to replicate human cognitive abilities, facilitating natural interaction and decision-making in complex scenarios. NLP enables machines to understand and generate human language, powering applications like chatbots, sentiment analysis, and language translation. These components drive advancements across industries, shaping the future of technology and human-machine interaction [11].

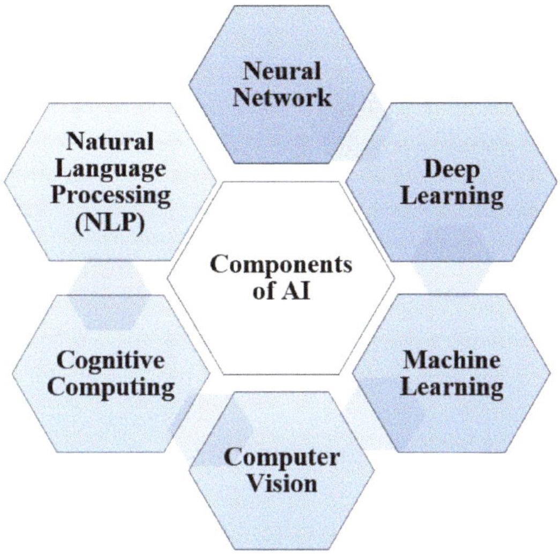

Figure 4.1 Components of artificial intelligence.

4.1.2 Importance of AI in neuroscience

AI has become an invaluable tool in neuroscience, revolutionizing the way researchers study the brain and its functions. The integration of AI in neuroscience has led to significant advancements in understanding brain structure and function, diagnosing neurological disorders, and developing innovative treatments. AI's ability to handle and analyze vast amounts of data with high precision and speed has opened new frontiers in neuroscience research.

The significance of AI in enhancing neuroscience is emphasized in Figure 4.2. AI methods such as machine learning, neural networks, and deep

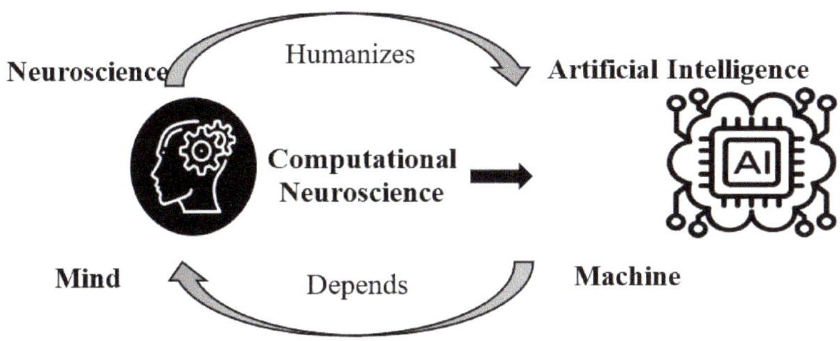

Figure 4.2 Importance of artificial intelligence in neuroscience.

learning have transformed the examination of complex neural data, providing accurate diagnoses and custom treatments for neurological illnesses. AI techniques are improving brain-computer interfaces, enhancing the analysis of neuroimaging, and helping in the detection of new biomarkers. These advancements have an important effect on research and clinical practices in the field of neuroscience.

4.1.3 Significant contributions of AI to neuroscience

AI has made significant contributions to neuroscience in recent years, revolutionizing the way researchers study and understand the human brain. Figure 4.3 shows the deep impact of AI on neuroscience, including advances in neuroimaging analysis, improvements in brain-computer interfaces, and the discovery of new biomarkers.

AI technologies have revolutionized the way researchers analyze complex brain imaging data, such as MRI scans, fMRI, and EEG data. These AI algorithms can identify patterns, abnormalities, and even predict neurological disorders like Alzheimer's disease and schizophrenia [12–13]. AI has been instrumental in identifying neuroimaging biomarkers associated with

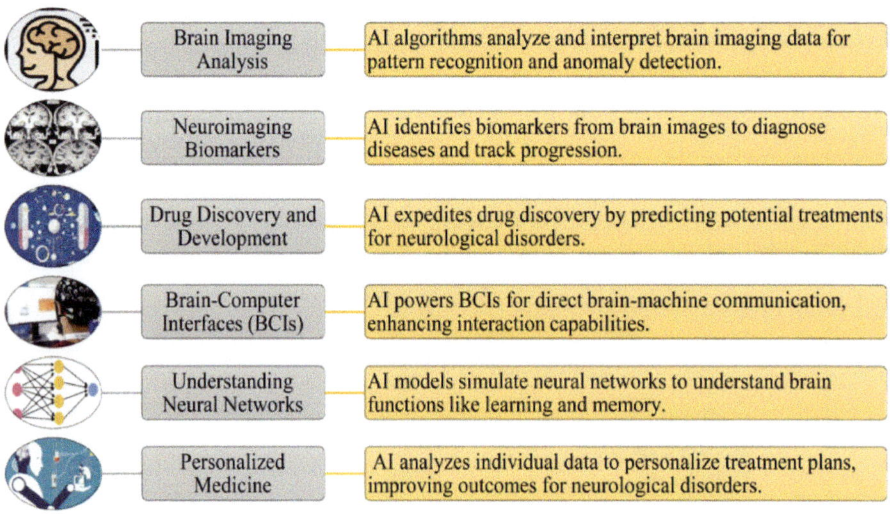

Figure 4.3 Some key ways in which artificial intelligence has impacted neuroscience.

specific neurological conditions, enabling early diagnosis, monitoring disease progression, and improving treatment strategies. In drug discovery and development, AI accelerates the process by analyzing vast biological and chemical data to pinpoint potential drug targets and predict effective treatments for brain disorders [14–15]. On the frontier of brain–machine interaction, AI powers the development of BCIs that establish direct communication between the brain and external devices. BCIs hold promise for individuals with neurological conditions to regain mobility, communication, and enhance their quality of life [16–18]. AI's ability to simulate neural networks and model brain functions using deep learning algorithms has deepened our understanding of how the brain processes information. By uncovering the intricate dynamics of neural circuits, AI contributes to computational neuroscience and sheds light on fundamental principles of brain function [19–21]. In the realm of personalized medicine, AI analyzes individual genetics [22], imaging [23], and clinical data to tailor treatment plans for patients with neurological disorders [24–25]. This personalized approach enhances treatment outcomes and minimizes adverse effects, paving the way for more effective patient care.

In conclusion, the collaboration between AI experts and neuroscientists is propelling groundbreaking research at the interface of AI and neuroscience. This synergy holds immense potential to advance our knowledge of the brain, improve diagnosis and treatment of neurological disorders, and ultimately enhance human health and well-being.

4.1.4 Historical context and evolution

The relationship between AI and neuroscience has evolved significantly over the past few decades. Initially, the interaction was largely theoretical, with early AI models inspired by basic principles of neural activity. However, with advances in technology and computational power, the practical applications of AI in neuroscience have expanded dramatically. Table 4.1 represents the history of AI in neuroscience.

The integration of AI into neuroscience is continually growing, driven by technological advancements and increasing interdisciplinary collaboration. This synergy promises to unlock new insights into the brain and pave the way for novel therapeutic approaches.

Table 4.1 A brief recent history of artificial intelligence in neuroscience.

Decade	Milestone	Impact	Key technologies
1950s–1960s	Theoretical foundations laid with the development of ANN inspired by the brain's neural structure.	Establishment of the basis for AI modeling after neural networks.	Early neural network models
1980s	Introduction of more sophisticated neural network models, such as backpropagation, which improved the ability to train multi-layer networks.	Enhanced the training and performance of neural networks.	Backpropagation algorithm
1990s	Growth in neuroimaging technologies (MRI, fMRI) provided large datasets for AI analysis, enhancing understanding of brain activity.	Enabled detailed analysis of brain structure and function using AI techniques.	MRI, fMRI
2000s	Advances in computational power and data storage enabled the handling of big data, leading to significant breakthroughs in AI applications in neuroscience.	Facilitated the processing and analysis of large neuroimaging datasets for research purposes.	Increased computational power and storage capabilities
2010s–Present	Emergence of deep learning and more complex AI models, resulting in major contributions to neuroimaging, diagnosis, and treatment of neurological disorders.	Revolutionized the field of neuroimaging and neurology through advanced AI algorithms.	Deep learning models, AI applications in neuroimaging and diagnosis of disorders

4.2 Fundamental concepts

4.2.1 Basics of neuroscience

Neuroscience is the scientific study of the nervous system, encompassing the brain, spinal cord, and peripheral nerves. It involves various disciplines, such as biology, psychology, chemistry, and medicine, to understand the structure, function, development, and disorders of the nervous system.

Several important domains that contribute to the scientific knowledge of the nervous system are indicated in Figure 4.4. Neuroanatomy is the scientific examination of the structure of the nervous system, encompassing the arrangement and interconnections of neuronal circuits. Neurophysiology

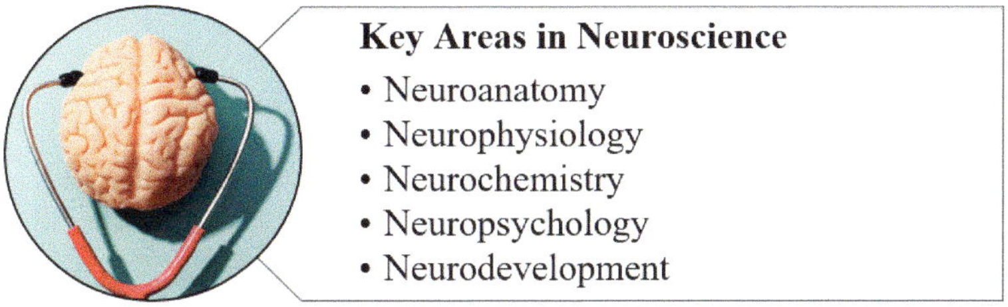

Figure 4.4 Key areas in neuroscience.

is the study of the operations and activities of the nervous system, specifically focusing on how neurons interact using electrical and chemical signals. Neurochemistry is a field of study that examines the chemical makeup and operations of the nervous system, specifically the actions of neurotransmitters. Neuropsychology researches the correlation between brain activity and behavior, specifically examining the impact of brain lesions or disorders on cognitive and behavioral processes. Neurodevelopment refers to the process of growth and maturation of the nervous system. Neurodevelopmental study encompasses the examination of the progression of the neural system from its initial embryonic phases to its mature state in adulthood. Neuropathology is the study and analysis of disorders affecting the neurological system, such as Alzheimer's disease, Parkinson's disease, and multiple sclerosis.

4.2.2 Introduction to AI and machine learning

AI is the field of computer science dedicated to creating systems capable of performing tasks that typically require human intelligence. These tasks include learning, reasoning, problem-solving, perception, and language understanding.

4.2.3 Key algorithms and techniques in AI

AI encompasses various algorithms and techniques, each suited for different tasks and applications. Here, we focus on some of the most impactful techniques relevant to neuroscience.

4.3 Neural Networks

4.3.1 Artificial neural networks (ANNs)

They are computational models inspired by the human brain, consisting of interconnected layers of nodes (neurons) that process information.

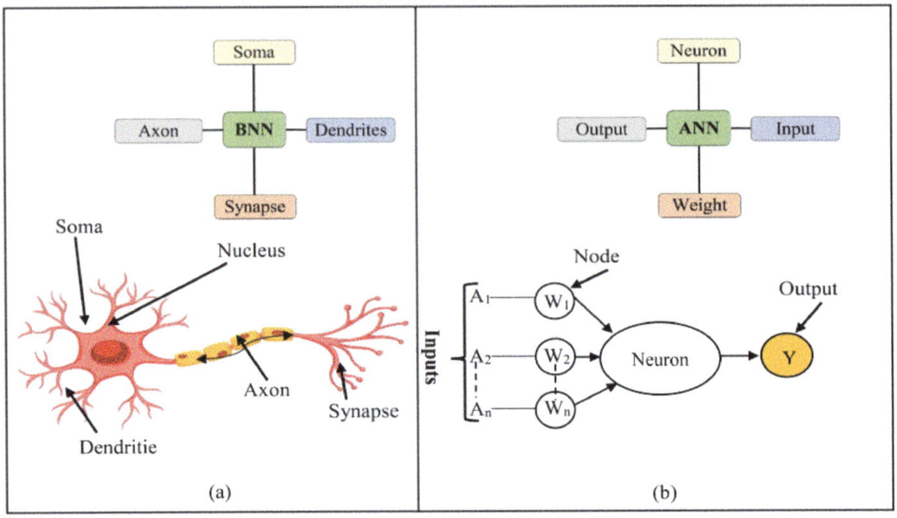

Figure 4.5 The structure of (a) ANN and (b) BNN.

Biological neuron network: Information is analyzed by nerve cells, which are unique biological cells. A large number of neurons – approximately 10^{11} with a large number of connections approximately 10^{15} are estimated.

In Figure 4.5 (a), dendrites resembling branches of a tree, dendrites receive information from neighboring neurons. It can additionally compare neurons to a neuron's receptors in a different sense. The soma, or cell body, of a neuron is in charge of processing information that comes from dendrites. Axons function similarly to cables such that they allow neurons to transmit data. Synapses are the points where an axon and other neuron dendrites link. An ANN can be defined as a machine duplicate of a biological neuron, with its many components carrying out different tasks, much like an original neuron shown in Figure 4.5 (b). ANNs come in two main varieties: feedforward and feedback.

4.3.2 Feedforward neural network

It is the basic type of neural network where connections between nodes do not form cycles. It is used for tasks like image recognition and classification.

In Figure 4.6, data is received by the neurons in input layer and transmitted to the next stages of the network. The number of neurons in the input layer and the feature or attribute numbers in the dataset must match. Hidden

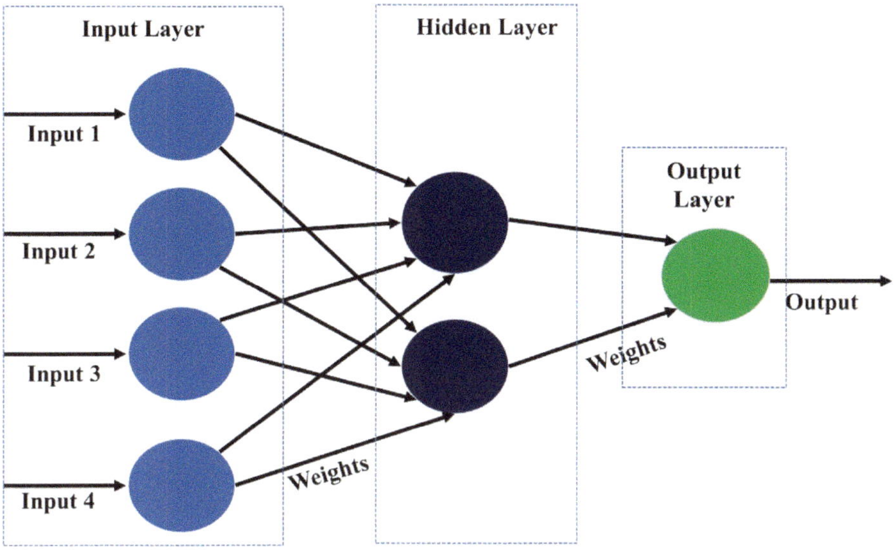

Figure 4.6 Structure of feedforward neural network.

layers are used to divide the input and output layers. There could be multiple hidden layers in a model, depending on its kind. The weight between neurons determines their strength or magnitude. It is also useful to compare input weights, much like with linear regression results. Weight usually has a value between 0 and 1, with a range of 0 to 1. Output layer displays the predicted feature based on the kind of model being constructed.

4.3.3 Recurrent neural networks (RNNs)

It is the type of neural network where connections between nodes form directed cycles, allowing them to maintain a "memory" of previous inputs. It is useful for sequential data like time series and NLP.

In Figure 4.7, the output of an RNN is controlled by the current input as well as the inputs that have already been received when the data cycles through the loop. After processing, the first input is passed to the middle layer by the input layer. Many hidden layers, each with its own functions for activation, weights, and biases, make up the middle layer. Because these settings are the same across the hidden layer, it will generate a single hidden layer and loop it over rather than producing several.

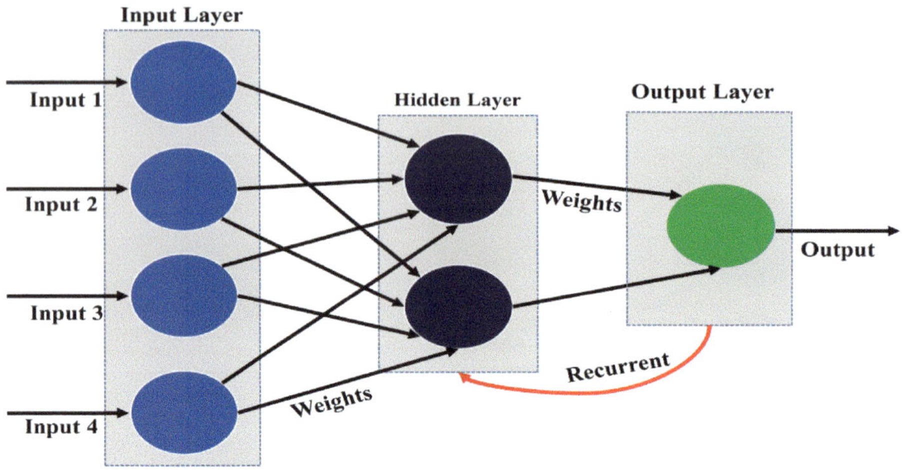

Figure 4.7 Structure of recurrent neural networks.

4.3.4 Convolutional neural networks (CNNs)

They are specialized neural networks for processing structured grid data like images. They use convolutional layers to automatically and adaptively learn spatial hierarchies of features.

Figure 4.8 highlights two significant components that contribute to a CNN design. The convolutional tool collects network features first. This

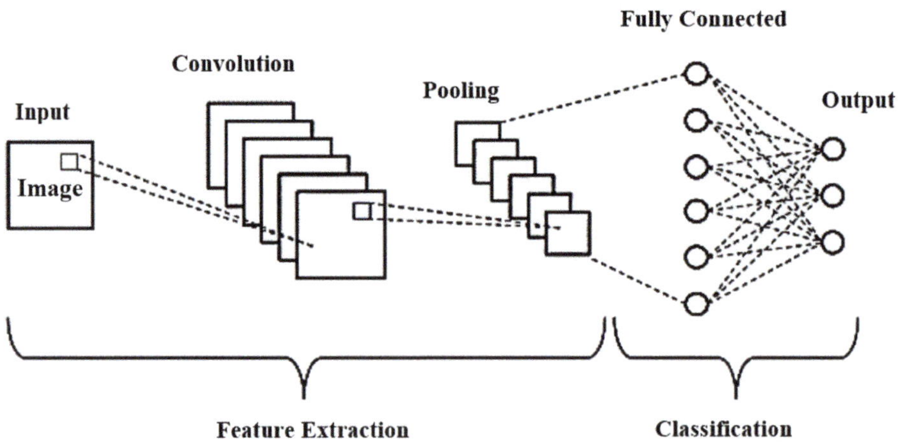

Figure 4.8 Basic CNN architecture.

image extraction network uses many pairs of convolutional and pooling layers to highlight and capture image features. Second, the fully connected layer uses the convolutional process output to predict the image class based on data extracted in previous stages. The CNN feature extraction model creates new features that summarize existing ones to minimize the number of features in a dataset. This multi-layer approach is shown in the CNN architecture diagram.

4.4 Deep Learning

4.4.1 Deep neural networks (DNNs)

DNNs are neural networks with multiple layers between input and output, enabling the modeling of complex patterns and representations.

Figure 4.9 shows the structure of deep neural networks, highlighting the fundamental elements that constitute their architecture. The input layer is the entry point for raw data into the network. After the input layer, there are other hidden layers, such as hidden layer 1 and hidden layer 2, where complex calculations and feature extraction take place. The output layer generates the network's predictions or classifications by utilizing the processed information from the hidden layers. The layers are composed of several neurons that are connected through weighted edges. These edges are modified during the training process in order to reduce errors in predictions.

DNNs are transforming the field of neuroscience by allowing for the precise processing of large and complex neural data with unparalleled precision.

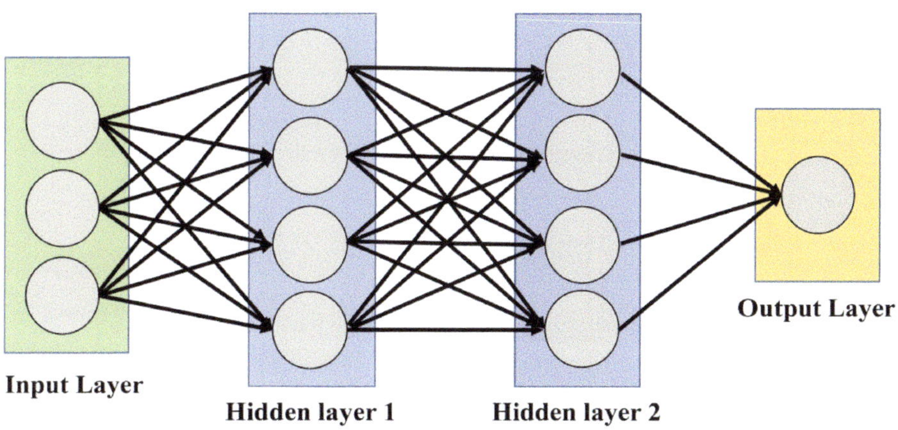

Figure 4.9 Structure of deep neural networks.

DNNs are very useful for analyzing neuroimaging data, such as MRI, fMRI, and PET scans. They can automatically detect patterns and abnormalities in these scans, which can help in the early detection of neurological disorders. DNNs enhance the analysis of neural signals in BCIs, enabling more efficient manipulation of prosthetic devices and communication aids for those with disabilities. In addition, DNNs contribute to the understanding of cognitive functions by duplicating neural processes and simulating brain activity, so offering a significant understanding of the interactions between various brain regions. The ability to process data with a large number of dimensions and identify complex patterns makes deep neural networks essential for improving both theoretical and applied neuroscience research.

4.4.2 Autoencoders

An autoencoder is a type of neural network designed for unsupervised learning that aims to learn efficient representations of input data.

Figure 4.10 shows the working process of autoencoders that consist of three main components: an encoder, a latent representation, and a decoder. The process begins with the input layer, which receives the original data. The encoder compresses this input into a lower-dimensional space, producing a latent representation that captures the essential features of the data. The decoder then reconstructs the data from this latent representation, aiming to recreate the input as accurately as possible. The final output of the autoencoder is a reconstruction of the input data. By minimizing the difference between the input and output during training, the autoencoder learns a compressed, abstract representation of the data, which can be useful for tasks like dimensionality reduction, denoising, and anomaly detection.

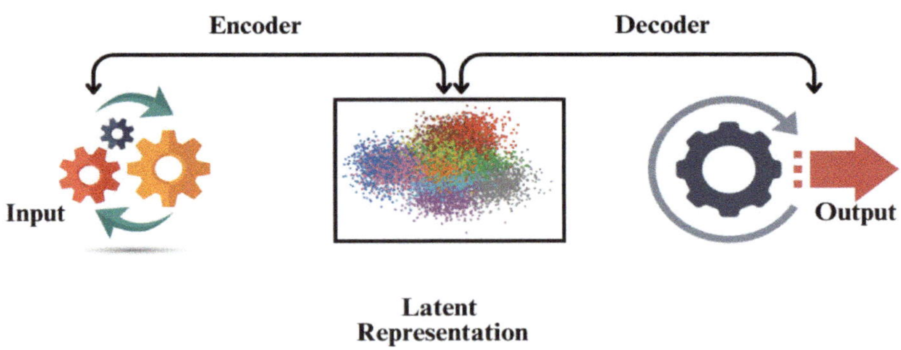

Figure 4.10 Structure of autoencoders.

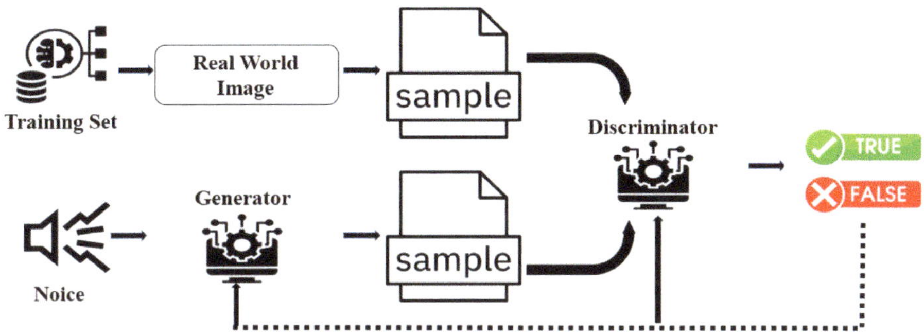

Figure 4.11 Workflow of generative adversarial networks (GANs)

4.4.3 Generative adversarial networks (GANs)

Generative adversarial network (GAN) consists of two neural networks, a generator and a discriminator, competing to produce realistic synthetic data. It is used in image generation, data augmentation, and more [26].

Figure 4.11 shows the process that makes GAN, which consists of a discriminator and a generator, operate. Real-world photographs make up the training set, serving as a discriminator sample set. The generator is fed random noise and outputs synthetic samples that approach actual visuals. The discriminator takes, assesses them, and decides whether they are true (actual) or false (fake). The generator gains more realistic picture production skills through iterative adversarial training, while the discriminator sharpens its capacity to discern between authentic and fraudulent images.

4.4.4 Support vector machines (SVMs)

Support vector machines (SVM) are a collection of supervised learning techniques employed for classification, regression, and the identification of outliers. SVMs are particularly efficient in high-dimensional spaces as they aim to identify the optimal hyperplane that may separate a dataset into various groups [27]. In Figure 4.12, the key concepts in this context are the hyperplane, which is a flat affine subspace in an N-dimensional space; support vectors, which are the data points that are closest to the hyperplane and have the most influence on its position and orientation; and the margin, which is the distance between the hyperplane and the nearest data points from either class. The goal is to maximize this margin.

SVMs are commonly employed in neuroscience to classify neuroimaging data, specifically to differentiate between various brain states using MRI or

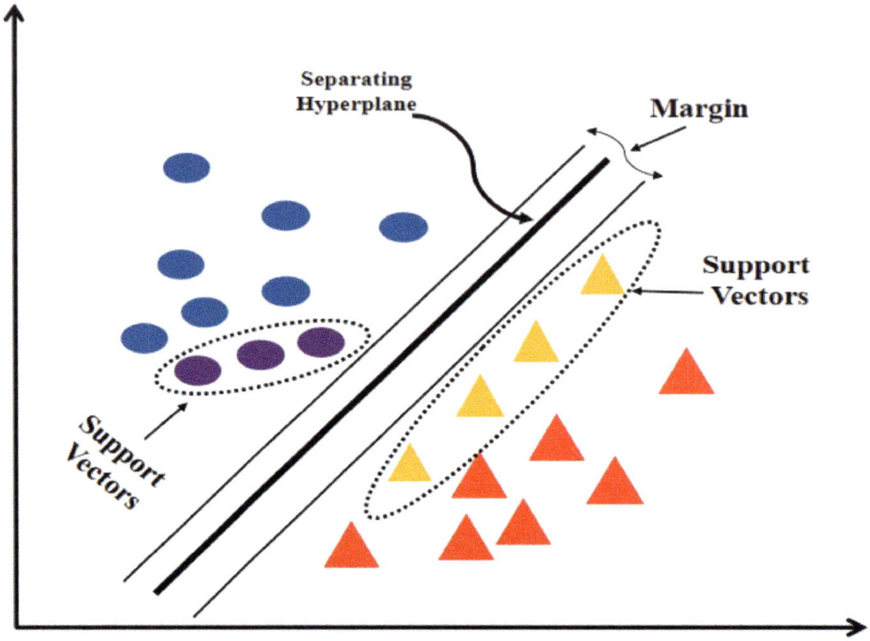

Figure 4.12 Supervised learning: Classification using SVM.

fMRI scans. They are also used to decode neural signals in brain–computer interfaces (BCIs) for the purpose of controlling prosthetic limbs or computer cursors. They have exceptional proficiency in pattern recognition, accurately detecting cognitive states or atypical brain processes, such as seizures in individuals with epilepsy, by analyzing EEG data. Moreover, SVMs are utilized to categorize functional connectivity patterns in the brain for the purpose of investigating neurological and psychiatric illnesses. For instance, SVMs can be employed to differentiate variations in brain network connection observed in individuals with autism spectrum disorders [28–30].

4.4.5 Decision trees and random forests

A decision tree is a supervised learning algorithm used for classification and regression tasks, where data is split into branches based on feature values to form a tree structure. The root node represents the initial decision point, decision nodes represent subsequent splits, and leaf nodes represent the final outcomes or classifications [31]. Random forest is an ensemble learning method that constructs multiple decision trees and combines their results to produce a more accurate and stable prediction, reducing overfitting, and improving generalization [32].

In Figure 4.13 (a), the work process of a decision tree begins with the root node, where data is split based on the feature that provides the best separation, as measured by criteria. This splitting continues at decision nodes until leaf nodes are reached, which represent the final predictions. From Figure 4.13 (b) in random forest, multiple decision trees are created using different subsets of the data and features. Each tree is independently trained, and the final prediction is derived by aggregating the results from all trees, through methods such as majority voting for classification or averaging for regression.

In neuroscience, decision trees and random forests are valuable for their clarity and effectiveness in handling complex data. They are used to analyze neuroimaging data, such as MRI or fMRI scans, to differentiate between healthy and diseased brain states. These algorithms are also applied in brain–computer interfaces to decode neural signals for controlling prosthetic devices. Additionally, they aid in pattern recognition tasks, such as identifying biomarkers for neurological disorders, and in examining functional connectivity patterns to study brain network alterations in conditions like autism and schizophrenia. By using decision trees and random forests, neuroscientists can gain deeper insights into brain function and pathology [33–34].

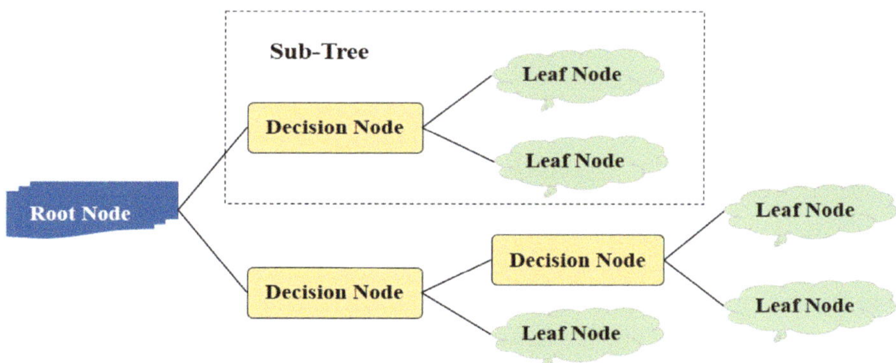

(a)

Figure 4.13 *Continued.*

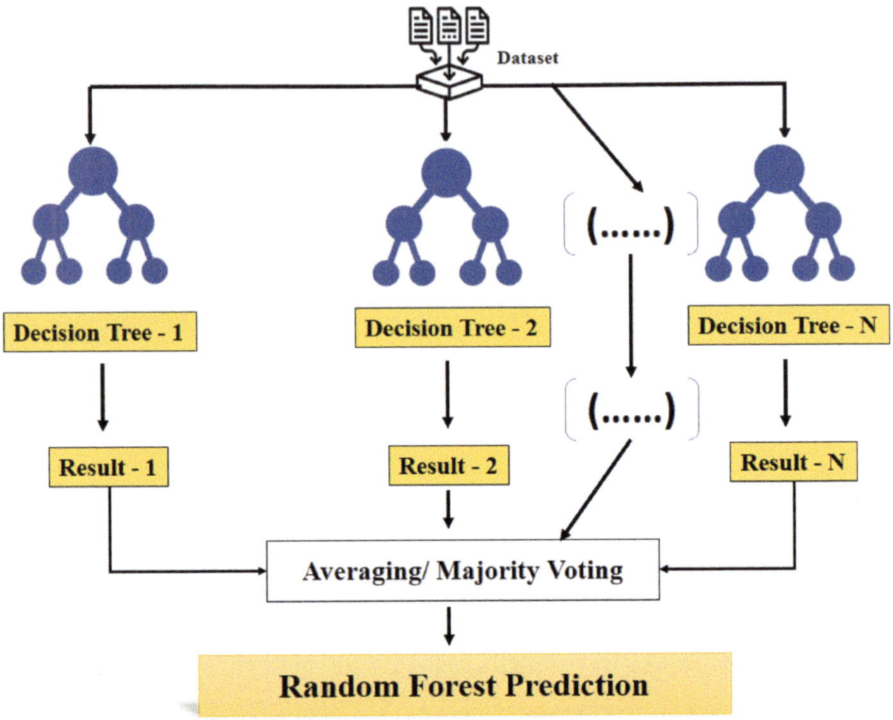

(b)

Figure 4.13 Illustration of (a) decision tree architecture, starting from the root node, progressing through decision nodes, and ending at leaf nodes representing final classifications, and (b) random forest architecture, composed of multiple decision trees built from different data subsets, aggregated to improve accuracy and reduce overfitting.

4.4.6 Clustering algorithms

Techniques like K-means, hierarchical clustering, and DBSCAN, are used for grouping similar data points into clusters without predefined labels. By understanding these fundamental concepts, we can better appreciate how AI and ML are applied to address complex challenges in neuroscience, ultimately leading to deeper insights and innovative solutions [35].

Figure 4.14 shows K-means hierarchical clustering and DBSCAN clusters of different variance. K-means algorithm starts by establishing K center points. It then assigns each data point to the midpoint that is closest to it. The algorithm continues to recalculate its centers repeatedly until they reach a stable state, which causes the formation of K clusters. Hierarchical clustering

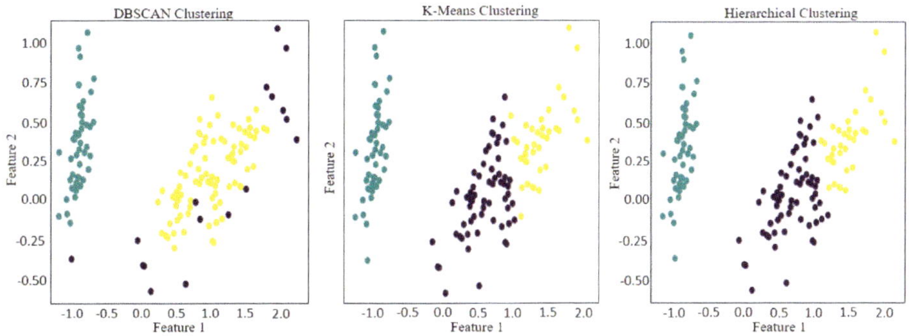

Figure 4.14 Example of clustering data with two clusters of different variance. DBSCAN clustering (left) detects the cluster with low variance and high point density (yellow) while discarding all other points as outliers (turquoise). K-means clustering (center) separates the clusters but misclassifies a few points in the middle, adding them to the purple cluster instead of the yellow one. Hierarchical clustering (right) effectively separates the clusters based on their variances.

develops a cluster hierarchy by either merging the nearest clusters or dividing the largest clusters, resulting in a tree diagram that can be divided at various levels. DBSCAN starts by selecting a random point and then expands it into a cluster if it has a sufficient number of nearby points within a specified radius (epsilon). Points that fail to satisfy this requirement are known as noise, while dense regions are accurately identified as clusters.

4.5 Applications of AI in Neuroscience

4.5.1 Neuroimaging analysis

Neuroimaging techniques provide critical insights into the structure and function of the brain. AI enhances the analysis of neuroimaging data, offering more precise and efficient methods for understanding brain activity and diagnosing neurological conditions. Figure 4.15 shows four important brain imaging techniques such as MRI, fMRI, PET, and CT.

AI algorithms improve the resolution and speed of MRI scans, facilitating detailed structural analysis of the brain. AI also aids in detecting anomalies and quantifying brain structures [36]. AI techniques, such as ML and deep learning, analyze the complex data from fMRI scans to map brain activity and understand functional connectivity [37]. AI enhances PET scan analysis by improving image reconstruction, noise reduction, and the quantification of metabolic processes [38]. AI helps in the detection and classification of

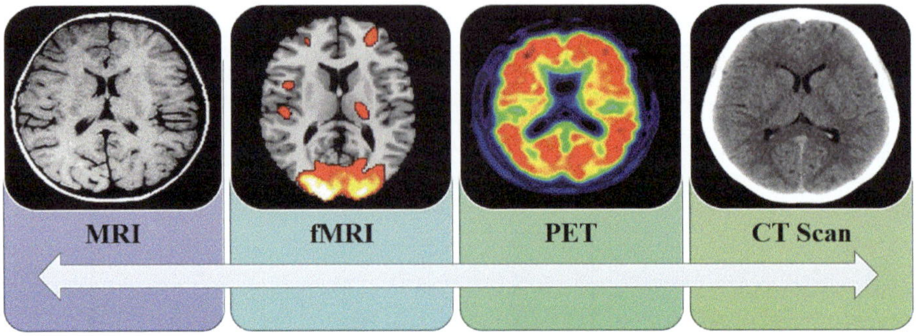

Figure 4.15 Neuroimaging: Four important brain imaging techniques.

lesions, tumors, and other abnormalities in CT images with high accuracy and speed [39], [40].

4.5.2 Brain–computer interfaces (BCIs)

Brain–computer interfaces (BCIs) enable direct communication between the brain and external devices, offering new possibilities for medical treatments and augmenting human capabilities.

Figure 4.16 illustrates the design and implementation process of BCIs. It typically involves the acquisition of neural signals through methods such as EEG, followed by signal processing to filter and extract features. Machine learning algorithms then interpret these signals, translating the processed data into commands for controlling external devices or software applications. AI algorithms play a critical role in processing brain signals, ensuring

Figure 4.16 Design and implementation of BCIs.

accurate and noise-free data. These methods identify relevant features from brain signals and classify them to interpret user intentions or commands. AI also ensures the real-time processing of brain signals, crucial for the responsive performance of BCIs in applications like neuroprosthetics and communication aids.

4.5.3 Neural data analysis

Neural data, obtained from various electrophysiological techniques, provides insights into brain function. AI plays a crucial role in analyzing this data to understand neural mechanisms and diagnose disorders.

Electrophysiological data from EEG and MEG offer valuable insights into brain activity. AI algorithms analyze EEG data to identify patterns related to brain activity, detect abnormalities, and diagnose conditions such as epilepsy and sleep disorders [41]. Similarly, AI methods process MEG data to map brain activity with high temporal and spatial resolution, enhancing the study of cognitive processes and neurological disorders. By leveraging these AI techniques, researchers and clinicians can achieve more accurate diagnoses and a deeper understanding of brain function, facilitating the development of targeted treatments and interventions.

Signal processing and interpretation of neural data are significantly enhanced by AI techniques. AI methods, including deep learning, improve the quality of neural signals by effectively filtering out noise and artifacts, leading to cleaner data for analysis [42]. Additionally, AI excels in feature extraction, identifying critical features from neural signals that correlate with specific brain functions or states. Furthermore, AI models are adept at pattern recognition, which aids in the detection of seizures, cognitive states, and other neurological events. By leveraging these AI capabilities, researchers and clinicians can gain more precise insights into neural activity, ultimately improving diagnostic accuracy and treatment outcomes for neurological conditions.

4.5.4 Cognitive neuroscience

AI contributes significantly to cognitive neuroscience by modeling cognitive functions and understanding the underlying neural mechanisms.

AI simulates cognitive processes such as perception, memory, decision-making, and learning, providing insights into how the brain performs these functions. Figure 4.17 presents a chart of understanding cognitive processes

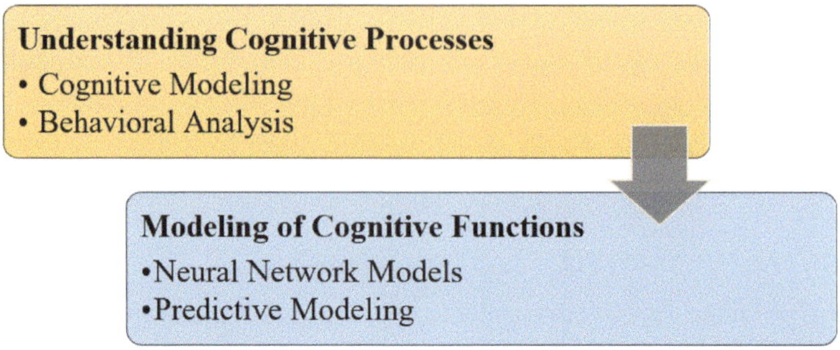

Figure 4.17 AI in understanding cognitive processes and modeling of cognitive functions.

and modeling of cognitive functions. AI analyzes behavioral data to correlate cognitive functions with neural activity, aiding the study of attention, emotion, and executive functions. AI leverages neural networks to model complex cognitive functions, exploring interactions and information processing between different brain regions. Additionally, AI predicts cognitive performance and potential impairments based on neural data, supporting early detection of cognitive decline and mental health conditions.

In these applications, neuroscience can achieve more accurate diagnoses, better understand brain functions, and develop innovative treatments for neurological and cognitive disorders.

4.6 AI in Neurodegenerative Diseases

AI is revolutionizing the field of neurodegenerative diseases by enhancing early diagnosis, improving treatment strategies, and providing insights into disease progression. Neurodegenerative diseases, such as Alzheimer's and Parkinson's, are complex and multifaceted, making them challenging to diagnose and treat. AI's ability to analyze vast datasets and identify patterns can significantly aid in addressing these challenges.

Table 4.2 summarizes the main areas where AI is making significant contributions to the understanding, diagnosis, and treatment of neurodegenerative diseases.

4.7 AI in Neuropsychiatric Disorders

AI is playing a transformative role in the field of neuropsychiatric disorders by improving diagnosis, personalizing treatment plans, and enhancing

Table 4.2 Key applications and examples of AI in neurodegenerative diseases.

Application	Description	Examples
Early diagnosis and prognosis [43]	AI algorithms analyze medical images (MRI, CT, PET) to detect early signs of neurodegeneration. Predicts disease progression using longitudinal data.	AI detects subtle brain changes before clinical symptoms, forecasting disease trajectory.
Treatment and management [44]	Develops personalized treatment plans by analyzing genetic, lifestyle, and clinical data. Monitors disease progression and treatment response in real time.	Personalized therapy recommendations for specific patients, real-time therapy adjustments.
Alzheimer's disease [45]	Predictive models identify high-risk individuals by analyzing genetic, clinical, and imaging data.	Early risk assessments based on comprehensive data analysis.
Parkinson's disease [46–47],	ML algorithms detect early signs through voice and movement pattern analysis.	Non-invasive diagnostic methods provide additional insights into disease progression.

our understanding of these complex conditions. Neuropsychiatric disorders, including depression, schizophrenia, and bipolar disorder, often present diagnostic challenges due to their multifaceted nature and the subjective assessment of symptoms. AI's ability to analyze large datasets, recognize patterns, and make predictions offers significant advantages in tackling these challenges. By integrating AI with clinical practice, researchers and clinicians can achieve more accurate diagnoses, develop tailored treatment strategies, and monitor patient progress more effectively.

Table 4.3 outlines the primary applications of AI in neuropsychiatric disorders, highlighting how AI enhances diagnostic processes, personalizes treatments, and provides real-world case examples to illustrate these advancements.

4.8 Future Directions

As AI continues to change, several emerging trends and technologies are controlled to further transform neuroscience and medicine.

There is a growing emphasis on developing AI models that are transparent and interpretable, allowing clinicians to understand and trust AI-driven decisions. Figure 4.18 represents a list of explainable AI, Edge AI, and biomarker discovery. AI processing is moving closer to data sources through

Table 4.3 Key applications and examples of AI in neuropsychiatric disorders.

Application	Description	Examples
Diagnosis and classification [48]	AI algorithms analyze various data sources to improve diagnostic accuracy and classify neuropsychiatric disorders.	ML models distinguish between types of depression and other mood disorders.
Personalized treatment plans [49]	AI develops tailored therapeutic strategies by analyzing individual patient data, including genetic profiles, lifestyle factors, and treatment responses.	AI-driven recommendations for medication and therapy adjustments based on patient-specific data.
Depression [50]	AI systems analyze patient data to identify biomarkers and predict treatment responses.	Predictive models for selecting the most effective antidepressant medications for individuals.
Schizophrenia [51]	AI helps in early diagnosis by detecting subtle patterns in speech and behavior, which are often missed by traditional methods.	Early detection through speech analysis and monitoring of behavioral changes using wearable devices.

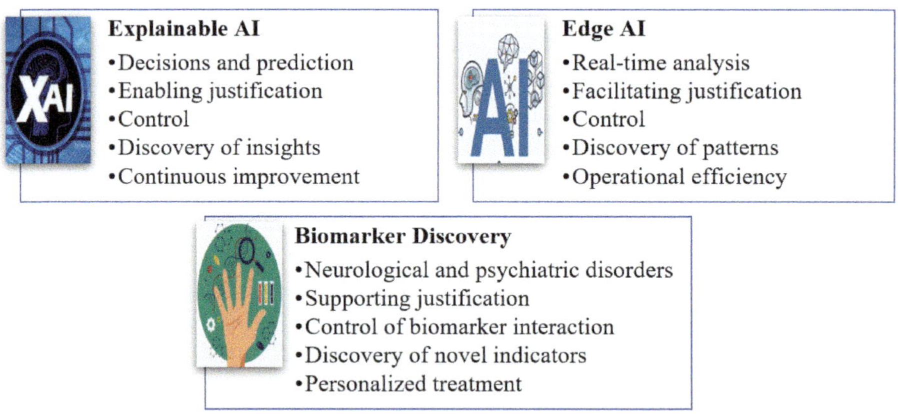

Figure 4.18 Emerging trends and technologies.

edge computing, enabling real-time analysis and decision-making in applications such as wearable devices and remote monitoring systems [48–49]. AI is increasingly used to identify novel biomarkers for various neurological and psychiatric disorders, improving early diagnosis and personalized treatment. Inspired by the brain's architecture, neuromorphic computing involves designing hardware that mimics neural networks, promising faster and more efficient AI processing for neuroscience applications. Quantum computing holds the potential to solve complex problems that are currently intractable for classical computers, such as simulating large-scale neural networks and optimizing personalized treatment plans. Advances in synthetic biology, combined with AI, could lead to the development of bioengineered tissues and organs, potentially offering new treatments for neurodegenerative diseases and injuries.

4.9 Conclusion

In conclusion, the integration of AI into neuroscience represents a transformative leap forward, promising profound impacts on both research and clinical practice. Throughout this chapter, we have explored various applications where AI is enhancing our understanding and management of brain-related disorders. AI's ability to analyze vast datasets in neuroimaging, such as MRI and PET scans, has revolutionized diagnostic accuracy. By detecting subtle patterns and anomalies that may evade human observation, AI enables earlier and more precise diagnoses of conditions like Alzheimer's disease and schizophrenia. Moreover, AI-driven biomarker discovery has opened new avenues for personalized medicine, identifying genetic and molecular signatures that guide tailored treatment strategies. The concept of explainable AI has emerged as a critical enabler in the adoption of AI technologies within clinical settings. By providing transparent explanations of its decision-making processes, AI builds trust among healthcare providers and patients, facilitating the integration of AI-driven insights into everyday medical practice. Looking forward, the future of AI in neuroscience holds immense promise. Emerging technologies such as Edge AI are pushing the boundaries by enabling real-time data analysis and decision-making directly at the point of care. This capability is particularly valuable in remote and underserved areas, where access to specialist healthcare may be limited. Collaborative research efforts between AI scientists, neuroscientists, and clinicians are essential for further advancements. By combining expertise in AI algorithms with deep domain knowledge in neuroscience, researchers can continue to innovate in areas such as brain-machine interfaces, computational modeling of neural circuits, and predictive analytics for neurodevelopmental disorders. While challenges such as data privacy, algorithm biases, and ethical considerations remain, the potential benefits of AI in neuroscience are undeniable. As technology continues to evolve and our understanding of brain function grows, AI stands poised to revolutionize neuroscientific research and enhance the quality of care for individuals affected by neurological and psychiatric conditions.

Acknowledgments

We would like to express our heartfelt appreciation to the editors for their invaluable contributions to this chapter. Their expertise and careful feedback were instrumental in refining our ideas and ensuring the highest standards in our work. Their guidance throughout the editorial process was invaluable, providing clarity and coherence to our research findings. We are deeply thankful for their unwavering support and commitment to excellence. This chapter would not have been possible without their dedication and expertise.

References

[1] F. Nagi et al., "Applications of Artificial Intelligence (AI) in Medical Education: A Scoping Review," Stud Health Technol Inform, vol. 305, pp. 648–651, Jun. 2023, doi: 10.3233/SHTI230581.

[2] F. Yasmin et al., "GastroNet: Gastrointestinal Polyp and Abnormal Feature Detection and Classification With Deep Learning Approach," IEEE Access, vol. 11, pp. 97605–97624, 2023, doi: 10.1109/ACCESS.2023.3312729.

[3] P. Ji, Y. Wang, T. Peron, C. Li, J. Nagler, and J. Du, "Structure and function in artificial, zebrafish and human neural networks," Phys Life Rev, vol. 45, pp. 74–111, Jul. 2023, doi: 10.1016/j.plrev.2023.04.004.

[4] "Refining neural network algorithms for accurate brain tumor classification in MRI imagery - PubMed." Accessed: Jun. 17, 2024. [Online]. Available: https://pubmed.ncbi.nlm.nih.gov/38773391/

[5] M. Tabassum, A. Al Suman, C. Russo, A. Di Ieva, and S. Liu, "A Deep Learning Framework for Skull Stripping in Brain MRI," Annu Int Conf IEEE Eng Med Biol Soc, vol. 2023, pp. 1–4, Jul. 2023, doi: 10.1109/EMBC40787.2023.10340846.

[6] C. L. Cheek, P. Lindner, and E. L. Grigorenko, "Statistical and Machine Learning Analysis in Brain-Imaging Genetics: A Review of Methods," Behav Genet, vol. 54, no. 3, pp. 233–251, May 2024, doi: 10.1007/s10519-024-10177-y.

[7] F. Yasmin et al., "PoxNet22: A Fine-Tuned Model for the Classification of Monkeypox Disease Using Transfer Learning," IEEE Access, vol. 11, pp. 24053–24076, 2023, doi: 10.1109/ACCESS.2023.3253868.

[8] Abdulaziz Albahr, M. A. Albahar, M. I. Thanoon, and M. Binsawad, "Computational Learning Model for Prediction of Heart Disease Using Machine Learning Based on a New Regularizer," Computational Intelligence and Neuroscience, vol. 2021, pp. 1–10, Nov. 2021, doi: 10.1155/2021/8628335.

[9] "Computer-vision based analysis of the neurosurgical scene - A systematic review - PubMed." Accessed: Jun. 17, 2024. [Online]. Available: https://pubmed.ncbi.nlm.nih.gov/38020988/

[10] B. F. Saway et al., "The evolution of neuromodulation for chronic stroke: From neuroplasticity mechanisms to brain-computer interfaces," Neurotherapeutics, vol. 21, no. 3, p. e00337, Apr. 2024, doi: 10.1016/j.neurot.2024.e00337.

[11] M. M. Nour and Q. J. M. Huys, "Natural Language Processing in Psychiatry: A Field at an Inflection Point," Biol Psychiatry Cogn Neurosci Neuroimaging, vol. 8, no. 10, pp. 979–981, Oct. 2023, doi: 10.1016/j.bpsc.2023.08.001.

[12] E. L. Schiffrin and J. C. Engert, "Hypertension, brain imaging phenotypes, and cognitive impairment: lessons from Mendelian randomization," Eur Heart J, vol. 44, no. 23, pp. 2126–2128, Jun. 2023, doi: 10.1093/eurheartj/ehad187.

[13] K. Sailunaz, S. Alhajj, T. Özyer, J. Rokne, and R. Alhajj, "A survey on brain tumor image analysis," Med Biol Eng Comput, vol. 62, no. 1, pp. 1–45, Jan. 2024, doi: 10.1007/s11517-023-02873-4.

[14] T. van Eimeren, K. Giehl, K. Reetz, C. Sampaio, and T. A. Mestre, "Neuroimaging biomarkers in Huntington's disease: Preparing for a new era of therapeutic development," Parkinsonism Relat Disord, vol. 114, p. 105488, Sep. 2023, doi: 10.1016/j.parkreldis.2023.105488.

[15] S. Nisar and M. Haris, "Neuroimaging genetics approaches to identify new biomarkers for the early diagnosis of autism spectrum disorder," Mol Psychiatry, vol. 28, no. 12, pp. 4995–5008, Dec. 2023, doi: 10.1038/s41380-023-02060-9.

[16] F. Bai, S. Li, and H. Li, "AI enhances drug discovery and development," Natl Sci Rev, vol. 11, no. 3, p. nwad303, Mar. 2024, doi: 10.1093/nsr/nwad303.

[17] M. Aljofan and A. Gaipov, "Drug discovery and development: the role of artificial intelligence in drug repurposing," Future Med Chem, vol. 16, no. 7, pp. 583–585, Apr. 2024, doi: 10.4155/fmc-2024-0048.

[18] V. Bijak et al., "The current role and evolution of X-ray crystallography in drug discovery and development," Expert Opin Drug Discov, vol. 18, no. 11, pp. 1221–1230, 2023, doi: 10.1080/17460441.2023.2246881.

[19] J. Chen et al., "fNIRS-EEG BCIs for Motor Rehabilitation: A Review," Bioengineering (Basel), vol. 10, no. 12, p. 1393, Dec. 2023, doi: 10.3390/bioengineering10121393.

[20] J. Peksa and D. Mamchur, "State-of-the-Art on Brain-Computer Interface Technology," Sensors (Basel), vol. 23, no. 13, p. 6001, Jun. 2023, doi: 10.3390/s23136001.

[21] Y. Qin, Y. Zhang, Y. Zhang, S. Liu, and X. Guo, "Application and Development of EEG Acquisition and Feedback Technology: A Review," Biosensors (Basel), vol. 13, no. 10, p. 930, Oct. 2023, doi: 10.3390/bios13100930.

[22] "Utilizing multimodal AI to improve genetic analyses of cardiovascular traits - PubMed." Accessed: Jun. 17, 2024. [Online]. Available: https://pubmed.ncbi.nlm.nih.gov/38562791/

[23] S. Hatamikia et al., "Ovarian cancer beyond imaging: integration of AI and multiomics biomarkers," Eur Radiol Exp, vol. 7, no. 1, p. 50, Sep. 2023, doi: 10.1186/s41747-023-00364-7.

[24] M. Hasan, F. Yasmin, and L. Deng, "Rule Mining of Early Diabetes Symptom and Applied Supervised Machine Learning and Cross Validation Approaches based on the Most Important Features to Predict Early-stage Diabetes," IJIRMPS - International Journal of Innovative Research in Engineering & Multidisciplinary Physical Sciences, vol. 11, no. 3, Jun. 2023, doi: 10.37082/IJIRMPS.v11.i3.230225.

[25] J. Guintivano et al., "Meta-Analyses of Genome-Wide Association Studies for Postpartum Depression," Am J Psychiatry, vol. 180, no. 12, pp. 884–895, Dec. 2023, doi: 10.1176/appi.ajp.20230053.

[26] S. Bengesi, H. El-Sayed, M. K. Sarker, Y. Houkpati, J. Irungu, and T. Oladunni, "Advancements in Generative AI: A Comprehensive Review of GANs, GPT, Autoencoders, Diffusion Model, and Transformers," IEEE Access, vol. 12, pp. 69812–69837, 2024, doi: 10.1109/ACCESS.2024.3397775.

[27] "Quantum machine learning for support vector machine classification | Evolutionary Intelligence." Accessed: Jul. 02, 2024. [Online]. Available: https://link.springer.com/article/10.1007/s12065-022-00756-5

[28] "Automatic sleep–wake classification and Parkinson's disease recognition using multifeature fusion with support vector machine - Shen - 2024 - CNS Neuroscience & Therapeutics - Wiley Online Library." Accessed: Jul. 02, 2024. [Online]. Available: https://onlinelibrary.wiley.com/doi/full/10.1111/cns.14708

[29] "Decoding fMRI data with support vector machines and deep neural networks - ScienceDirect." Accessed: Jul. 02, 2024. [Online]. Available: https://www.sciencedirect.com/science/article/pii/S0165027023002236

[30] A. Ororbia, A. Mali, A. Kohan, B. Millidge, and T. Salvatori, "A Review of Neuroscience-Inspired Machine Learning." arXiv, Feb. 16, 2024. doi: 10.48550/arXiv.2403.18929.

[31] "A decision tree algorithm to identify predictors of post-stroke complex regional pain syndrome | Scientific Reports." Accessed: Jul. 02, 2024. [Online]. Available: https://www.nature.com/articles/s41598-024-60597-3

[32] "Remote Sensing | Free Full-Text | Comparison of Random Forest and XGBoost Classifiers Using Integrated Optical and SAR Features for Mapping Urban Impervious Surface." Accessed: Jul. 02, 2024. [Online]. Available: https://www.mdpi.com/2072-4292/16/4/665

[33] "Frontiers | Random forest analysis of midbrain hypometabolism using [18F]-FDG PET identifies Parkinson's disease at the subject-level." Accessed: Jul. 02, 2024. [Online]. Available: https://www.frontiersin.org/journals/computational-neuroscience/articles/10.3389/fncom.2024.1328699/full

[34] "Neuromarketing and decision-making: Classification of consumer preferences based on changes analysis in the EEG signal of brain regions - ScienceDirect." Accessed: Jul. 02, 2024. [Online]. Available: https://www.sciencedirect.com/science/article/abs/pii/S1746809423009023

[35] F. Yasmin et al., "AQIPred: A Hybrid Model for High Precision Time Specific Forecasting of Air Quality Index with Cluster Analysis," Hum-Cent Intell Syst, vol. 3, no. 3, pp. 275–295, Sep. 2023, doi: 10.1007/s44230-023-00039-x.

[36] Z. Cui, J. Yang, and Y. Qiao, "Brain MRI segmentation with patch-based CNN approach," in 2016 35th Chinese Control Conference (CCC), Jul. 2016, pp. 7026–7031. doi: 10.1109/ChiCC.2016.7554465.

[37] "Study design in fMRI: Basic principles - ScienceDirect." Accessed: May 18, 2024. [Online]. Available: https://www.sciencedirect.com/science/article/abs/pii/S0278262605001752

[38] "Higher SNR PET image prediction using a deep learning model and MRI image - IOPscience." Accessed: Jun. 06, 2024. [Online]. Available: https://iopscience.iop.org/article/10.1088/1361-6560/ab0dc0/meta

[39] D. Berman et al., "Machine and deep learning models for accurate detection of ischemia and scar with myocardial blood flow positron emission tomography imaging," Journal of Nuclear Cardiology, vol. 32, p. 101797, Feb. 2024, doi: 10.1016/j.nuclcard.2024.101797.

[40] "HTTU-Net: Hybrid Two Track U-Net for Automatic Brain Tumor Segmentation | IEEE Journals & Magazine | IEEE Xplore." Accessed: Jun. 17, 2024. [Online]. Available: https://ieeexplore.ieee.org/document/9103502

[41] I. Ahmad, M. Zhu, G. Li, D. Javeed, P. Kumar, and S. Chen, "A Secure and Interpretable AI for Smart Healthcare System: A Case Study on Epilepsy Diagnosis Using EEG Signals," IEEE Journal of Biomedical and Health Informatics, vol. 28, no. 6, pp. 3236–3247, Jun. 2024, doi: 10.1109/JBHI.2024.3366341.

[42] "Advancing Audio Fingerprinting Accuracy with AI and ML: Addressing Background Noise and Distortion Challenges | IEEE Conference Publication | IEEE Xplore." Accessed: Jun. 19, 2024. [Online]. Available: https://ieeexplore.ieee.org/abstract/document/10475656

[43] R. Bharti et al., "Prediction of Heart Disease Using a Combination of Machine Learning and Deep Learning.," Computational Intelligence and Neuroscience, vol. 2021, p. 8387680, Jul. 2021, doi: 10.1155/2021/8387680.

[44] M. Y. Sikkandar et al., "Hidden Markov Model based Predicting of Alzheimer's Disease with graph cut segmentation using MR Diffusion Tensor Imaging (DTI)," Journal of Intelligent & Fuzzy Systems, vol. 46, no. 2, pp. 4277–4289, Jan. 2024, doi: 10.3233/JIFS-234613.

[45] N. Yamanakkanavar, J. Y. Choi, and B. Lee, "MRI Segmentation and Classification of Human Brain Using Deep Learning for Diagnosis of Alzheimer's Disease: A Survey," Sensors, vol. 20, no. 11, Art. no. 11, Jan. 2020, doi: 10.3390/s20113243.

[46] R. Gupta, S. Kumari, A. Senapati, R. K. Ambasta, and P. Kumar, "New era of artificial intelligence and machine learning-based detection, diagnosis, and therapeutics in Parkinson's disease," Ageing Res Rev, vol. 90, p. 102013, Sep. 2023, doi: 10.1016/j.arr.2023.102013.

[47] V. Purrer et al., "Artificial-intelligence-based MRI brain volumetry in patients with essential tremor and tremor-dominant Parkinson's disease," Brain Commun, vol. 5, no. 6, p. fcad271, 2023, doi:10.1093/braincomms/fcad271.

[48] P. Kamboj, A. Banerjee, V. L. Boerwinkle, and S. K. S. Gupta, "The expert's knowledge combined with AI outperforms AI alone in seizure onset zone localization using resting state fMRI," Front. Neurol., vol. 14, Jan. 2024, doi: 10.3389/fneur.2023.1324461.

[49] "Explainable AI: A Brief Survey on History, Research Areas, Approaches and Challenges | SpringerLink." Accessed: Jun. 17, 2024. [Online]. Available: https://link.springer.com/chapter/10.1007/978-3-030-32236-6_51

[50] C. Lin et al., "Predicting suicidality in late-life depression by 3D convolutional neural network and cross-sample entropy analysis of resting-state fMRI," Brain and Behavior, vol. 14, no. 1, p. e3348, 2024, doi: 10.1002/brb3.3348.

[51] "JPM | Free Full-Text | Neuroinflammation and Schizophrenia: New Therapeutic Strategies through Psychobiotics, Nanotechnology, and Artificial Intelligence (AI)." Accessed: Jun. 19, 2024. [Online]. Available: https://www.mdpi.com/2075-4426/14/4/391

5

Brain Networks in Neuroscience: Tailoring Treatments with AI Insights

Chander Prabha

Chitkara University Institute of Engineering and Technology,
Chitkara University, India
E-mail: prabhanice@gmail.com

Abstract

The human brain is a highly complex network of interconnected neurons, whose intricate dynamics underpin cognitive functions, behaviors, and neurological health. In recent years, the advent of advanced neuroimaging techniques and large-scale data analysis has enabled the detailed mapping of both structural and functional brain networks. Artificial intelligence (AI) is revolutionizing this field by providing powerful tools for processing and interpreting vast amounts of brain data, thus unveiling new dimensions of neural connectivity and functionality. Personalized neuroscience, driven by AI, tailors diagnostic and therapeutic strategies to the unique neural profiles of individuals, enhancing the precision of interventions for neurological disorders. AI algorithms excel in identifying subtle patterns in neuroimaging data, predicting disease progression, and optimizing treatment plans, leading to more effective and individualized patient care. Applications range from early detection of Alzheimer's disease and personalized epilepsy treatment to tailored rehabilitation programs for stroke recovery. This personalized approach not only improves clinical outcomes but also advances our understanding of brain network dynamics in health and disease. Ethical considerations, such as data privacy, algorithmic bias, and accessibility, are paramount in integrating AI into neuroscience. As AI continues to evolve, its synergy with neuroscience promises to unlock deeper insights into brain function and foster innovative solutions for brain health, heralding a new era of personalized and precise neurological care.

Keywords: Neurons, cognitive functions, brain network, genetics, diseases.

5.1 Introduction

Neuroscience is the scientific study of the nervous system, encompassing its structure, function, development, genetics, biochemistry, physiology, pharmacology, and pathology. As an interdisciplinary field, neuroscience integrates knowledge from biology, psychology, medicine, and other disciplines to understand the mechanisms underlying behavior, cognition, and neurological disorders. The ultimate goal of neuroscience is to decipher how the brain and nervous system work, how they develop, how they can be repaired when damaged, and how their functions can be enhanced. Recent years have seen significant advancements in neuroscience, driven by new technologies and interdisciplinary approaches. These approaches are:

- **Brain mapping initiatives**: Projects like the Human Connectome Project and the BRAIN Initiative aim to map the intricate connectivity of the human brain, providing comprehensive blueprints of neural circuits.
- **Artificial intelligence (AI) and machine learning (ML)**: AI and ML algorithms are increasingly used to analyze vast amounts of neuroimaging and genetic data, uncovering patterns and insights that were previously inaccessible.
- **Neuroprosthetics and brain–computer interfaces (BCIs)**: Advances in neuroprosthetics and BCIs are enabling direct communication between the brain and external devices, offering new possibilities for restoring function in individuals with paralysis or other neurological impairments.
- **Neurogenomics**: The integration of genomics with neuroscience is revealing how genetic variations contribute to neural diversity and susceptibility to neurological disorders, paving the way for personalized medicine.

Personalized neurology is an emerging approach that tailors diagnostic and therapeutic strategies to the unique neurological profile of each individual. This paradigm shift from a one-size-fits-all methodology to a more customized approach holds great promise for improving the efficacy of treatments and enhancing patient outcomes. Advances in neuroimaging, genetic analysis, and artificial intelligence (AI) are driving this personalization, allowing for more precise and targeted interventions [1]. Key components of personalized neurology include:

- **Neuroimaging**: Techniques such as MRI, fMRI, DTI, and PET scans provide detailed images of brain structure and function. These images help identify specific patterns of connectivity and activity associated with different neurological conditions.
- **Genetic analysis**: Genomic studies reveal genetic variations that contribute to neurological disorders. Identifying these variations can inform risk assessments and guide the selection of targeted therapies.
- **Clinical data integration**: Combining neuroimaging and genetic data with clinical history and behavioral assessments provides a comprehensive picture of an individual's neurological health.

Neurology is concerned with the human brain and the human brain is a complex organ, comprised of billions of neurons forming intricate networks. These neurons form complex networks that underpin every cognitive process, sensation, emotion, and action we experience [2]. Understanding these networks is crucial for diagnosing and treating neurological disorders, enhancing cognitive function, and personalizing medicine. The advent of AI has revolutionized neuroscience, enabling unprecedented insights into brain networks and fostering personalized approaches to brain health. Neuroscience, the study of the nervous system, leverages various techniques to map and analyze these networks, revealing insights into how the brain operates in health and disease. They can be broadly classified into structural networks (physical connections between brain regions) and functional networks (patterns of synchronized activity) shown in Figure 5.1. Table 5.1 shows the neuroimaging techniques used by these networks along with their key metrics.

The chapter organization is as follows. The introduction section briefly explored the concept of neuroscience, personalized neurology and the structure of brain networks. Section 5.2 presents its importance. Section 5.3 deals

Table 5.1 Neuroimaging techniques and key metrics of brain networks.

Factor brain network	Neuroimaging techniques	Key metrics
Structural networks	MRI, DTI	Connectivity strength White matter integrity
Functional networks	fMRI, EEG, MEG	Coherence Synchronicity Functional connectivity

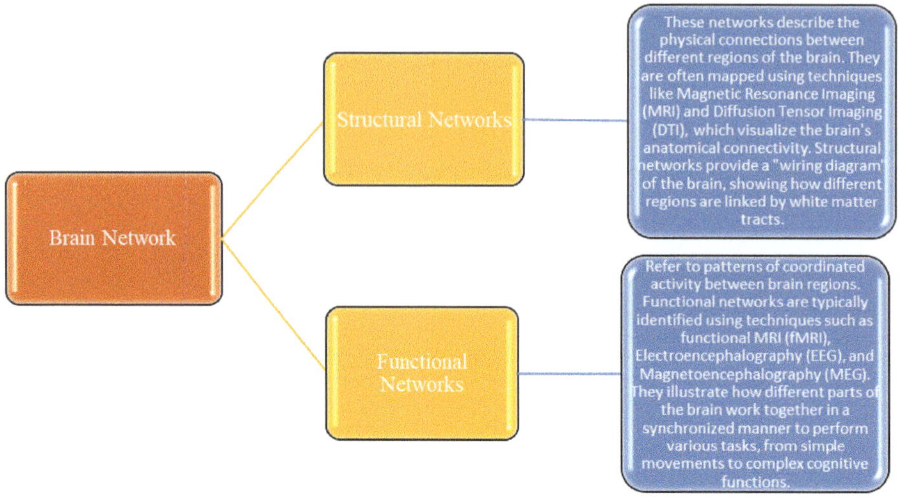

Figure 5.1 Types of brain network.

with personalization in neuroscience via AI-driven analysis. Section 5.4 deals with the role of AI in brain network analysis. Some of the key roles that AI plays in advancing neuroscience are also a part of this section. Section 5.5 deals with some of the case studies and their applications. Finally, Section 5.6 concludes the chapter along with future research directions.

5.2 Importance of Brain Networks

Brain networks refer to the interconnected pathways through which neurons communicate. These are fundamental to understanding how the brain functions in both health and disease. These interconnected systems of neurons and their interactions underpin every aspect of our cognitive and behavioral capabilities (every thought, emotion, and action). The study of brain networks is crucial for several reasons, ranging from insights into normal brain function to advancements in diagnosing and treating neurological disorders [3]. Understanding brain networks is essential for several reasons. These are depicted in Figure 5.2. The normal brain's key aspects include:

- **Cognitive processes**: Brain networks support a wide range of cognitive functions such as memory, attention, language, and executive functions. For example, the default mode network (DMN) is active during restful introspection and mind-wandering, while the central executive network (CEN) is involved in goal-directed behavior and problem-solving.

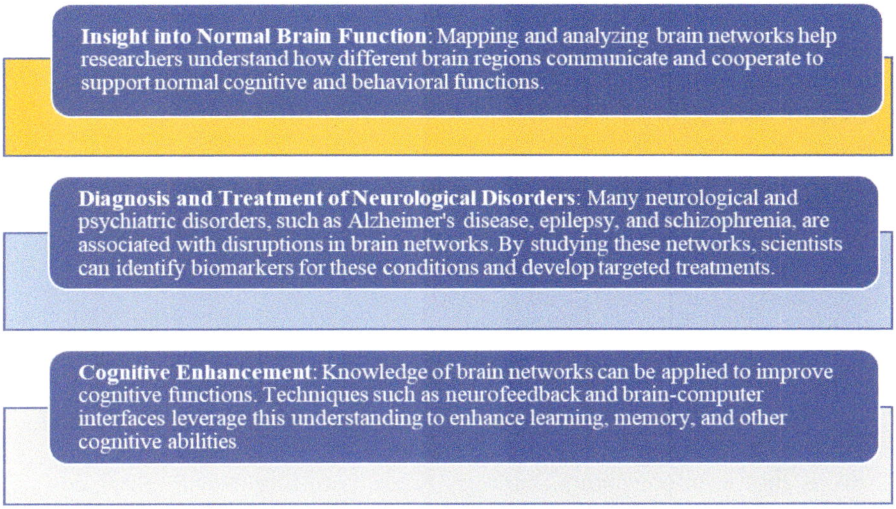

Figure 5.2 Brain network importance.

- **Sensorimotor integration**: Networks such as the sensorimotor network are responsible for processing sensory inputs and coordinating motor outputs, enabling activities ranging from simple reflexes to complex voluntary movements.
- **Emotional regulation**: The limbic network, including structures like the amygdala and prefrontal cortex, plays a crucial role in processing emotions and regulating mood. Understanding these networks can shed light on how emotions are generated and controlled.

Many neurological and psychiatric disorders are characterized by disruptions in brain networks [4]. Studying these networks can provide valuable insights into the pathophysiology of these conditions and inform clinical practice.

- **Neurodegenerative diseases**: Conditions like Alzheimer's disease is associated with specific patterns of network degeneration. For example, the spread of tau pathology in Alzheimer's can be traced along network connections, aiding in early diagnosis and monitoring disease progression.
- **Psychiatric disorders**: Disorders such as schizophrenia and depression involve abnormalities in functional connectivity. Identifying these disruptions can help in diagnosing these conditions and tailoring therapeutic interventions. For instance, altered connectivity in the DMN and CEN is often observed in depression.

- **Epilepsy**: Network analysis can help localize epileptogenic zones, which are areas of the brain responsible for generating seizures. This information is crucial for planning surgical interventions in drug-resistant epilepsy.

Knowledge of brain networks can be leveraged to improve cognitive function and support recovery from brain injuries [4].

- **Neurofeedback**: This technique involves training individuals to regulate their brain activity. By providing real-time feedback on brain network activity, neurofeedback can help enhance cognitive functions such as attention and memory.
- **Brain–computer interfaces**: BCIs can interpret brain signals and translate them into commands for external devices. This technology holds promise for restoring communication and mobility in individuals with severe motor impairments, such as those resulting from spinal cord injuries or stroke.
- **Rehabilitation**: Tailored rehabilitation programs can be developed based on an individual's specific network disruptions. For example, stroke rehabilitation can be enhanced by targeting network plasticity to restore lost functions.

5.3 Personalization in Neuroscience

AI-driven analysis of brain networks is paving the way for personalized neuroscience. By tailoring interventions to the unique network characteristics of individuals, researchers and clinicians can develop more effective diagnostic and therapeutic strategies [5]. Personalized approaches are particularly promising in the fields of:

- **Precision diagnostics**: By analyzing an individual's brain network patterns, AI can improve the accuracy of early diagnosis for neurological disorders. Personalized diagnostic criteria can lead to earlier and more effective interventions.
- **Customized treatment**: Treatments can be tailored to an individual's unique brain network profile. For example, personalized neurostimulation protocols, such as transcranial magnetic stimulation (TMS) or deep brain stimulation (DBS), can be developed to target specific network disruptions.
- **Optimization of cognitive interventions**: Personalized cognitive training programs can be designed based on an individual's brain network

characteristics, enhancing the effectiveness of interventions aimed at improving cognitive functions.

5.3.1 Advancing neuroscience research

The study of brain networks is also driving advances in basic neuroscience research. By mapping the intricate web of connections in the brain, researchers are uncovering fundamental principles of brain organization and function [6]. This knowledge is crucial for developing new theories of brain function and for creating more sophisticated models of brain activity.

5.4 Role of AI in Brain Network Analysis

AI is transforming the field of neuroscience by enabling the analysis of complex brain data, uncovering new insights into brain function, and enhancing the diagnosis and treatment of neurological disorders. AI encompasses various techniques, including machine learning, deep learning, and natural language processing, which are being applied to a wide range of neuroscientific problems. Further AI, particularly ML and DL has significantly advanced the analysis of brain networks. By processing vast amounts of neuroimaging data, AI can uncover patterns and relationships that are beyond human capability to discern. The complexity of brain networks and the vast amount of data generated by neuroimaging techniques presents significant challenges for traditional analytical methods [7]. AI has emerged as a powerful tool in neuroscience, capable of processing and interpreting large datasets with high accuracy.

- **Machine learning**: AI algorithms can identify patterns and relationships within neuroimaging data that might be imperceptible to human researchers. ML models can classify different brain states, predict disease progression, and uncover biomarkers for various conditions. Figure 5.3 shows ML's role in neuroscience.
- **Deep learning (DL)**: This subset of ML involves neural networks that mimic the structure and function of the human brain. DL models can integrate multimodal data, such as imaging, genetic, and clinical information, providing a comprehensive understanding of brain networks. Figure 5.4 shows DL's role in neuroscience.

Some of the key roles that AI plays in advancing neuroscience are:

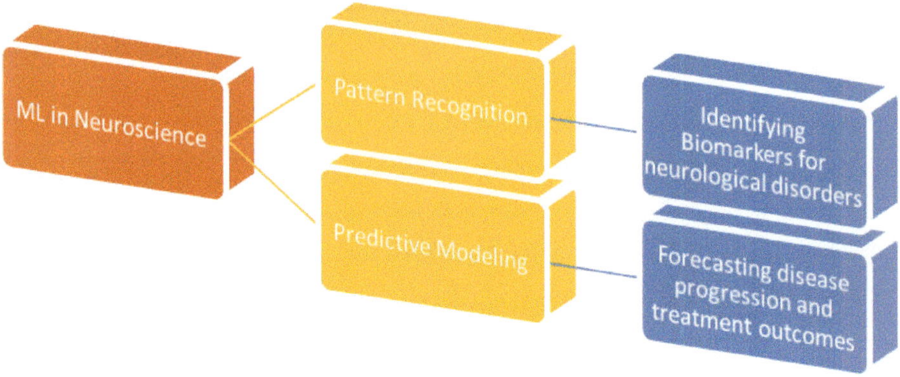

Figure 5.3 ML in neuroscience.

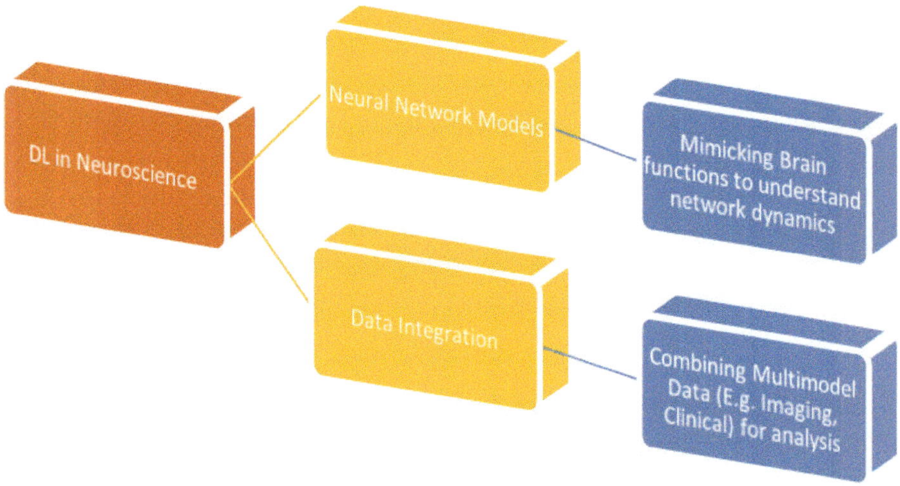

Figure 5.4 DL in neuroscience.

5.4.1 AI in neuroimaging

AI is revolutionizing neuroimaging by improving the acquisition, processing, and interpretation of brain images. Techniques such as MRI, fMRI, PET, and CT scans generate vast amounts of data that are challenging to analyze using traditional methods [8].

- **Image enhancement**: AI algorithms enhance image quality by reducing noise and correcting artifacts, leading to clearer and more accurate images.

- **Automated segmentation**: ML models can automatically segment brain regions, tumors, and lesions from neuroimaging data, improving the efficiency and accuracy of analysis.
- **Pattern recognition**: DL models identify patterns in brain images that may be indicative of neurological conditions, enabling early detection and diagnosis of diseases like Alzheimer's, Parkinson's, and multiple sclerosis.

5.4.2 AI in functional brain analysis

Functional neuroimaging techniques such as fMRI, EEG, and MEG capture dynamic brain activity. AI helps in analyzing these complex data sets to understand brain function and connectivity [9].

- **Functional connectivity analysis**: AI algorithms analyze synchronized activity across different brain regions, revealing functional networks involved in various cognitive and behavioral processes.
- **Brain state classification**: ML models classify different brain states, such as wakefulness, sleep stages, and cognitive load, based on neural activity patterns.
- **Brain–computer Interfaces**: AI enhances BCIs by decoding brain signals in real time, enabling direct communication between the brain and external devices for applications in neuroprosthetics and assistive technologies.

5.4.3 AI in genetic and molecular neuroscience

The integration of genomics with neuroscience is providing deeper insights into the genetic basis of brain function and neurological disorders [10].

- **Genomic data analysis**: AI processes large-scale genomic data to identify genetic variations associated with neurological diseases, leading to better understanding and potential therapeutic targets.
- **Gene expression profiling**: Machine learning models analyze gene expression data to uncover how different genes are regulated in the brain and their roles in neural development and function.

5.4.4 AI in clinical neuroscience

AI is enhancing clinical practice by improving the diagnosis, treatment, and management of neurological and psychiatric disorders [11].

- **Early diagnosis**: AI algorithms analyze clinical and neuroimaging data to detect early signs of neurological conditions, enabling timely intervention and improved outcomes.
- **Personalized treatment**: Machine learning models predict individual responses to treatments based on genetic, neuroimaging, and clinical data, allowing for personalized therapeutic strategies.
- **Predictive modeling**: AI models forecast disease progression and treatment outcomes, helping clinicians make informed decisions about patient care.

5.4.5 AI in cognitive and behavioral neuroscience

AI aids in the study of cognitive processes and behaviors by analyzing complex data from experiments and simulations [12].

- **Cognitive modeling**: AI helps create computational models of cognitive processes, such as decision-making, memory, and attention, providing insights into the underlying neural mechanisms.
- **Behavioral analysis**: Machine learning algorithms analyze behavioral data to understand the neural basis of behaviors and to identify patterns associated with different cognitive states and disorders.

AI enables the customization of brain health interventions based on individual network characteristics. This personalized approach is transforming several areas as shown in Table 5.2.

5.5 Case Studies and Applications

5.5.1 Alzheimer's disease

- **Early detection**: AI models analyze fMRI and PET scans to identify early biomarkers of Alzheimer's disease, allowing for early diagnosis and intervention [18].
- **Treatment personalization**: Machine learning algorithms predict patient responses to different therapeutic options, enabling personalized treatment plans.

5.5.2 Epilepsy

- **Seizure prediction**: AI analyzes EEG data to predict the onset of seizures, allowing for timely intervention and improved management of epilepsy.

Table 5.2 Personalization areas via AI.

Areas	Approach	Response
Diagnosis	**AI algorithms:** Early detection of disorders like Alzheimer's, epilepsy [13] **Biomarker identification:** AI algorithms analyze neuroimaging and genetic data to identify biomarkers for early diagnosis of conditions such as Alzheimer's disease and Parkinson's disease. Early detection allows for timely intervention, which can slow disease progression and improve quality of life [14]. **Predictive modeling:** Machine learning models predict the course of neurological diseases based on initial patient data. This helps clinicians anticipate future challenges and plan appropriate interventions.	**Precision diagnostics:** Tailoring diagnostic criteria to individual brain network patterns
Treatment	**Neurostimulation personalization:** Techniques like transcranial magnetic stimulation (TMS) and deep brain stimulation (DBS) can be optimized using AI to target specific brain regions and neural circuits involved in disorders such as depression and epilepsy. **Pharmacogenomics:** AI-driven analysis of genetic data enables the customization of medication regimens. By predicting how individual patients will respond to different drugs, AI helps in selecting the most effective treatments with the least side effects [15].	**Pharmacotherapy:** Customizing drug regimens based on network analysis
Cognitive enhancement	**Brain–computer interfaces:** Adaptive systems for learning and rehabilitation [16] **Customized rehabilitation programs**: AI helps design personalized rehabilitation strategies for patients recovering from stroke or brain injuries by analyzing their unique brain network disruptions. Tailored programs can enhance motor and cognitive recovery by focusing on the most affected neural pathways [17].	**Neurofeedback and brain–computer interfaces:** AI-driven neurofeedback and BCIs offer personalized training to help patients with motor impairments, enabling them to regain control over lost functions. These systems adapt to the individual's neural signals, providing more effective and customized therapeutic experiences.

- **Surgical planning**: Machine learning models help to identify epileptogenic zones in the brain, guiding surgical interventions to treat drug-resistant epilepsy.

5.5.3 Stroke rehabilitation

- **Recovery prediction**: AI models predict recovery trajectories based on initial clinical assessments and neuroimaging data, helping clinicians design personalized rehabilitation programs.
- **Neurofeedback**: AI-driven neurofeedback systems train patients to regulate their brain activity, enhancing cognitive and motor recovery after stroke.

Table 5.3 further presents some examples and outcomes of case studies in different areas of neuroscience.

5.5.4 Ethical considerations

AI-enhanced personalized treatments are in great demand due to their precision; however, several ethical issues must be addressed while doing these treatments. Some of these are:

- **Data privacy**: Ensuring the confidentiality and security of sensitive patient data is paramount, given the extensive use of neuroimaging and genetic information [21].
- **Algorithmic bias**: AI models must be carefully designed and validated to avoid biases that could lead to unequal treatment across different populations.
- **Accessibility and equity**: The benefits of AI-driven personalized treatments should be made accessible to all patients, regardless of socioeconomic status or geographic location, to prevent disparities in healthcare.

Table 5.3 Case studies examples and outcomes.

Case studies	Example	Outcome
Personalized medicine [19]	AI-driven analysis of fMRI data to tailor cognitive behavioral therapy (CBT) for depression.	Improved efficacy and reduced treatment duration.
Neurorehabilitation	Custom neurofeedback programs for stroke recovery.	Enhanced motor function and cognitive recovery.
Ethical and future considerations [20]	Data privacy: Ensuring the confidentiality of sensitive neuroimaging data.	Bias and fairness: Avoiding biases in AI algorithms that could lead to unequal treatment.

5.6 Discussion

The integration of AI into the study of brain networks in neuroscience represents a transformative shift towards more personalized and effective treatments for neurological disorders. Brain networks, the intricate systems of interconnected neurons, are crucial for understanding cognitive functions and behaviors. Disruptions in these networks often underlie various neurological and psychiatric conditions. AI's ability to process and analyze complex datasets has become instrumental in mapping these networks and deriving actionable insights for clinical practice.

AI algorithms excel in identifying subtle patterns within neuroimaging data that may be imperceptible to human analysts. This capability is particularly beneficial in the early detection of neurodegenerative diseases such as Alzheimer's and Parkinson's. By analyzing structural and functional MRI scans, AI can pinpoint early biomarkers, enabling interventions at a stage where they can be most effective. This proactive approach not only delays the progression of diseases but also improves the overall quality of life for patients.

Tailoring treatments to individual patients is one of the most significant advantages of incorporating AI insights into neuroscience. For example, in epilepsy treatment, AI can analyze EEG data to predict seizures, allowing for timely medical interventions. Furthermore, AI-driven models can help identify the precise brain regions responsible for epileptic activity, guiding surgical procedures to maximize efficacy while minimizing collateral damage to healthy brain tissue.

Pharmacogenomics, another area benefiting from AI, involves customizing drug therapies based on a patient's genetic makeup. AI algorithms can predict how different patients will respond to specific medications, thereby optimizing treatment plans to enhance efficacy and reduce adverse effects. This personalized approach is particularly beneficial in managing psychiatric conditions like depression, where treatment responses can vary widely among individuals.

AI's role extends to rehabilitation, especially for stroke recovery. By analyzing a patient's unique brain network disruptions, AI can develop personalized rehabilitation protocols that target the most affected neural pathways, thereby accelerating recovery. Additionally, AI-enhanced neurofeedback systems can provide real-time monitoring and adjustment of therapeutic activities, helping patients regain lost functions more effectively.

The integration of AI in neuroscience is not without challenges. Data privacy is a significant concern, given the sensitive nature of neuroimaging and genetic information. Ensuring that AI models are free from biases that could lead to unequal treatment is also critical. Moreover, making these advanced treatments accessible to all patients, regardless of socioeconomic status, remains a crucial goal to prevent disparities in healthcare.

Thus, AI is revolutionizing the field of neuroscience by providing deep insights into brain networks and enabling highly personalized treatment strategies. From early diagnosis and targeted therapies to optimized rehabilitation protocols, AI enhances our ability to treat neurological disorders effectively. As we continue to refine these technologies and address ethical considerations, the potential for AI to transform neuroscience and improve patient outcomes is immense.

5.7 Conclusion and Future Scope

The study of brain networks is at the forefront of neuroscience research, offering profound insights into the workings of the human brain. The integration of AI into this field is revolutionizing one's ability to analyze complex brain data, leading to more personalized and effective approaches in diagnosis, treatment, and cognitive enhancement. As AI technology continues to advance, it holds the potential to unlock even deeper understanding and innovative solutions for brain health. This introduction sets the stage for a deeper exploration of how AI and personalized approaches are transforming our understanding and management of brain networks in neuroscience. Thus, the integration of AI into neuroscience is unveiling new dimensions of brain network analysis, leading to personalized approaches in diagnosis, treatment, and cognitive enhancement. As AI technology continues to evolve, its application in brain health promises to revolutionize the understanding and interaction with the brain, paving the way for a future of personalized and precise neuroscience. Future research should focus on Integrative Models and real-time analysis. In integrative modeling one can combine AI with other technologies (e.g., genomics, proteomics) and real-time deals with developing systems for real-time monitoring and intervention.

Acknowledgments

We are deeply grateful to the editors for their invaluable assistance in writing this chapter. Their insights and expertise played a crucial role in shaping the content. Their guidance ensured both the accuracy and clarity of the information presented. Without their support, this work would not have reached its full potential.

References

[1] Bassett, D. S., & Sporns, O. (2017). Network neuroscience. Nature Neuroscience, 20(3), 353-364.

[2] Esteva, A., et al. (2019). A guide to deep learning in healthcare. Nature Medicine, 25(1), 24-29.

[3] Poldrack, R. A., et al. (2020). The importance of standards for sharing brain-imaging data. Nature Neuroscience, 23(12), 1614-1623.

[4] Gerloff, C., Konrad, K., Bzdok, D., Büsing, C., & Reindl, V. (2022). Interacting brains revisited: A cross-brain network neuroscience perspective. *Human Brain Mapping*, *43*(14), 4458–4474. doi:10.1002/hbm.25966

[5] Bilek, E., Zeidman, P., Kirsch, P., Tost, H., Meyer-Lindenberg, A., & Friston, K. (2022). Directed coupling in multi-brain networks underlies generalized synchrony during social exchange. NeuroImage, 252, 119038.

[6] Maas, D. A., & Douw, L. (2023). Multiscale network neuroscience in neuro-oncology: How tumors, brain networks, and behavior connect across scales. *Neuro-Oncology Practice*, *10*(6), 506–517. doi:10.1093/nop/npad044

[7] Ardesch DJ, Scholtens LH, De Lange SC, et al. Scaling principles of white matter connectivity in the human and nonhuman primate brain. Cereb Cortex. 2022;32(13):2831–2842

[8] Barbey AK. Network neuroscience theory of human intelligence. Trends Cogn Sci. (2018) 22:8–20. doi: 10.1016/j.tics.2017.10.001

[9] VanRullen, R., and Reddy, L. (2019). Reconstructing faces from fMRI patterns using deep generative neural networks. *Commun. Biol.* 2, 1–10. doi: 10.1038/s42003-019-0438-y

[10] Yang, Z., Nasrallah, I. M., Shou, H., Wen, J., Doshi, J., Habes, M., et al. (2021). A deep learning framework identifies dimensional representations of Alzheimer's disease from brain structure. *Nat. Commun.* 12, 1–15. doi: 10.1038/s41467-021-26703-z

[11] Gong, C., Jing, C., Chen, X., Pun, C. M., Huang, G., Saha, A., ... Wang, S. (2023). Generative AI for brain image computing and brain network computing: a review. *Frontiers in Neuroscience*, *17*. doi:10.3389/fnins.2023.1203104

[12] Ibanez, A., Kringelbach, M. L., & Deco, G. (2024). A synergetic turn in cognitive neuroscience of brain diseases. *Trends in Cognitive Sciences*, *28*(4), 319–338. doi:10.1016/j.tics.2023.12.006

[13] Ibanez, A., & Zimmer, E. R. (2023). Time to synergize mental health with brain health. *Nature Mental Health*, *1*(7), 441–443. doi:10.1038/s44220-023-00086-0

[14] Venkadesh, S., & Van Horn, J. D. (2021). Integrative models of brain structure and dynamics: Concepts, challenges, and methods. *Frontiers in Neuroscience*, *15*. doi:10.3389/fnins.2021.752332

[15] Deshpande, G., & Jia, H. (2020). Multi-level clustering of dynamic directional brain network patterns and their behavioral relevance. *Frontiers in Neuroscience*, *13*. doi:10.3389/fnins.2019.01448

[16] Li, Q., Calhoun, V. D., & Iraji, A. (2024). Revealing complex functional topology brain network correspondences between humans and marmosets. *Neuroscience Letters*, *822*(137624), 137624. doi:10.1016/j.neulet.2024.137624

[17] Ienca, M., & Ignatiadis, K. (2020). Artificial intelligence in clinical neuroscience: Methodological and ethical challenges. *AJOB Neuroscience*, *11*(2), 77–87. doi:10.1080/21507740.2020.1740352

[18] Pereira, A., Jr, Garcia, J. W., & Muotri, A. (2023). Neural stimulation of brain organoids with dynamic patterns: A Sentiomics approach directed to Regenerative Neuromedicine. *NeuroSci*, *4*(1), 31–42. doi:10.3390/neurosci4010004

[19] Kaushal, C., Islam, M. K., Singla, A., & Al Amin, M. (2022, January 10). An IoMT-based smart remote monitoring system for healthcare. *IoT-Enabled Smart Healthcare Systems, Services and Applications*, pp. 177–198. doi:10.1002/9781119816829.ch8

[20] Prabha, C., Singh, J., Agarwal, S., Verma, A., & Sharma, N. (2022). Introduction to computational intelligence in healthcare. In *Computational Intelligence in Healthcare* (pp. 1–15). Boca Raton: CRC Press.

[21] Verma, K., Bhardwaj, S., Arya, R., Islam, M. S. U., Bhushan, M., Kumar, A., & Samant, P. (2019). Latest tools for data mining and machine learning. *International Journal of Innovative Technology and Exploring Engineering*, *8*(9S), 18–23. doi:10.35940/ijitee.L2195.1089S219

6

Brain Network Dynamics: Implications for Health and Disease

Tamanna Haque Ritu[1], Hirak Mondal[2], Anindya Nag[3], and Riya Sil[4]

[1]Pharmacy Discipline, Khulna University, Bangladesh
[2]Department of Computer Science and Engineering, North Western University, Bangladesh
[3]Computer Science and Engineering Discipline, Khulna University, Bangladesh
[4]Computer Science and Engineering, Presidency University, India
E-mail: tamannahaqueritu@gmail.com; hirakmondal.cse@gmail.com; anindyanag@ieee.org; riyasil1802@gmail.com

Abstract

Understanding both healthy and sick brain states is greatly enhanced by the study of brain network dynamics. Graph theory is used to comprehend structural and functional networks as this chapter explores the complex neural network dynamics within the healthy human brain. The complexity and interconnection of the brain are shown by the connection between the anatomical and functional connections of various networks. We scrutinize further the changes in these dynamics in different illness states, concentrating on Parkinson's disease (PD), schizophrenia, and autism spectrum disorder (ASD). Changes in brain connection during tasks and rest in ASD are revealed by graph theoretical methods as unique patterns and architectures. Because schizophrenia affects neuronal networks so deeply, it causes particular changes in brain dynamics. Both non-motor and motor symptoms of PD are based on particular network abnormalities and include bradykinesia and tremors. Furthermore, we address the possible therapeutic insights provided by the modification of brain networks by PD therapies. These results taken together highlight how crucial it is to comprehend brain network dynamics to create successful therapies and interventions. Using this thorough investigation, we aim to progress our knowledge of brain network changes in various diseases and advancing into more focused and efficient therapeutic strategies. This review lays the groundwork for the next studies in this developing area by highlighting the important function of brain network dynamics in both health and disease.

Keywords: Brain network, Parkinson's disease, autism spectrum disorder, graph theory, fMRI, schizophrenia.

6.1 Introduction

The brain is responsible for regulating several physiological functions such as cognition, memory, emotion, sensory perception, motor control, visual processing, respiration, thermoregulation, and appetite. The human brain is thought to contain 80–100 billion nerve cells as well as there are 100 trillion connections among them. A complex network of connections and interactions exists between different brain regions for the brain to function. One well-known hypothesis of brain function in the past, which dates back to the late nineteenth century, concentrated on giving discrete brain regions – which frequently include millions of neurons unique, distinct tasks. It has become more widely recognized in recent years that these specialized brain regions interact with one another to support more sophisticated mental activities rather than functioning alone. The concept that the brain is a network of interconnected and interactive parts is gaining popularity in contemporary research. It necessitates an examination of both functional integrations, or the synthesis of information from different brain regions, and segregation, or the localized processing of information. Over the last decade, there has been a significant advancement in non-invasive neuroimaging methods, such as diffusion spectrum imaging (DSI) and functional magnetic resonance imaging (fMRI). These methods have enabled the investigation of patterns of interconnections in the living human brain. The concept of the human brain being a network composed of numerous interconnected elements has gained considerable support as a result of these findings. The development of techniques to extract significant fundamental concepts from intricately interconnected systems is necessary to provide answers to issues like in what precise way can individual brain regions interact dynamically to generate brain networks or systems at a broad scale and in the end, how these networks facilitate human behavior and cognition. Utilizing the mathematical framework of complex networks, the interactions' patterns can be succinctly described as brain graphs. In these graphs, each brain region is depicted as a node, and every connection is illustrated as an edge. Through these methods, it has been demonstrated that human brain networks display features like a small-world structure, potentially bolstering cognitive functions directly. Figure 6.1 illustrates the structure and function of the human brain.

Furthermore, an increasing body of research suggests that these traits undergo modifications during periods of sickness. These changes might potentially serve as useful biomarkers for neurological and mental disorders, while also aiding our comprehension of the underlying processes behind cognitive impairments [1]. The key topics explored in this chapter are as follows.

- This chapter delves into the topic of healthy brain network dynamics by utilizing graph theory to provide a thorough understanding.
- This chapter investigates changes in brain network dynamics in individuals with autism, schizophrenia, and Parkinson's disease.
- This chapter employs sophisticated analytical methods to uncover complex patterns in brain connectivity.
- This chapter provides valuable insights into the treatment implications that arise from network changes in Parkinson's disease.

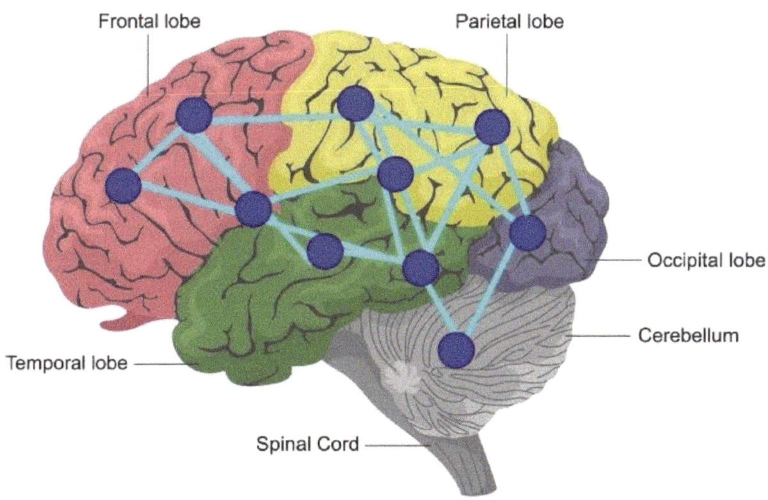

Figure 6.1 Structure and function of the human brain.

The subsequent portion of this chapter is structured in such a manner. Section 6.2 presents an analysis of the neural network dynamics in the healthy human brain. This includes topics such as graph theory, dynamic networks, structural brain networks, functional networks, and the relationship between structural and functional connectivity. Section 6.3 discusses brain network alterations in various illness states, such as autism, schizophrenia, and Parkinson's disease. It specifically focuses on brain network dynamics in these conditions. Section 6.4 presents the conclusion.

6.2 Neural Network Dynamics in the Healthy Human Brain

Network-based methodologies may be used to investigate neuroscientific topics across several spatial dimensions to comprehend the patterns of interaction among genes, proteins, neurons, cortical columns, and extensive brain regions.

6.2.1 Graph theory

A particular method for identifying, evaluating, and visualizing network design is based on graph theory and provides an exceptionally broad range of tools [2]. Graph theory is a branch of mathematics that has traditionally focused on the analysis and investigation of intricate networks. A set of nodes V, often referred to as vertices, linked by connections E, sometimes referred to as edges, is commonly known as a network or graph. An adjacency matrix A is the most basic representation of a network, where the strength of the connection between nodes *i* and *j* is denoted by the *ij*th element. It is possible for edges to be undirected ($A_{ij} = A_{ij}$) or directed ($A_{ij} \neq A_{ij}$). Weighted or unweighted edges are another option. When edges are present in an unweighted (or binary) network, their weight is 1, and when they are absent, their weight is 0. Any integer weight assigned to an edge in a weighted network indicates how strongly linked the two nodes it connects are to one another. Although there are other kinds of networks (such as ones with many types of edges or vertices), these are the most prevalent network components applied to neuroscience [1].

6.2.2 Dynamic network

Studies of the intuitive ideas put out have typically been carried out in the setting of static networks, or networks with a topology that remains constant across time. However, because functional brain networks fluctuate in configuration across timescales spanning from milliseconds to years, they are dynamic rather than static [2]. The brain is constantly active, even while we're not consciously doing anything, as scientists have known since Hans Berger's revolutionary recordings of brain activity. Initially, researchers employed methods such as EEG and MEG to analyze brain activity. They considered the continuous fluctuations in the brain's blood flow signals, detected by fMRI, as noise that needed to be filtered out and concentrated on reactions to certain stimuli or activities. This idea was called into question, nevertheless, when organized patterns in these spontaneous brain impulses were found. Researchers discovered that specific brain regions remained persistently active even while the body was at rest; this pattern of activity was dubbed the "default mode" [3]. This network came to be linked to mental vacancies and daydreaming. Additional investigation uncovered additional attention-related and cognitive control-related networks, each with unique patterns of activity. Remarkably, several networks had a propensity to remain active during quiet periods, implying intricate interactions among distinct brain regions. With the development of technology, scientists were able to use different analysis methods to dissect these resting-state brain activities into distinct networks. They discovered that these networks may be divided into a small number of distinct roles. While some networks process complicated stimuli and activities, others respond primarily to motor or sensory tasks. This change in knowledge has fundamentally changed our perspective of the brain, enabling us to perceive its intrinsic activity as an essential component of its function rather than only considering it as a reaction to external stimuli [4].

Researchers in the fields of physics, engineering, mathematics, and other related disciplines are now working on extending the universal network characteristics to multilayer networks in order to better understand and represent the dynamics of these networks [5]. In a temporal multilayer network, an adjacency tensor is formed by connecting each node in the network to itself in the neighboring time frames. Adjacency matrices, which represent the state of a network at a certain instant, are connected by this. Several traditional network diagnostic measures, such as route length and clustering coefficient, were first established based on multilayer networks and continue to be widely used in various academic fields. The recognition of the hypothesis that the brain functions as a dynamic network is growing [6].

6.2.3 Structural brain network

When constructing structural brain networks using diffusion MRI (dMRI) data, it is common to use a standard brain atlas to define the network nodes.

The definition of network edges is based on a white matter microstructure measure, like the average fractional anisotropy or the number of streamlines along a tract that connects nodes *i* and *j* [7]. The imaging modality being utilized has a significant impact on how nodes and edges are defined. Prior to developing a structural network using neuroimaging data, it is essential to establish the areas of interest (network nodes) and interaction measurements (network edges). Because the definition of morphometric networks' edges is dependent on cross-subject correlations between surface, cortical thickness, and grey matter volume, among other regional morphometric parameters, they differ from structural networks. As low as 3% of the potential edges are present in healthy structural brain networks, based on the latest diffusion imaging MRI procedures [8]. These networks have poor long-range connectivity and dense local connectivity, which are characteristics of tiny world networks. The average anatomical connection distance is less than that of randomly connected physical networks, indicating a tendency to minimize wiring expenses [9]. The aforementioned discoveries have given rise to the concept that the structure of the brain may be comprehended as the result of evolution, aiming to improve network efficiency while limiting the expenses connected to wiring, conduction speed, and cytoplasmic volume [10].

These networks have a highly ordered design consisting of modules of closely linked nodes that are nested or arranged hierarchically [11]. This architecture probably offers the structural elements required for quick adaptability to a changing environment across several functional dimensions, serving as a vital support system for cognitive function. A limited selection of brain areas exhibits a particularly high degree of hub status within these modules. The hub nodes display a notable correlation with each other, serving as the central structural core or affluent club of the network, facilitating the majority of the shortest pathways. According to some, the central hub areas of the brain may serve as the communication center [12].

6.2.4 Functional networks

The blood oxygen level-dependent (BOLD) signal, which is an indicator of neuronal activity, is used in functional MRI (fMRI) investigations to indirectly evaluate the activity of a specific brain area by monitoring changes in oxygenation. Subjects can be asked to complete a cognitive activity while undergoing fMRI data collection, or they can lie still in the scanner and do nothing; this is known as the "resting state" [13]. Even though the concept of the "resting state" is cognitively unconstrained (a fact that originally sparked a heated debate regarding its applicability to brain function research [13]), resting brain fluctuations and resting-state networks generate topographical patterns that are typically uniform among people [14], scanning sessions [15, 16], and imaging facilities [17]. The general structure of these patterns remains stable across global state transitions, such as waking and sleeping [18], or other states of consciousness. However, particular functional connections may undergo changes influenced by experiences, such as specific sensorimotor training [19, 20]. Resting-state networks (RSNs) are not exclusive to humans; studies have shown their presence in the brains of rodents and macaque monkeys as well [21, 22].

Usually, the nodes of networks created from fMRI correspond to large-scale areas that are specified by the atlases. While fMRI boasts outstanding spatial resolution, detecting BOLD activity with precision down to 1–2 mm, EEG and MEG offer superior temporal resolution, capturing electromagnetic activity at sampling frequencies of up to 2000 Hz. Traditionally, nodes in MEG or EEG data were detected by placing sensors on the scalp. However, in recent times, nodes inside the cortical volume have also been detected using source localization methods [23]. The temporal patterns of regional activation in fMRI, EEG, and MEG networks are compared using methods such as synchronization or a Pearson correlation coefficient. This is done to determine the strength of connections between these nodes. There is a substantial and expanding body of research that consistently shows many important topological characteristics in working brain networks. One of these architectures is a small-world architecture [24], which allows for fast information flow and synchronization while limiting the need for wiring, and also achieves a balance between local processing and global integration [25]. Functional brain networks have a hierarchical modular arrangement, similar to structural networks, especially during periods of rest [26]. The large-scale modules consist of brain regions that are strongly functionally coupled and correspond to well-established subnetworks responsible for certain cognitive tasks [27, 28]. These modules are found in several task domains. It is unknown to what extent functional network architecture can be used cognitively. According to preliminary research, verbal memory scores [29] and IQ [30] are two cognitive factors that are linked to network architecture during rest.

6.2.5 Relationship of structural and functional connectivity

The brain's structural connection patterns, or connectome, have an anatomical foundation, as suggested by the consistency of RSN topography [31]. This theory was tested using neurocomputational models; the proposal suggests a connection between the emerging patterns of functional connectivity, which arise from the anatomical coupling matrix of inter-regional projections, and the spontaneous dynamics of the brain [32]. Empirical research conducted on nonhuman primates revealed significant overlap in resting-state functional linkages and anatomical projections identified by tract-tracing experiments [3, 34]. A significant statistical correlation was discovered between the structural and functional connections in human studies that combined diffusion imaging/tractography with resting-state fMRI recordings from the same group of individuals. Additional research has investigated the anatomical linkages concerning functional RSNs and found that they are related to functionally coherent networks [35, 36]. Figure 6.2 depicts the use of EEG/dMRI combination to examine the correlation between structural and functional connectivity.

An extensive network study of both structural and functional connectivity revealed an intricate and intricate correlation between structural and

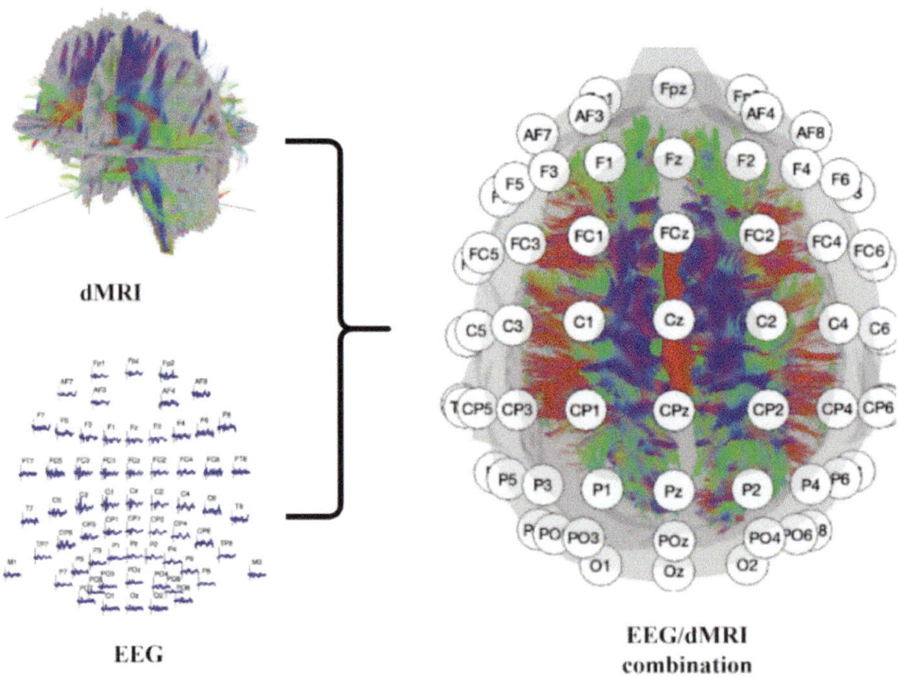

Figure 6.2 EEG and dMRI together to look into the connection between anatomical and functional connectivity.

functional connections, even when functional connectivity mirrored the underlying structural networks. For instance, a substantial amount of meaningful functional connections were discovered between pairs of areas that lacked a direct structural connection. Network models indicate that each functional link in the network is likely the result of several dynamic effects that travel through it via various channels, many of which are indirect and need multiple intermediary steps.

6.3 Brain Network Alterations in Disease States

Human brain networks have been developed and analyzed in many modified states, including mental diseases, neurological ailments, and other forms of damage. The field of "connectomics," which focuses on studying these networks, provides a novel framework for comprehending these altered states [37]. A significant benefit of human connectomics is that it will contribute to a better knowledge of the molecular foundations, including the genetic basis, of brain and mental diseases [38]. Mapping patterns of structural brain connectivity and determining how these relate to new patterns of brain dynamics are the main goals of human connectomics [39]. It has been demonstrated that almost all mental and brain diseases, as well as brain damage and recovery, are related to disrupted interactions among different brain regions [40].

Finding genetic risk factors and comprehending the biochemical underpinnings of the majority of common brain and mental diseases are highly challenging due to the genetic foundation of these disorders and their heterogeneous clinical manifestations. The study of intermediate phenotypes – phenotypes that fall somewhere between genetic variants and clinical symptoms – is one possible strategy to address this complexity. By illuminating the distinct genetic or biological elements that underlie each category, these intermediate phenotypes may aid in the classification of diverse illnesses into more logical subgroups. One interesting option for such intermediate phenotypes is the connectome, which includes the complex network of connections in the brain and its dynamic activity patterns. It combines a variety of environmental and genetic factors and provides a wide range of possible biomarkers to help describe brain illnesses. There is increasing interest in creating metrics that describe both normal variations and anomalies in brain anatomy and function as research on brain networks advances. These measurements could reveal elements of environmental and genetic illness causes as well as act as objective indicators for prognosis and treatment planning [2]. The objective of extensive initiatives like the Human Connectome Project is to analyze variations in the structural organization of brain networks across people without any health issues, which serves as a guide for comprehending abnormalities connected to disease. Like many other common diseases, it is proposed that brain and mental problems are quantitative deviations from normal rather than qualitative differences. These illnesses most likely arise from the combined effects of multiple genetic variants, constituting the extremes of continuous phenotypic distributions, rather than being driven by a single genetic component. It is still not quite clear, yet, how various brain network measurements connect to these phenotypic features. With the increased availability of information on both healthy and disordered brain networks, there is a chance to investigate how connection measures might transform how prevalent mental diseases are classified and understood [2].

6.3.1 Brain network dynamics in autism

Autism spectrum disorders (ASDs), also known as heritable neurodevelopmental conditions, present a diverse range of behavioral symptoms that can vary greatly in severity [41]. The main behavioral impairments associated with ASD, as defined by the American Psychiatric Association (2013), include challenges in social communication and interaction, repetitive behaviors, narrow interests, and reduced responsiveness to external stimuli [42]. While these deficits serve as primary diagnostic criteria for ASD, it's important to acknowledge that not all individuals on the autism spectrum exhibit every symptom. Moreover, there is a wide spectrum of social, emotional, and cognitive impairments among those diagnosed with ASD. Nevertheless, because of the diversity among individuals with ASD, it is very difficult to identify a singular underlying cause in the nervous system and create successful therapies. Therefore, a primary objective of ASD research is to get a comprehensive understanding of the neurological foundations of ASD from several viewpoints.

6.3.1.1 Brain connectivity in autism spectrum disorder (ASD)

Examining how brain areas interact with one another via social brain networks to complete complicated tasks instead of focusing on activity within specific brain regions can help explain functional brain abnormalities in autism. Early disruptions in the physical structure and functional connections of local circuits in persons with ASD may affect the development of large-scale brain networks necessary for sophisticated cognitive processing [43]. These anatomical and functional anomalies might hinder the normal process of brain circuitry reconfiguration, which is necessary for the development of fully integrated networks and the comprehension and induction of complex social behavior [44].

6.3.1.2 Connectivity in task-related brain networks

Research on task-based functional connectivity in individuals with ASD has revealed alterations in connections within multiple brain networks responsible for processing intricate social and emotional information. In their study, Just et al. [45] (2004) discovered that individuals diagnosed with ASD had a decrease in connectivity across several high-level association brain areas during a phrase comprehension test. The authors hypothesized that cognitive impairments in individuals with ASD may result from a widespread lack of connection between brain areas that are crucial for integrating information [46]. Several prior task-based connectivity studies have shown underconnectivity in brain regions linked to tasks in persons with ASD, therefore supporting this notion. It encompasses fronto-parietal connections in working memory tasks [47], in theory of mind [48], in visuomotor coordination tasks [49], in visual imagery tasks [50], in executive functioning tasks [51], and in response inhibition tasks [52].

Significantly, new functional and structural neuroimaging research has shown a connection between behavioral abnormalities in ASD and changes in brain connections. Abrams et al. [53] investigated the functional connectivity of the bilateral posterior superior temporal sulcus in children with ASD when they were in a state of rest. In neurotypical people, the posterior superior temporal sulcus plays a vital role in the perception of human speech. However, people with ASD do not exhibit activation in this specific location. Children diagnosed with ASD had diminished connectivity across many reward-related brain regions, including the posterior superior temporal sulcus, insula, orbitofrontal cortex, nucleus accumbens, and ventromedial prefrontal cortex [54]. The researchers discovered a significant association between decreased connectivity in the posterior superior temporal sulcus and reward circuitry, and an increase in communication difficulties. These results imply that children with ASD may find the human voice less intrinsically appealing, which could have a detrimental effect on language development.

6.3.1.3 Resting state connectivity fMRI

The fact that task-based functional MRI investigations need participant engagement and are therefore usually restricted to older, higher-functioning autistic children is one of their main limitations. Later-life brain alterations might be a result of aberrant social development rather than its cause, and results might not apply to children who are non-verbal, have lower functioning levels, or have more severe social deficiencies. In addition, it is crucial for routine fMRI investigations that aim to compare the brain activity of individuals with ASD and those without ASD to consider and address any existing differences in baseline performance on the task. Additionally, it is important to ensure that the test used is carefully constructed to specifically address the problems related to the neurobiology of ASD. Many of these concerns are allayed by a relatively recent method of understanding functional brain connection: investigating the connections between brain regions during rest rather than during task execution. Resting state-functional connectivity refers to the measurement of the correlation or synchronization of brain activity across different regions of the brain when the individual is in a resting state, without any specific task or stimulation. MRI is a technique that utilizes fMRI to discover and analyze spontaneous fluctuations in brain activity at low frequencies (<0.1 Hz) without the need for a specific task. It is also used to identify brain regions that operate together, known as intrinsic functional brain networks [55]. The functional brain architecture is believed to consist of several interacting large-scale neural networks, as evidenced by studies that reveal synchronized activity across the brain during rest and task conditions. Recently, research using neuroimaging to define ASD has started to describe the functional connections both within and between brain networks. Figure 6.3 shows Abnormalities in resting-state and task-dependent activity, as well as functional connectivity, have been seen in the regions of the brain responsible for self-processing in individuals with ASD.

Multiple extensively replicated brain networks have been revealed by rs-fcMRI investigations in neurotypical persons. We are going to focus on networks related to the etiology of ASD. The "default mode network" (DMN) is a functional connectivity network that is crucial for social cognition. It includes the inferior parietal lobule, hippocampus formation, lateral temporal cortex, medial prefrontal cortex (MPFC), lateral temporal cortex and posterior cingulate cortex (PCC). This network has been extensively studied. Owing to its inactivation during goal-directed tasks and its inverse relationship with the "attentional control network," the DMN has been connected to inwardly focused cognition [56]. Research consistently suggests that individuals with ASD, including toddlers, adolescents, and adults, may have decreased connectivity between nodes in the DMN. This is consistent with the recognized role of certain nodes in the DMN during social cognitive activities, such as witnessing social interactions. It also relates to the observed

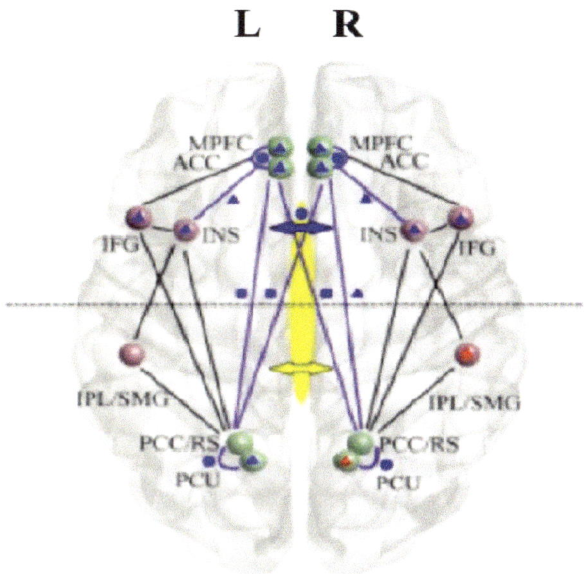

Figure 6.3 Abnormalities in resting-state and task-dependent activity, as well as functional connectivity, have been seen in the regions of the brain responsible for self-processing in individuals with ASD.

behavioral impairments often seen in individuals with ASD, such as improper theory of mind (TOM) processing and difficulties in social relationships. Nevertheless, considering the dynamic nature of its interactions with other brain systems, it is questionable if the DMN is the only functional network affected by ASD.

The "salience network" has lately gained significant interest in ASD research [57]. It involves identifying the most significant information in one's surroundings, including social signals. The primary components of the salience network are the anterior insula and anterior cingulate cortex. Neuroimaging studies suggest that these regions may play a role in the experience of social isolation and cognitive control. A recent study examined the links within extensive brain networks of children with ASD and compared them to a group of individuals without ASD who were similar in characteristics. Studies have shown that several brain networks, such as the DMN and the Salience Network, exhibit hyperconnectivity. Crucially, the connection values of the salience network exhibited the highest accuracy in classifying individuals without ASD from those with ASD. These measures were also found to correlate with assessments of the severity of restricted interests and repetitive behaviors. Another investigation into the connectivity of the salience network in both adults and adolescents revealed reduced connections with the medial temporal lobe network, including the amygdala. Reduced functional connectivity between networks may suggest impaired social-emotional integration of information across various brain regions in individuals with ASD.

6.3.1.4 Graph theoretical approaches: Unveiling patterns and structures

Although some neuroimaging studies reveal disparities in regional network connections in individuals with ASD, the implications of these results for more intricate system-level dysfunctions in the brains of ASD patients remain uncertain. Thus far, there have been only a limited number of graph theory investigations undertaken on ASD, and the published results have been inconclusive. Except for a higher "betweenness centrality" in a specific area of the DMN in individuals with ASD, Redcay et al. [58] discovered minimal distinctions between adolescents with ASD and neurotypical controls in a comprehensive examination of network characteristics in four functional brain networks. This signifies the degree of centrality of a node inside the network or the frequency with which it traverses across it. Keown et al. [59] discovered heightened local functional connectivity in the temporo-occipital regions in a separate investigation including adolescents with ASD. This finding was associated with more severe ASD symptoms. Rudie et al. [60] used graph theory methods to investigate the functional and structural connections in a group of children and adolescents with ASD. These infants exhibited alterations in both local and global network parameters, such as modularity and local efficiency, within their brain networks. In general, studies on ASD utilizing graph theory suggest that there are significant disruptions in the communication of brain networks. This indicates that in individuals with autism, critical networks are less separated and less organized into distinct modules. These difficulties may occur inside networks as well as across different networks. Despite being at an early stage of development, further research may uncover more complex connections between large brain networks in individuals with ASD.

6.3.2 Brain dynamics in schizophrenia

Schizophrenia is a very debilitating mental condition that is mostly inherited and may negatively impact social interactions, emotional control, and cognitive abilities by disrupting the integration of several domains of cognition and mental functioning. Ever since Eugen Bleuler used the word "disconnection" in 1925, it has been widely acknowledged that schizophrenia involves a disturbance or "disconnection" of neural connections in the brain, since the disorder seems to impair "the numerous associative pathways that govern our cognitive processes." Instead of experiencing a decrease in connections, the disorder is now commonly believed to be linked to "dysconnectivity," which refers to an aberrant pattern of connections between different sections of the brain. This pattern may include both the strengthening and weakening of pathways, leading to changes in how the brain functions as a whole. Recent investigations have used various imaging and electrophysiological methods to demonstrate distinct and widespread abnormalities in both the structure and function of brain connections. An often disrupted structural route is the link between regions of the frontal and temporal lobes [10].

Studies have shown a noticeable disruption in the efficient communication between the parietal and prefrontal areas in both control subjects and those with schizophrenia, particularly during working memory tasks. Several whole-brain connectivity experiments have shown that schizophrenia disrupts extensive brain networks, affecting more than just individual brain areas or pathways. Resting-state fMRI investigations of patients diagnosed with schizophrenia demonstrate distinct disturbances in the functional connectivity of the default mode network. Furthermore, other research has demonstrated specific deficits in the functional communication between different components of the brain's resting-state networks that are responsible for cognitive control. Additionally, there are widespread patterns of abnormal functional connections that are unique to certain brain regions, such as stronger or weaker connections between the dorsolateral prefrontal cortex and other areas of the brain. dMRI and tractography have shown that deficiencies in connectivity between frontal and temporal brain areas lead to a decrease in the importance of key brain hubs and a less integrated network structure. Studies have shown that people with schizophrenia have decreased structural connectivity and abnormalities in the centrality of hub regions, such as the medial frontal and left temporal lobe. Both investigations revealed a decline in the efficacy of the global network, suggesting insufficient functional integration. Recent network research on structural brain connections has shown that some locations that are disrupted in their pathways form what is known as the brain's "wealthy club." This refers to a group of nodes that are densely and strongly interconnected. Impairments in the connections between brain clubs, which are crucial for communication, are expected to result in functional anomalies in the processing of integrated neurons [12].

6.3.3 Parkinson's disease

Parkinson's disease (PD), the second most common neurodegenerative disorder, is recognized as a complex condition affecting multiple systems. It presents with motor symptoms like tremors, akinesia, and stiffness, alongside non-motor symptoms including cognitive impairments. The loss of nigrostriatal dopamine in PD impacts various neuronal circuits spread throughout the brain, leading to abnormalities in brain networks that contribute to the disease's neuropathology. The neurophysiological processes behind Parkinsonian symptoms remain uncertain. Enhancing our comprehension of the disruption of intricate brain networks caused by PD would aid in bridging the divide between well-established pathological alterations and clinically observed symptoms of PD. Neuroimaging tools, including fMRI, enable the assessment of functional connectivity in the brain networks of persons with PD. fMRI refers to a collection of magnetic resonance imaging (MRI) techniques that may identify alterations in brain activity. These techniques include diffusion imaging, perfusion imaging, and BOLD contrast imaging. However, BOLD fMRI is the most often linked technique to fMRI, since it measures variations in blood oxygen saturation levels. The following is a summary of ongoing studies on the breakdown of functional networks in PD and its impact on both motor and non-motor symptoms:

6.3.3.1 Motor symptoms-related network changes

- **Bradykinesia:** Multiple neuroimaging studies have investigated the network connections established by individuals with Parkinson's disease (PD) during the performance of different motor activities. In individuals with PD, there is a reduced functional link between the striatum and cortical motor areas, including the premotor cortex (PMC), primary motor cortex (M1) and supplementary motor area (SMA), during self-initiated movement. Furthermore, there is a disruption in the link between the prefrontal cortex, PMC, and SMA. Both voluntary movement initiation and the processes of planning and making decisions about movement rely on the SMA. The SMA serves as a primary recipient of the basal ganglia motor circuit. Evidence demonstrates that levodopa treatment may enhance motor function in individuals with Parkinson's disease by partly reinstating the functionality of the SMA. Dysfunction of the SMA has been associated with motor deficits. One important aspect that likely leads to bradykinesia in Parkinson's disease is the degeneration of the nigrostriatal dopamine pathway, leading to disconnection of the striato-SMA route. Motor automaticity is proposed as the potential cause of bradykinesia. Automaticity refers to the capacity to perform a physical action without the need for conscious consideration of its specific elements. The neural network has superior efficiency compared to the attentional network when it comes to processing motor automaticity in individuals who are in good health. The motor program is presumably stored in the sensorimotor striatum (specifically the posterior putamen), and it remains unaffected by external factors. Individuals with PD often experience a decline in the speed and range of their everyday activities, such as akinesia (lack of movement), decreased arm swing, freezing of gait (FOG), and micrographia (small handwriting). These motor issues, which are mostly caused by bradykinesia (slowness of movement), are closely linked to a reduction in motor automaticity.

The neural network studies suggest that reduced motor automaticity in PD can be attributed to several neural mechanisms. These include ineffective neural coding of movement, difficulty in transferring motor skills from the brain to the sensorimotor striatum, instability of the automatic mode within the striatum, and the need for attentional control to perform actions that are typically done automatically in healthy individuals. Bradykinesia is the result of people with PD losing automatic abilities they have previously learned and finding it difficult to learn new ones or regain lost motor skills [61].

- **Tremor:** Due to the potential disruption of fMRI signals by tremor, there has been far less research on the connectivity of networks associated to tremor compared to bradykinesia. Helmich et al. carried out an extensive study to evaluate the functional connectivity between the basal ganglia nuclei and the cerebellothalamic circuit. They used electromyography to detect tremors while doing fMRI scanning, as stated in their paper [62]. A study revealed that there was a short activation of the basal ganglia nuclei during the onset of the tremor, which differed from the activity seen in the cerebellothalamic circuit associated with the intensity of the tremor. The functional linkages between the cerebellothalamic circuit and the putamen and internal globus pallidus have been enhanced. The ensuing investigation provides support for the hypothesis that Parkinsonian tremors may arise from an interruption in the connection between the basal ganglia and the cerebellothalamic circuit. Functional connectivity studies have also been used to investigate several other motor symptoms associated with Parkinson's disease. For instance, Tessitore et al. found that individuals with PD had reduced functional connectivity within frontoparietal networks, which are responsible for attention-related tasks. Functional neuroimaging research indicates that disruptions in the frontal cortical regions, basal ganglia, and midbrain locomotor region may be responsible for the FoG. Neurological features of different subtypes of Parkinson's disease may be differentiated using a network connection.

6.3.3.2 Non-motor symptoms

The majority of people with PD have symptoms that are not motor in nature as well, such as emotional, cognitive, or olfactory abnormalities. Recently, there has been an increased emphasis on describing the brain network responsible for these non-motor symptoms. PD patients often have cognitive impairments. Individuals who suffer from both dementia and PD have a distinct disturbance in their corticostriatal connection. Furthermore, research has shown that individuals with PD who have cognitive deficits have altered connections within the DMN. The DMN consists of the medial prefrontal cortex, precuneus, anterior cingulate cortex, inferior parietal lobe and posterior cingulate cortex. It frequently exhibits deactivation during task performance. The DMN is believed to improve cognitive function by directing neuronal resources to important brain areas. An increased correlation between the rostral anterior cingulate cortex and the dorsal caudate was linked to a decline in cognitive function, specifically in memory and visuospatial abilities. On the other hand, DMN impairment was connected to a more rapid progression of cognitive decline. These results indicate that the dysfunction of the DMN may be the underlying cause of the impairments in executive function seen in PD.

Depression is the most prevalent mental disorder among individuals with Parkinson's disease. Research has shown that individuals with depression in Parkinson's disease have atypical connections within the prefrontal-limbic network. Interrupted operational depression is the most common mental health disorder associated with those who have PD. Research has shown that people with depression in PD have abnormal connections in the prefrontal-limbic network. Depression in individuals with PD is linked to a disturbance in the functional connections between the precuneus, prefrontal cortex, and cerebellum. Integrating multimodal information is essential for affective, sensorimotor, and cognitive functions, and the cingulate cortex facilitates this process. Moreover, certain cognitive processes connected to emotions, including the recognition and expectation of pain, introspective thinking about oneself, negative conditioning, and meditation, seem to be influenced by the median cingulate cortex. Individuals with Parkinson's disease may have damage to networks that are related to the median cingulate cortex.

Research has shown a correlation between network connectivity and non-motor symptoms such as apathy, delusions, and olfactory impairment. Patients with PD who have a reduced sense of smell have reduced communication between the posterior cingulate cortex and certain parts of the brain, including the right frontal areas, right parietal areas, and both sides of the major sensory areas. Additionally, these patients exhibit increased connectivity between the striatum and the cortex, compared to individuals with normal sense of smell. Apathetic individuals with PD had reduced functional connectivity, primarily in the frontal and limbic striatal regions. Additionally, there was a reduction in connectivity between the limbic division of the left striatum and the ipsilateral frontal cortex, as well as the remaining portions of the left striatum. Patients with PD had a much more robust occipital-cortico-striatal connection as compared to those who did not have visual hallucinations [63].

6.3.3.3 Intervention-related network changes

Functional connectivity can be utilized to explore the neural mechanisms underlying the effects of anti-Parkinsonian medications. Levodopa therapy has been shown to enhance motor function by restoring the connection between the motor pathways in the striatum and cortex, and by normalizing the function of the motor pathways in the basal ganglia (for example, by increasing neuronal activity in the striatum). Levodopa is the preferred medicine for managing the symptoms of PD. Nevertheless, a number of people have levodopa-induced dyskinesia (LID), characterized by daily variations in mobility and involuntary movements. There is still a lot to uncover about the neurological connections in the development of LID. A recent research discovered that the link between the putamen and M1 in individuals increased after the administration of levodopa and the development of LID. The strengthened link between the striatum and the cortex that occurs in response to levodopa may have ramifications for the development of LID. Another study discovered a reduction in connectivity between the inferior frontal cortex and M1, along with an increase in connectivity between the inferior frontal cortex and the putamen in patients with LID.

This discovery indicates that the neuronal network focused on the inferior frontal lobe may potentially play a role in the pathological mechanisms that lead to LID. Despite its effectiveness in treating PD, the specific neuronal mechanism of action for DBS remains unclear. Research has demonstrated that deep brain stimulation (DBS) applied to the subthalamic nucleus (STN) can alter the connections within the striato-thalamo-cortical and STN-cortical pathways, leading to symptomatic improvements. The pedunculopontine nucleus (PPN) is the central hub for resolving issues pertaining to posture and walking. Research has shown that PPN-DBS has the ability to restore normalcy to PPN connections that are aberrant. Table 6.1 provides a concise summary of the research on brain connections in PD.

Table 6.1 Brain functional connectivity of PD research summary.

Authors	**Task**	**Type of connectivity**	**Findings**
Herz et al. [64]	Resting-state	Effective connectivity dynamic causal modeling (DCM)	Patients who developed levodopa-induced dyskinesias exhibited increased connectivity between the putamen and primary motor cortex after taking levodopa, even while suppressing movement.
Yao et al. [65]	Resting-state	Functional connectivity	PD patients experiencing visual hallucinations have increased occipital-corticostriatal connection.
Manza et al. [66]	Resting-state	Functional connectivity	The study identified a significant correlation between cognitive impairment and heightened connectivity between the rostral anterior cingulate cortex and the dorsal caudate. Conversely, motor deficits were linked to decreased connectivity between the anterior putamen and the midbrain.
Wu et al. [67]	Visuomotor association task	Functional connectivity	Both chronic and worsening impairments of the basal ganglia motor circuit were observed. Additionally, associations were identified between disconnections in the cerebellum, rostral cingulate motor area, and pre-SMA, and the development of progressive micrographia.
Kahan et al. [68]	Resting-state	Effective connectivity DCM	DBS enhanced the cortico-striatal, thalamo-cortical, and direct pathways, while diminishing the effectiveness of the subthalamic nucleus's incoming and outgoing connections.

6.4 Conclusion

The study of brain network dynamics provides valuable insights into both healthy and pathological brain states. By utilizing graph theory and advanced analytical methods, we can gain a deeper understanding of the intricate relationship between anatomical and functional connections in the healthy brain. This insight is critical when studying brain network changes linked with autism spectrum disorder, schizophrenia, and Parkinson's disease. Each illness exhibits distinct patterns of connection disruption, emphasizing the significance of personalized approaches in diagnosis and treatment. The insights gained from researching these dynamics can help to generate more effective therapies and therapeutic procedures, thereby improving patient outcomes. This review emphasizes the importance of continuous study into brain network dynamics to better understand and treat diverse neurological and psychiatric diseases. Future studies should focus on better understanding the particular mechanisms behind brain network changes in diverse disorders. Furthermore, establishing new analytical methodologies and targeted interventions based on these findings is critical for improving therapeutic outcomes.

References

[1] Braun, U., Muldoon, S. F., & Bassett, D. S. (2015). On human brain networks in health and disease. Encyclopedia of Life Sciences, 1–9. https://doi.org/10.1002/9780470015902.a0025783

[2] Sporns, O. (2013b). Structure and function of Complex Brain Networks. Dialogues in Clinical Neuroscience, 15(3), 247–262. https://doi.org/10.31887/dcns.2013.15.3/osporns

[3] Raichle, M. E., MacLeod, A. M., Snyder, A. Z., Powers, W. J., Gusnard, D. A., & Shulman, G. L. (2001). A default mode of brain function. Proceedings of the National Academy of Sciences, 98(2), 676–682. https://doi.org/10.1073/pnas.98.2.676

[4] Thomas Yeo, B. T., Krienen, F. M., Sepulcre, J., Sabuncu, M. R., Lashkari, D., Hollinshead, M., Roffman, J. L., Smoller, J. W., Zöllei, L., Polimeni, J. R., Fischl, B., Liu, H., & Buckner, R. L. (2011). The organization of the human cerebral cortex estimated by intrinsic functional connectivity. Journal of Neurophysiology, 106(3), 1125–1165. https://doi.org/10.1152/jn.00338.2011

[5] Power, J. D., Cohen, A. L., Nelson, S. M., Wig, G. S., Barnes, K. A., Church, J. A., Vogel, A. C., Laumann, T. O., Miezin, F. M., Schlaggar, B. L., & Petersen, S. E. (2011). Functional Network Organization of the human brain. Neuron, 72(4), 665–678. https://doi.org/10.1016/j.neuron.2011.09.006

[6] Calhoun, V. D., Kiehl, K. A., & Pearlson, G. D. (2008). Modulation of temporally Coherent Brain Networks estimated using ICA at rest and during cognitive tasks. Human Brain Mapping, 29(7), 828–838. https://doi.org/10.1002/hbm.20581

[7] Fox, M. D., Zhang, D., Snyder, A. Z., & Raichle, M. E. (2009). The global signal and observed anticorrelated resting State Brain Networks. Journal of Neurophysiology, 101(6), 3270–3283. https://doi.org/10.1152/jn.90777.2008

[8] Hagmann, P., Cammoun, L., Gigandet, X., Meuli, R., Honey, C. J., Wedeen, V. J., & Sporns, O. (2008). Mapping the structural core of human cerebral cortex. PLoS Biology, 6(7). https://doi.org/10.1371/journal.pbio.0060159

[9] Kaiser, M., & Varier, S. (2011). Evolution and development of brain networks: From caenorhabditis elegans to homo sapiens. Network: Computation in Neural Systems, 22(1–4), 143–147. https://doi.org/10.3109/0954898x.2011.638968

[10] Bullmore, E., & Sporns, O. (2012). The economy of Brain Network organization. Nature Reviews Neuroscience, 13(5), 336–349. https://doi.org/10.1038/nrn3214

[11] Bassett, D. S., Brown, J. A., Deshpande, V., Carlson, J. M., & Grafton, S. T. (2011). Conserved and variable architecture of human white matter connectivity. NeuroImage, 54(2), 1262–1279. https://doi.org/10.1016/j.neuroimage.2010.09.006

[12] van den Heuvel, M. P., Kahn, R. S., Goñi, J., & Sporns, O. (2012). High-cost, high-capacity backbone for Global Brain Communication. Proceedings of the National Academy of Sciences, 109(28), 11372–11377. https://doi.org/10.1073/pnas.1203593109

[13] Morcom, A. M., & Fletcher, P. C. (2007). Does the brain have a baseline? why we should be resisting a rest. NeuroImage, 37(4), 1073–1082. https://doi.org/10.1016/j.neuroimage.2006.09.013

[14] Damoiseaux, J. S., Rombouts, S. A., Barkhof, F., Scheltens, P., Stam, C. J., Smith, S. M., & Beckmann, C. F. (2006). Consistent resting-state networks across healthy subjects. Proceedings of the National Academy

of Sciences, 103(37), 13848–13853. https://doi.org/10.1073/pnas.060 1417103

[15] Shehzad, Z., Kelly, A. M., Reiss, P. T., Gee, D. G., Gotimer, K., Uddin, L. Q., Lee, S. H., Margulies, D. S., Roy, A. K., Biswal, B. B., Petkova, E., Castellanos, F. X., & Milham, M. P. (2009). The Resting Brain: Unconstrained yet reliable. Cerebral Cortex, 19(10), 2209–2229. https://doi.org/10.1093/cercor/bhn256

[16] Honey, C. J., Sporns, O., Cammoun, L., Gigandet, X., Thiran, J. P., Meuli, R., & Hagmann, P. (2009). Predicting human resting-state functional connectivity from structural connectivity. Proceedings of the National Academy of Sciences, 106(6), 2035–2040. https://doi.org/10.1073/pnas.0811168106

[17] Biswal, B. B., Mennes, M., Zuo, X.-N., Gohel, S., Kelly, C., Smith, S. M., Beckmann, C. F., Adelstein, J. S., Buckner, R. L., Colcombe, S., Dogonowski, A.-M., Ernst, M., Fair, D., Hampson, M., Hoptman, M. J., Hyde, J. S., Kiviniemi, V. J., Kötter, R., Li, S.-J., … Milham, M. P. (2010). Toward discovery science of human brain function. Proceedings of the National Academy of Sciences, 107(10), 4734–4739. https://doi.org/10.1073/pnas.0911855107

[18] Larson-Prior, L. J., Zempel, J. M., Nolan, T. S., Prior, F. W., Snyder, A. Z., & Raichle, M. E. (2009). Cortical network functional connectivity in the descent to sleep. Proceedings of the National Academy of Sciences, 106(11), 4489–4494. https://doi.org/10.1073/pnas.0900924106

[19] Lewis, C. M., Baldassarre, A., Committeri, G., Romani, G. L., & Corbetta, M. (2009). Learning sculpts the spontaneous activity of the resting human brain. Proceedings of the National Academy of Sciences, 106(41), 17558–17563. https://doi.org/10.1073/pnas.0902455106

[20] Tambini, A., Ketz, N., & Davachi, L. (2010). Enhanced brain correlations during rest are related to memory for recent experiences. Neuron, 65(2), 280–290. https://doi.org/10.1016/j.neuron.2010.01.001

[21] Vincent, J. L., Patel, G. H., Fox, M. D., Snyder, A. Z., Baker, J. T., Van Essen, D. C., Zempel, J. M., Snyder, L. H., Corbetta, M., & Raichle, M. E. (2007). Intrinsic functional architecture in the anaesthetized Monkey Brain. Nature, 447(7140), 83–86. https://doi.org/10.1038/nature05758

[22] White, B. R., Bauer, A. Q., Snyder, A. Z., Schlaggar, B. L., Lee, J.-M., & Culver, J. P. (2011). Imaging of functional connectivity in the Mouse Brain. PLoS ONE, 6(1). https://doi.org/10.1371/journal.pone.0016322

[23] Maldjian, J. A., Davenport, E. M., & Whitlow, C. T. (2014). Graph theoretical analysis of resting-state MEG data: Identifying interhemispheric connectivity and the default mode. NeuroImage, 96, 88–94. https://doi.org/10.1016/j.neuroimage.2014.03.065

[24] Bassett, Danielle Smith, & Bullmore, E. (2006). Small-world brain networks. The Neuroscientist, 12(6), 512–523. https://doi.org/10.1177/1073858406293182

[25] Sporns, O. (2013a). Network attributes for segregation and integration in the human brain. Current Opinion in Neurobiology, 23(2), 162–171. https://doi.org/10.1016/j.conb.2012.11.015

[26] Achard, S., Salvador, R., Whitcher, B., Suckling, J., & Bullmore, E. (2006). A resilient, low-frequency, small-world human brain functional network with highly connected association cortical hubs. The Journal of Neuroscience, 26(1), 63–72. https://doi.org/10.1523/jneurosci.3874-05.2006

[27] Power, J. D., Cohen, A. L., Nelson, S. M., Wig, G. S., Barnes, K. A., Church, J. A., Vogel, A. C., Laumann, T. O., Miezin, F. M., Schlaggar, B. L., & Petersen, S. E. (2011a). Functional Network Organization of the human brain. Neuron, 72(4), 665–678. https://doi.org/10.1016/j.neuron.2011.09.006

[28] Cole, M. W., Bassett, D. S., Power, J. D., Braver, T. S., & Petersen, S. E. (2014). Intrinsic and task-evoked network architectures of the human brain. Neuron, 83(1), 238–251. https://doi.org/10.1016/j.neuron.2014.05.014

[29] Lynall, M.-E., Bassett, D. S., Kerwin, R., McKenna, P. J., Kitzbichler, M., Muller, U., & Bullmore, E. (2010). Functional connectivity and brain networks in Schizophrenia. Journal of Neuroscience, 30(28), 9477–9487. https://doi.org/10.1523/jneurosci.0333-10.2010

[30] van den Heuvel, M. P., Stam, C. J., Kahn, R. S., & Hulshoff Pol, H. E. (2009). Efficiency of functional brain networks and intellectual performance. Journal of Neuroscience, 29(23), 7619–7624. https://doi.org/10.1523/jneurosci.1443-09.2009

[31] Honey, Christopher J., Thivierge, J.-P., & Sporns, O. (2010). Can structure predict function in the human brain? NeuroImage, 52(3), 766–776. https://doi.org/10.1016/j.neuroimage.2010.01.071

[32] Honey, Christopher J., Kötter, R., Breakspear, M., & Sporns, O. (2007). Network structure of cerebral cortex shapes functional connectivity on multiple time scales. Proceedings of the National Academy of Sciences, 104(24), 10240–10245. https://doi.org/10.1073/pnas.0701519104

[33] Adachi, Y., Osada, T., Sporns, O., Watanabe, T., Matsui, T., Miyamoto, K., & Miyashita, Y. (2011). Functional connectivity between anatomically unconnected areas is shaped by collective network-level effects in the macaque cortex. Cerebral Cortex, 22(7), 1586–1592. https://doi.org/10.1093/cercor/bhr234

[34] Adachi, Y., Osada, T., Sporns, O., Watanabe, T., Matsui, T., Miyamoto, K., & Miyashita, Y. (2011a). Functional connectivity between anatomically unconnected areas is shaped by collective network-level effects in the macaque cortex. Cerebral Cortex, 22(7), 1586–1592. https://doi.org/10.1093/cercor/bhr234

[35] van den Heuvel, M. P., Mandl, R. C. W., Kahn, R. S., & Hulshoff Pol, H. E. (2009). Functionally linked resting-state networks reflect the underlying structural connectivity architecture of the human brain. Human Brain Mapping, 30(10), 3127–3141. https://doi.org/10.1002/hbm.20737

[36] Greicius, M. D., Supekar, K., Menon, V., & Dougherty, R. F. (2008). Resting-state functional connectivity reflects structural connectivity in the default mode network. Cerebral Cortex, 19(1), 72–78. https://doi.org/10.1093/cercor/bhn059

[37] Fornito, A., & Bullmore, E. T. (2015). Connectomics: A new paradigm for understanding brain disease. European Neuropsychopharmacology, 25(5), 733–748. https://doi.org/10.1016/j.euroneuro.2014.02.011

[38] Bassett, Danielle S, & Bullmore, E. T. (2009). Human brain networks in health and disease. Current Opinion in Neurology, 22(4), 340–347. https://doi.org/10.1097/wco.0b013e32832d93dd

[39] Hulshoff Pol, H., & Bullmore, E. (2013). Neural networks in psychiatry. European Neuropsychopharmacology, 23(1), 1–6. https://doi.org/10.1016/j.euroneuro.2012.12.004

[40] Tost, H., Bilek, E., & Meyer-Lindenberg, A. (2012). Brain connectivity in Psychiatric Imaging Genetics. NeuroImage, 62(4), 2250–2260. https://doi.org/10.1016/j.neuroimage.2011.11.007

[41] Bailey, A., Le Couteur, A., Gottesman, I., Bolton, P., Simonoff, E., Yuzda, E., & Rutter, M. (1995). Autism as a strongly genetic disorder: Evidence from a british twin study. Psychological Medicine, 25(1), 63–77. https://doi.org/10.1017/s0033291700028099

[42] Abrams, D. A., Lynch, C. J., Cheng, K. M., Phillips, J., Supekar, K., Ryali, S., Uddin, L. Q., & Menon, V. (2013). Underconnectivity between voice-selective cortex and reward circuitry in children with autism. Proceedings of the National Academy of Sciences, 110(29), 12060–12065. https://doi.org/10.1073/pnas.1302982110

[43] Belmonte, M. K., Allen, G., Beckel-Mitchener, A., Boulanger, L. M., Carper, R. A., & Webb, S. J. (2004). Autism and abnormal development of brain connectivity: Figure 1. The Journal of Neuroscience, 24(42), 9228–9231. https://doi.org/10.1523/jneurosci.3340-04.2004

[44] Courchesne, E., & Pierce, K. (2005). Why the frontal cortex in autism might be talking only to itself: Local over-connectivity but long-distance disconnection. Current Opinion in Neurobiology, 15(2), 225–230. https://doi.org/10.1016/j.conb.2005.03.001

[45] Just, M. A. (2004). Cortical activation and synchronization during sentence comprehension in high-functioning autism: Evidence of underconnectivity. Brain, 127(8), 1811–1821. https://doi.org/10.1093/brain/awh199

[46] Mundy, P., Sullivan, L., & Mastergeorge, A. M. (2009). A parallel and distributed-processing model of joint attention, social cognition and autism. Autism Research, 2(1), 2–21. https://doi.org/10.1002/aur.61

[47] Koshino, H., Carpenter, P. A., Minshew, N. J., Cherkassky, V. L., Keller, T. A., & Just, M. A. (2005). Functional connectivity in an fmri working memory task in high-functioning autism. NeuroImage, 24(3), 810–821. https://doi.org/10.1016/j.neuroimage.2004.09.028

[48] Kana, R. K., Keller, T. A., Cherkassky, V. L., Minshew, N. J., & Just, M. A. (2009). Atypical frontal-posterior synchronization of theory of mind regions in autism during mental state attribution. Social Neuroscience, 4(2), 135–152. https://doi.org/10.1080/17470910802198510

[49] Villalobos, M. E., Mizuno, A., Dahl, B. C., Kemmotsu, N., & Müller, R.-A. (2005). Reduced functional connectivity between V1 and inferior frontal cortex associated with visuomotor performance in autism. NeuroImage, 25(3), 916–925. https://doi.org/10.1016/j.neuroimage.2004.12.022

[50] Kana, R. K. (2006). Sentence comprehension in autism: Thinking in pictures with decreased functional connectivity. Brain, 129(9), 2484–2493. https://doi.org/10.1093/brain/awl164

[51] Just, Marcel Adam, Cherkassky, V. L., Keller, T. A., Kana, R. K., & Minshew, N. J. (2006). Functional and anatomical cortical underconnectivity in autism: Evidence from an fmri study of an executive function task and corpus callosum morphometry. Cerebral Cortex, 17(4), 951–961. https://doi.org/10.1093/cercor/bhl006

[52] Kana, Rajesh K., Keller, T. A., Minshew, N. J., & Just, M. A. (2007). Inhibitory control in high-functioning autism: Decreased activation and

underconnectivity in inhibition networks. Biological Psychiatry, 62(3), 198–206. https://doi.org/10.1016/j.biopsych.2006.08.004

[53] Belin, P., Zatorre, R. J., Lafaille, P., Ahad, P., & Pike, B. (2000). Voice-selective areas in human auditory cortex. Nature, 403(6767), 309–312. https://doi.org/10.1038/35002078

[54] Gervais, H., Belin, P., Boddaert, N., Leboyer, M., Coez, A., Sfaello, I., Barthélémy, C., Brunelle, F., Samson, Y., & Zilbovicius, M. (2004). Abnormal cortical voice processing in autism. Nature Neuroscience, 7(8), 801–802. https://doi.org/10.1038/nn1291

[55] Biswal, B., Zerrin Yetkin, F., Haughton, V. M., & Hyde, J. S. (1995a). Functional connectivity in the motor cortex of resting human brain using echo-Planar MRI. *Magnetic Resonance in Medicine*, *34*(4), 537–541. https://doi.org/10.1002/mrm.1910340409

[56] Stevens, M. C., Pearlson, G. D., & Calhoun, V. D. (2009). Changes in the interaction of resting-state neural networks from adolescence to adulthood. *Human Brain Mapping*, *30*(8), 2356–2366. https://doi.org/10.1002/hbm.20673

[57] Seeley, W. W., Menon, V., Schatzberg, A. F., Keller, J., Glover, G. H., Kenna, H., Reiss, A. L., & Greicius, M. D. (2007). Dissociable intrinsic connectivity networks for salience processing and executive control. *The Journal of Neuroscience*, *27*(9), 2349–2356. https://doi.org/10.1523/jneurosci.5587-06.2007

[58] Redcay, E., Moran, J. M., Mavros, P. L., Tager-Flusberg, H., Gabrieli, J. D., & Whitfield-Gabrieli, S. (2013). Intrinsic functional network organization in high-functioning adolescents with autism spectrum disorder. *Frontiers in Human Neuroscience*, *7*. https://doi.org/10.3389/fnhum.2013.00573

[59] Keown, C. L., Shih, P., Nair, A., Peterson, N., Mulvey, M. E., & Müller, R.-A. (2013). Local functional overconnectivity in posterior brain regions is associated with symptom severity in autism spectrum disorders. *Cell Reports*, *5*(3), 567–572. https://doi.org/10.1016/j.celrep.2013.10.003

[60] Rudie, J. D., Brown, J. A., Beck-Pancer, D., Hernandez, L. M., Dennis, E. L., Thompson, P. M., Bookheimer, S. Y., & Dapretto, M. (2013). Altered functional and Structural Brain Network organization in autism. *NeuroImage: Clinical*, *2*, 79–94. https://doi.org/10.1016/j.nicl.2012.11.006

[61] Wu, T., Hallett, M., & Chan, P. (2015). Motor automaticity in parkinson's disease. *Neurobiology of Disease*, *82*, 226–234. https://doi.org/10.1016/j.nbd.2015.06.014

[62] Surdhar, I., Gee, M., Bouchard, T., Coupland, N., Malykhin, N., & Camicioli, R. (2012). Intact limbic-prefrontal connections and reduced amygdala volumes in parkinson's disease with mild depressive symptoms. *Parkinsonism & Related Disorders*, *18*(7), 809–813. https://doi.org/10.1016/j.parkreldis.2012.03.008

[63] Gorges, M., Müller, H.-P., Lulé, D., Pinkhardt, E. H., Ludolph, A. C., & Kassubek, J. (2015). To rise and to fall: Functional connectivity in cognitively normal and cognitively impaired patients with parkinson's disease. *Neurobiology of Aging*, *36*(4), 1727–1735. https://doi.org/10.1016/j.neurobiolaging.2014.12.026

[64] Herz, D. M., Haagensen, B. N., Nielsen, S. H., Madsen, K. H., Løkkegaard, A., & Siebner, H. R. (2016). resting-state connectivity predicts levodopa-induced dyskinesias in parkinson's disease. *Movement Disorders*, *31*(4), 521–529. https://doi.org/10.1002/mds.26540

[65] Yao, N., Pang, S., Cheung, C., Chang, R. S., Lau, K. K., Suckling, J., Yu, K., Mak, H. K.-F., McAlonan, G., Ho, S.-L., & Chua, S. (2015). Resting activity in visual and corticostriatal pathways in parkinson's disease with hallucinations. *Parkinsonism & Related Disorders*, *21*(2), 131–137. https://doi.org/10.1016/j.parkreldis.2014.11.020

[66] Manza, P., Zhang, S., Li, C. R., & Leung, H. (2015). Resting-state functional connectivity of the striatum in early-stage pArkinson's disease: Cognitive Decline and motor symptomatology. *Human Brain Mapping*, *37*(2), 648–662. https://doi.org/10.1002/hbm.23056

[67] Wu, T., Zhang, J., Hallett, M., Feng, T., Hou, Y., & Chan, P. (2015). Neural correlates underlying micrographia in parkinson's disease. *Brain*, *139*(1), 144–160. https://doi.org/10.1093/brain/awv319

[68] Kahan, J., Urner, M., Moran, R., Flandin, G., Marreiros, A., Mancini, L., White, M., Thornton, J., Yousry, T., Zrinzo, L., Hariz, M., Limousin, P., Friston, K., & Foltynie, T. (2014). Resting state functional MRI in parkinson's disease: The impact of deep brain stimulation on 'effective' connectivity. *Brain*, *137*(4), 1130–1144. https://doi.org/10.1093/brain/awu027

7

Neural Dynamics: Unraveling the Complexity of Brain Activity

Rezuana Haque[1] and Md Tanvir Islam[2]

[1]Computer Science and Engineering, BGC Trust University, Bangladesh
[2]Department of Software, SungkyunKwan University, South Korea
E-mail: rezuanahaque27@gmail.com; tanvirislamnwu@gmail.com

Abstract

Understanding how the brain works is important for both neuroscience and artificial intelligence. This chapter looks at neural dynamics, focusing on how neurons interact and work together over different times and areas. It explains how neural networks are inspired by the brain, how they process information. By studying the structure and signaling of neurons, we can see similarities between artificial and biological systems. The chapter also discusses the importance of brain rhythms and techniques for measuring brain activity, such as EEG, MEG, and fMRI. This study aims to connect neuroscience and machine learning, providing insights into how complex brain functions can be modeled and understood using computational methods.

Keywords: Brain rhythms, EEG, fMRI, MEG, neural dynamics.

7.1 Introduction

Neural dynamics refers to the activity and interactions of individual neurons or small groups of neurons, observed across various spatial and temporal scales. Studying these dynamics helps us understand how the brain normally functions and how it changes in different disorders [1, 2]. Different parts of the brain have different functions. For instance, the occipital lobe mainly handles visual processing, while the frontal lobe is responsible for decision-making and problem-solving. Understanding spatiotemporal activity enables us to identify and associate specific brain functions with particular regions [3]. The brain's spatiotemporal activity is incredibly complex, involving millions of neurons interacting in intricate ways. Each person's brain is unique, and these patterns can vary widely between individuals.

The brain is organized on multiple levels, from individual neurons and microscopic molecules to extensive brain networks and local circuits. These dynamic processes involve highly interconnected networks. Brain connectivity is classified as neuroanatomical (structural), functional, or effective [4]. Functional connectivity shows how different parts communicate, and effective connectivity shows causal influences between areas. Structural connections are the physical links between neurons and regions [4].

Neural activity is the electrical and chemical actions within neurons. Neural activity allows neurons to communicate and perform various functions. It is always adapting to internal states and external stimuli, showcasing the brain's neuroplasticity. This activity involves multiple brain processes happening simultaneously in different areas, with signals converging and diverging to allow for complex integration and dissemination of information. Neuromodulators such as serotonin, dopamine, and acetylcholine have a crucial impact on brain activity, influencing various brain regions and affecting functions like mood, arousal, and cognition [5].

The complexity of brain dynamics includes emergent properties, where complex behaviors and cognitive functions arise from the interactions of neurons. For example, consciousness results from the dynamic interactions of large-scale brain networks rather than the activity of individual neurons alone. Theories like the free energy principle and synergetics explain how the brain maintains stability and organizes itself through these interactions.

Researchers use computational models to simulate brain activity and understand how different parts of the brain work together to create complex behaviors. These models incorporate data from various levels of brain organization, linking molecular changes that occur in milliseconds to long-term processes that span years. Understanding the complexity of the brain requires both holistic (top-down) and detailed (bottom-up) methods. Top-down methods focus on the brain as a whole and bottom-up approaches study small parts of the brain. Modern brain imaging techniques and big data analysis are key tools that help scientists study the vast and dynamic activity of the brain.

This chapter addresses several key questions:

1. What is neural dynamics and why is it important?
2. What is a neuron and how do neurons transmit signals across the brain?
3. What are brain rhythms and the different types of rhythms?
4. What techniques are used for measuring brain activities?
5. How does the human brain inspire artificial neural networks?
6. How do feedforward networks and backpropagation work?

7.2 The Basic Structure and Function of Neurons

Neurons are special cells that send signals throughout the body. They are unique because they can receive and send information. Neurons use both electrical and chemical methods to transmit information. Neurons link to one another via synapses, and they also connect to muscles or glands through neuroeffector junctions. Typically, neurons are made up of an axon, dendrites, and a cell body. The cell body is often known as the soma. The axon transmits messages away from the cell, whereas dendrites receive signals [6]. Figure 7.1 represents the structure of a single neuron.

A neuron is identified by its long processes that radiate outward from the cell body, also called the soma. Axons are responsible for sending outgoing signals, whereas dendrites are responsible for receiving incoming signals. The incoming signals are called afferent signals. The outgoing signals are called

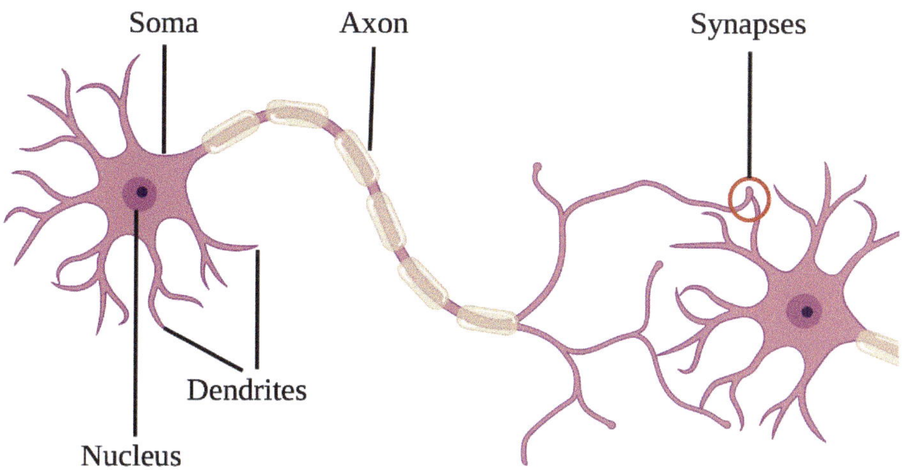

Figure 7.1 Illustration of a neuron, showing its main parts: soma (cell body), nucleus, dendrites, axon, and synapses.

efferent signals. Neurons come in different shapes and types, including multipolar, bipolar, pseudounipolar, and anaxonic neurons. Multipolar neurons have many dendrites and one axon, bipolar neurons have one dendrite and one axon, pseudounipolar neurons have a single extension that splits into two, and anaxonic neurons have no distinct axon. The cell body of a neuron contains the nucleus and other essential organelles. A neuron can have one or several dendrites depending on its function and location. Dendrites receive signals and also help in protein synthesis and signaling. When axons reach their terminals, they release chemicals like neurotransmitters, neuromodulators, or neurohormones. These chemicals allow neurons to communicate with other cells. They convert electrical signals to chemical signals that can cross synapses or neuromuscular junctions. The movement of materials inside axons is called axonal transport. Proteins like kinesin and dynein help in this process. Neurons send messages using voltage-gated ion channels. Potassium, sodium, and chloride ions are essential for maintaining the neuron's membrane potential, which is around −70 mV. As soon as a stimulus hits a certain level, it sends a consistent electrical signal that travels down the axon. Graded potentials vary in strength and diminish as they travel. Synapses are the junctions where neurons communicate with each other or with other types of cells. They allow the transmission of signals throughout the nervous system and enable complex processes like movement, sensation, and thought.

Neurons interact in complex ways to perform various functions in the body. Mature neurons cannot divide, so damage to them can cause neurological problems. But some parts of the brain, like the subependymal zone and the dentate gyrus in rats, have neural stem cells that can help make new neurons. This process is called neurogenesis.

7.3 Signal Transmission Process

Signal transmission happens when neurons communicate with each other. The presynaptic neuron sends the signal and releases neurotransmitters. The postsynaptic neuron receives the signal through receptors on its membrane, continuing the signal through the nervous system. Signal transmission happens in four steps:

1. **Action potential arrival:** An electrical signal called an action potential travels down the axon to the end of the presynaptic neuron, reaching the presynaptic terminal.

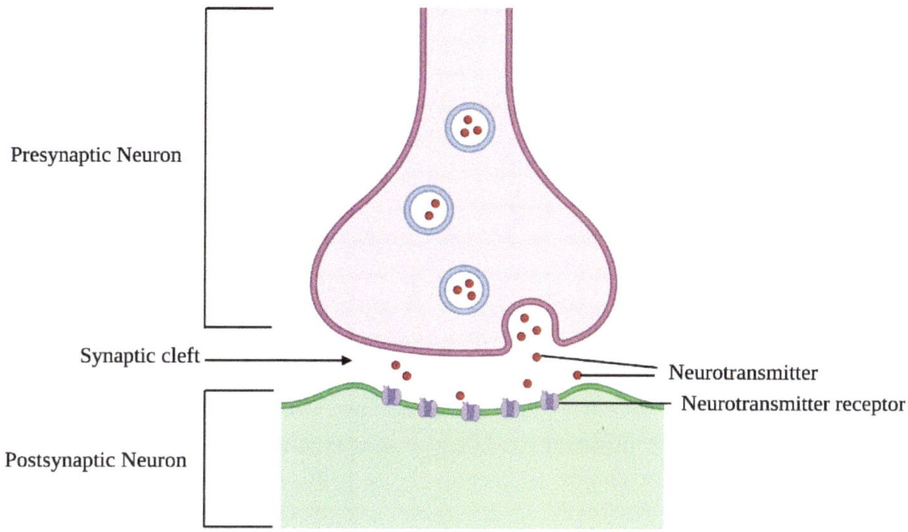

Figure 7.2 Diagram of a synapse showing the presynaptic neuron, synaptic cleft, and postsynaptic neuron. Neurotransmitters are released from the presynaptic neuron, cross the synaptic cleft, and bind to receptors on the postsynaptic neuron.

2. **Neurotransmitter release:** The action potential opens calcium channels in the presynaptic terminal. Calcium ions enter, causing neurotransmitters to be released into the synaptic cleft. The synaptic cleft is the gap between neurons.
3. **Neurotransmitter binding:** Neurotransmitters cross the synaptic cleft and bind to specific receptors on the postsynaptic membrane. Postsynaptic membrane is the receiving end of the next neuron.
4. **Postsynaptic response:** The postsynaptic response happens when neurotransmitters attach to receptors on the postsynaptic neuron. This action causes ion channels to either open or close. This changes the neuron's electrical charge, generating a response and continuing the signal. Figure 7.2 illustrates the process of synapses.

7.4 Brain Rhythms

Brain rhythms, also known as brain waves, are patterns of neural activity that appear as oscillations in the brain's electrical activity. Neuronal rhythms are prevalent in brain dynamics and are strongly connected with cognitive processing [7]. Brain rhythms are the macroscopic result of synchronized neural activity across brain regions. These rhythms facilitate cognitive and physiological processes by efficiently communicating and integrating information. Figure 7.3 represents the common brain rhythms with the respective ranges. These rhythms are measured using electroencephalography (EEG) and are classified based on their frequency:

- **Infra-slow oscillations (ISO) (<0.5 Hz):** These very slow waves help the brain grow early on and are mostly found in babies who were born before their due dates [8]. It influences brain connections during non-REM sleep. It appears in slow EEG responses during seizures [9].
- **Delta waves (0.5–4 Hz):** They occur in the frontal lobe during deep sleep. Pathological delta waves may be signs of brain dysfunction when they are seen in conscious persons. There are a few specific types of delta waves that are noticeable. These patterns include FIRDA, which is rhythmic delta activity in the front of the brain often seen in adults. OIRDA occurs in the back of the brain and is common in children. TIRDA happens in the temporal lobe and is frequently found in people with epilepsy [10, 11].
- **Theta waves (4–7 Hz):** They show up in the drowsy and early phases of sleep. As drowsiness increases, they tend to shift backward, and they are

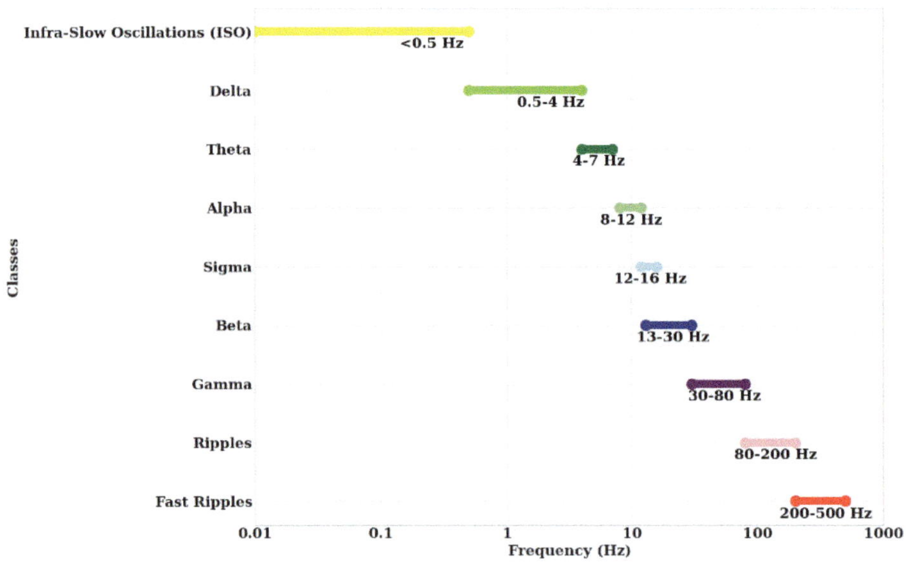

Figure 7.3 This chart displays different classes of brain waves and their corresponding frequency ranges, from infra-slow oscillations (ISO) to fast ripples.

most evident in the frontal region of the brain. Emotional situations may also cause an increase in theta waves, and if detected in waking states, this may be an indication of malfunction in the brain.

- **Alpha waves (8–12 Hz):** In a comfortable wakeful state, alpha waves are visible in the back portion of the brain. They begin at the age of three and are constant throughout a person's life; they are called the typical background rhythm in adulthood. When alpha waves are slow or altered, it may be a sign of cerebral malfunction [12]. Alpha waves are triggered by mental exertion and eye opening. This group also contains the Mu rhythm, which occurs in central brain regions and fades with motor activity [13].
- **Sigma waves (12–16 Hz):** Sleep spindles, or sigma waves, are seen during N2 sleep. They are mostly found in the front of the brain and can be slow (12–14 Hz) or fast (14–16 Hz) [14]. Spindle coma is one of the disorders that may be seen in patients with pathological spindles.
- **Beta waves (13–30 Hz):** They are common in both healthy adults and kids. They are most prominent in the frontal and center parts of the brain. Their amplitude increases during sleepiness and decreases throughout deeper phases of sleep. Sedative medications tend to enhance beta waves, while focal decreases in beta activity can indicate brain injury [15].
- **High-frequency oscillations (HFOs) (>30 Hz):** These consist of ripples (80–200 Hz), fast ripples (200–500 Hz), and gamma waves (30–80 Hz). Gamma waves are associated with sensory processing and cognitive functions. HFOs, particularly fast ripples are linked to epilepsy and can indicate epileptic brain tissue.

7.5 Techniques for Measuring Brain Activity

Understanding brain activity is important for detecting and treating a variety of neurological and psychiatric disorders. There are some advanced methods used for measuring and analyzing brain activity, and each has its own benefits and uses. Here are some of the most commonly used techniques:

- **Magnetoencephalography (MEG):** Magnetoencephalography (MEG) is a technique for measuring the magnetic fields created by brain activity. It helps researchers understand how different areas of the brain function and communicate. When neurons communicate, they generate tiny magnetic fields. Highly sensitive sensors called superconducting quantum interference devices (SQUIDs) detect these fields [16]. These measurements are taken in a magnetically shielded room (MSR) to eliminate external noises [17]. The process involves detecting magnetic fields, converting the signals into digital data, and analyzing the data to locate brain activity. MEG is used in treating epilepsy, stroke, brain injuries, and mental illnesses such as depression, schizophrenia, and autism, as well as chronic pain [17].

Figure 7.4 shows how magnetic fields are produced by neurons during communication and how MEG uses these fields to measure brain activity. The figure's left portion shows the magnetic fields that MEG detected throughout the brain. The right portion describes how these fields are created at the synaptic level during neural transmission.

- **Electroencephalography (EEG):** Electroencephalography (EEG) is a widely used method for studying the brain's electrical activity. It was first developed by Hans Berger over 50 years ago. Berger used radio equipment to record the electrical signals produced by the brain [18]. EEG measures the electrical activity produced by synchronized firing of neurons, especially pyramidal cells in the brain. Different parts of

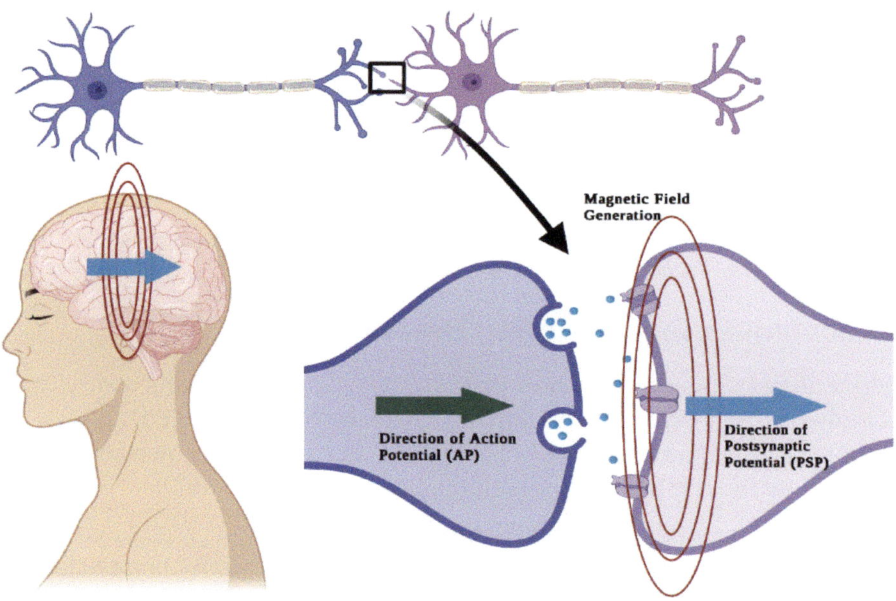

Figure 7.4 Illustration of neural magnetic field generation and detection by magnetoencephalography (MEG).

these cells have different electric charges. The dendrites are negatively charged, while the rest of the cell is positively charged. These neurons create electric dipoles, and the combined activity of many such cells can be detected by electrodes placed on the scalp. EEG electrodes can detect and show this combined activity as a series of positive and negative waves. Using the international 10–20 system, electrodes are placed on the head to record EEG. This system has four main reference points: the inion, the nasion, and two particular regions. Electrodes are attached to the scalp with conductive paste to record different brain waves. These include brain waves categorized as delta (0.5–4 Hz), theta (4–8 Hz), alpha (8–13 Hz), beta (13–30 Hz), and gamma (over 30 Hz) [19]. These waves reflect various states of brain activity. EEG can also capture evoked potentials during tasks, helping to analyze neural processes involved in perception and cognition. Clinically, EEG is used to diagnose and study conditions like epilepsy, sleep disorders, brain injuries, neurodegenerative diseases, and more, making it an important tool in both research and medical fields [20]. A study utilized EEG data to identify epileptic seizures [21]. Figure 7.5 illustrates an EEG setup recording brain activity from a subject.

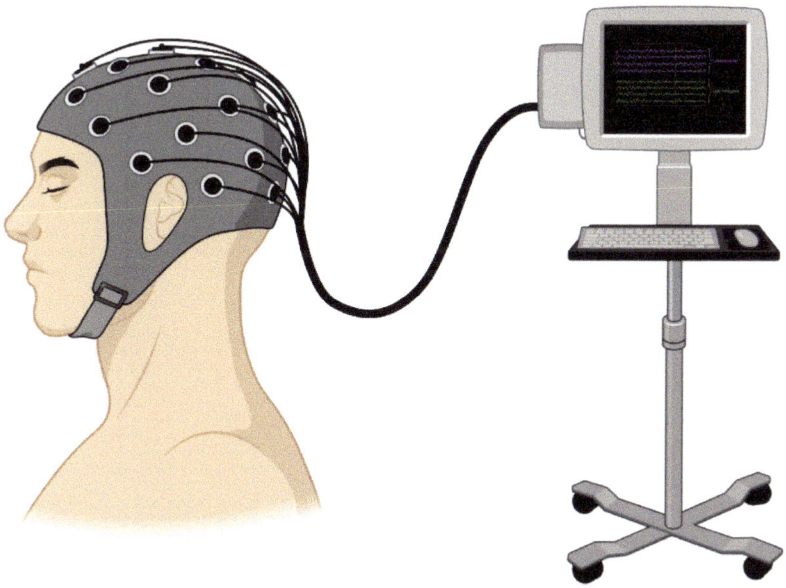

Figure 7.5 An EEG setup showing electrodes attached to a participant's scalp to measure brain activity.

- **Functional magnetic resonance imaging (fMRI):** Functional magnetic resonance imaging (fMRI) is a tool for studying brain activity without any invasive procedures. It works by detecting changes in blood oxygen levels. This is known as the blood oxygen level-dependent (BOLD) effect [22]. Hemoglobin in red blood cells behaves differently based on whether it carries oxygen or not. Oxyhemoglobin is non-magnetic, while deoxyhemoglobin is slightly magnetic. This difference helps fMRI create images showing where oxygen levels change in the brain. When a part of the brain becomes active, it uses more oxygen. Blood flow to that area increases to meet the demand. This changes the ratio of oxyhemoglobin to deoxyhemoglobin. These changes can be detected by fMRI. This shows up as either increases or decreases in the signal. By using this method, researchers can create maps of brain activity.
  fMRI doesn't measure brain activity directly. Instead, it measures the changes in blood flow and oxygenation that occur in response to neural activity. This relationship between neural activity and blood flow is called neurovascular coupling. When neurons are activated, they release neurotransmitters, which are then taken up by astrocytes (a type of brain cell). This process triggers changes in blood vessels, increasing blood flow to the active areas. Figure 7.6 shows fMRI results indicating regions of brain activity.

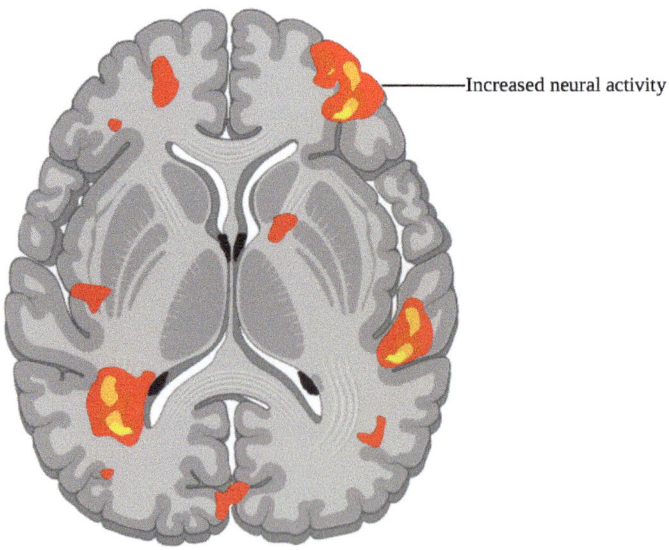

Figure 7.6 An fMRI scan highlighting active brain regions during different tasks.

Task-related fMRI studies specific brain functions by having people perform tasks during scanning. Resting-state fMRI studies brain activity when the person is not doing any tasks. This helps understand how different brain regions communicate in both healthy individuals and those with conditions like brain tumors, multiple sclerosis, Alzheimer's disease, epilepsy, and psychiatric disorders. Resting-state fMRI is useful for patients who cannot follow task instructions. MRI (Magnetic Resonance Imaging) is also used to create a detailed image of an organ, including the brain. Many research has used MRI images to detect brain tumors [23].

7.6 Neural Networks and Brain

Artificial neural networks are inspired by the human brain's structure and functioning. In the brain, neurons are the cells that process information. Each neuron connects to other neurons through synapses. Synapses allow them to communicate and form networks. Similarly, in neural networks, nodes are connected by links that can strengthen or weaken based on the data they process. In both cases, learning happens by adjusting these connections. Repeated events in the brain strengthen synapses, making it easier for neurons to communicate. Our brains have special parts for different jobs, like seeing and hearing. In neural networks, instead of parts, there are layers of connected neurons that change pictures or sounds into information. It's like our brains, but simpler and more like a computer.

An artificial neuron takes in multiple inputs through its dendrites, each multiplied by a weight that represents its importance. These weighted inputs are then summed together, with a bias applied to the total. This sum is then sent to an activation function, which decides whether the neuron "fires" (sends out a signal) or stays inactive. If the activation function's threshold exceeds, the neuron fires; otherwise, it remains silent. When the output is produced, it is sent to other neurons in the network and contributes to overall computation.

An MLP is a multi-layer artificial neural network (ANN). MLPs have input, hidden, and output layers. Each layer contains multiple artificial neurons, also known as nodes or units.

- **Input layer**: The input layer contains neurons that receive input data, such as image features or text words. This layer has the same number of neurons as input features. Each neuron represents one feature.

- **Hidden layers**: The computation takes place in the hidden layers. Neurons in each hidden layer receive inputs from neurons in the input layer or another hidden layer. Hidden layer neurons use weighted connections and activation functions to process these inputs and produce outputs for the following layer of neurons.
- **Output layer**: The neurons of the output layer generate the final output. This layer's neurons depend on the nature of the task the network is designed to solve. For example, if there is more than one class in a classification task, there will be one neuron for each class. Each neuron will indicate how likely it is that the person is in that class.

During the training process of a multi-layer perceptron (MLP), the network uses feedforward to make predictions for input data. Error is calculated by comparing predictions to target outputs. Backpropagation sends this error backward over the network. Backpropagation updates the weights of the connections to reduce the error. By repeating the steps of feedforward and backpropagation, the network gradually learns to make better predictions or classifications. It does this by adjusting its weights based on the training data.

In a feedforward neural network, the network takes input values x and produces an output y using a function,

$$y = f(x;\phi), \tag{7.1}$$

where φ represents the parameters learned during training. For a given input x, the network adjusts these parameters to lower the error between its predictions $y^\wedge$ and the true outputs y. To do this, a loss function $(y^\wedge,\ y)$ calculates the difference between the predicted and true values. One commonly used loss function for binary classification is the cross-entropy loss. The formula is,

$$\text{Cross}-\text{entropy loss} = -\frac{1}{M}\sum_{i=1}^{M}(y_i\log(\widetilde{y}_i) + (1 - y_i)\log(1 - \widetilde{y}_i)), \tag{7.2}$$

where y_i denotes the actual value for the ith point, and $y^{\sim}{}_i$ signifies the predicted value for the ith data point and M is the total number of data points.

When training a machine learning model, it starts with random values for its parameters. Trying random values each time would take forever to find the best ones. Optimization algorithms help the model quickly find the best parameters by making small adjustments based on the data. Gradient descent is a method to minimize a cost function by adjusting parameters. It starts at a random point and moves towards the best point by considering the slope (gradient) of the function. The next position is calculated by subtracting the gradient multiplied by a learning rate from the current position. The formula for moving to the next point is,

$$\theta_{\text{new}} = \theta_{\text{old}} - \alpha \cdot \nabla \mathrm{E}\left(\theta_{\text{new}}\right), \tag{7.3}$$

where θ_{new} represents the next parameter value, while θ_{old} is the current parameter value. The symbol α represents learning rate, and $\nabla E(\theta_{\text{old}})$ refers to the gradient of the error function concerning the current parameter value.

Backpropagation calculates the cost function gradient for each network parameter. It has two steps. The two phases are forward and backward passes. In the forward pass, the input data x is passed through the network to compute the output $y^\wedge$. The network's predictions $y^\wedge$ are then compared to the actual outputs y to calculate the loss. The difference between the predicted and actual outputs is passed back through the network during the backward pass. This involves computing the gradient of the loss function for each network parameter, including weights and biases. The error for the ith neuron in the kth layer is denoted as $\delta^{(k)}$. The gradient of the cost function with respect to a weight $w_{ij}{}^{(k)}$ connecting the jth neuron in layer (k – 1) to the ith neuron in layer k is given by,

$$\frac{\partial J}{\partial w_{ij}^{(k)}} = \delta_i^{(k)} a_j^{(k-1)}, \tag{7.4}$$

where $a_j^{(k-1)}$ is the activation of the jth neuron in the (k – 1)th layer, and $\delta_i^{(k)}$ is the error term for the ith neuron in the kth layer. J denotes the cost function that calculates the error between predicted outputs and target values. The weights are updated using the gradient descent method. The gradient descent update rule is,

$$w_{ij}^{(l)} = w_{ij}^{(l)} - \eta \frac{\partial J}{\partial w_{ij}^{(l)}}, \tag{7.5}$$

where η is the learning rate, $w_{ij}^{(k)}$ is the current weight, and $\frac{\partial J}{\partial w_{ij}^{(k)}}$ is the gradient of the cost function with respect to the weight. By repeating the forward and backward passes and updating the weights and biases, the network minimizes the cost function and learns to make accurate predictions.

Figure 7.7 shows how data from different brain imaging methods (EEG, MRI, fMRI) are processed using a neural network to diagnose brain disorders or understand brain activity patterns. Here's a step-by-step explanation:

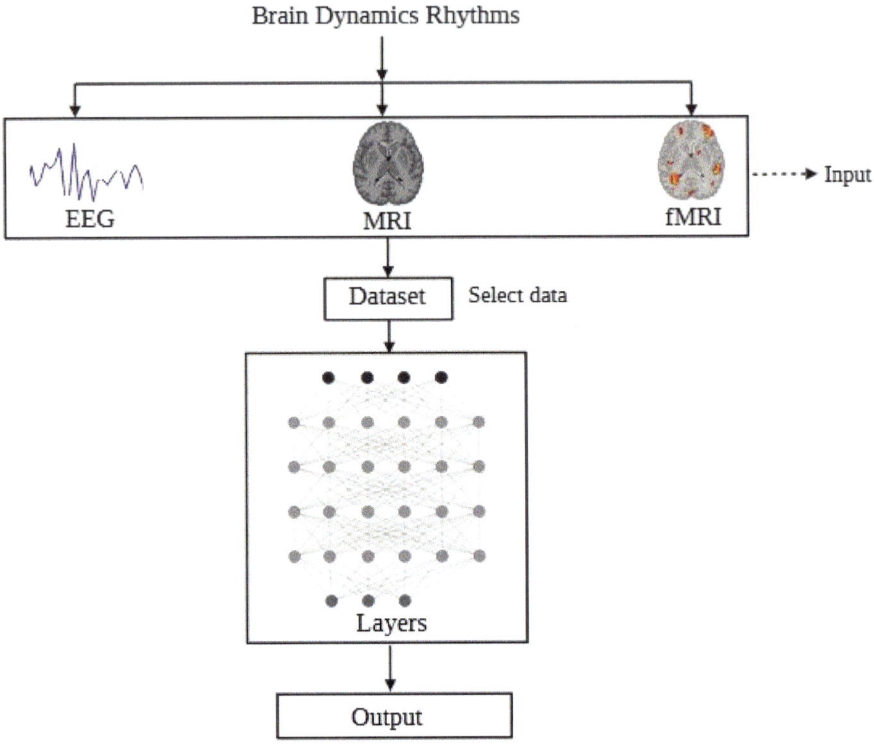

Figure 7.7 Workflow of diagnosing brain diseases using neural networks.

7.6.1 Input data

EEG, MRI, and fMRI are the three most popular methods which are used by professionals to diagnose various brain diseases.

- **Electroencephalography (EEG):** Records electrical activity of the brain.
- **Magnetic resonance imaging (MRI):** Provides detailed images of brain structures.
- **Functional MRI (fMRI):** Measures brain activity by detecting changes in blood flow.

Dataset:

- The EEG, MRI, and fMRI data are collected and organized into a dataset. This dataset contains all the necessary information from the different brain imaging techniques.

Data selection:

- Specific data is selected from the dataset to be used in the analysis. This step ensures that only relevant information is processed.

Layers (neural network):

- The selected data is passed through multiple layers of a neural network. These layers process the data to extract meaningful patterns and features.

Output:

- The neural network generates an output based on the processed data. This output could be a diagnosis of a brain condition, identification of brain patterns, or other useful information.

Pseudo code

---

1. **Select data type:**
    - **selected_data_type = select_data_type() # Choose from 'EEG', 'MRI', 'fMRI'**

2. **Load data:**
    - **If selected_data_type is 'EEG':**
        - **data = load_eeg_data()**
    - **Else if selected_data_type is 'MRI':**
        - **data = load_mri_data()**
    - **Else if selected_data_type is 'fMRI':**
        - **data = load_fmri_data()**
3. **Preprocess data:**
    - **data = preprocess_data(data)**
4. **Create dataset:**
    - **labels = get_labels()**
5. **Split data:**
    - **train_data, test_data, train_labels, test_labels = split_data(data, labels)**

6. **Build neural network:**
   - **model = build_neural_network(input_shape)**
7. **Train neural network:**
   - **model.train(train_data, train_labels)**
8. **Evaluate neural network:**
   - **accuracy = model.evaluate(test_data, test_labels)**
   - **print("Model accuracy:", accuracy)**
9. **Make predictions:**
   - **new_data = load_new_data(selected_data_type) # Load new data of the selected type**
   - **predictions = model.predict(new_data)**
   - **print("Predictions:",predictions)**

7.7 Discussion

Neural networks have changed AI by replicating how neurons interact. Future AI research should create more complex networks and better algorithms. Combining neuroscience and machine learning can improve models and help us understand the brain better. Hybrid models that include features like synaptic plasticity and neurogenesis can make AI systems more effective. Neurological illnesses like brain tumors, epilepsy, Alzheimer's disease etc. can be diagnosed using machine learning [21], [23–25]. Brain-inspired computing systems that replicate the brain's capacity for information processing and experience-based learning can be created using machine learning algorithms. The brain's efficient processing and organized information can also inspire new AI methods.

7.8 Conclusion

The brain is one of the most important organs in the human body. Its role is to process information, control functions, and allow complex behaviors. We showed how the brain processes information biologically in this research. We focused on the structure and function of neurons, which communicate through electrical signals. We also highlighted brain rhythms and described EEG, MEG, and fMRI. These tools help us study brain function. Neural networks are inspired by the human brain. These are artificial systems that replicate the biological processes of human brain. We discussed how neural networks process information and improve their predictions using methods like gradient descent and backpropagation. This study aims to connect neuroscience and machine learning, showing how understanding the brain can help us create better artificial intelligence. By learning from both fields, we can develop more advanced neural networks that work like the brain, leading to new discoveries in both neuroscience and AI.

Acknowledgments

We would like to express our heartfelt gratitude to editors for guiding us in writing this study.

References

[1] Vince D. Calhoun, R. Miller, G. Pearlson, and T. Adalı, "The Chronnectome: Time-Varying Connectivity Networks as the Next Frontier in fMRI Data Discovery," Neuron, vol. 84, no. 2, pp. 262–274, Oct. 2014, doi: https://doi.org/10.1016/j.neuron.2014.10.015.

[2] J. Sui, R. Jiang, J. Bustillo, and V. Calhoun, "Neuroimaging-based Individualized Prediction of Cognition and Behavior for Mental Disorders and Health: Methods and Promises," Biological Psychiatry, Feb. 2020, doi: https://doi.org/10.1016/j.biopsych.2020.02.016.

[3] R. M. Hutchison et al., "Dynamic functional connectivity: Promise, issues, and interpretations," NeuroImage, vol. 80, pp. 360–378, Oct. 2013, doi: https://doi.org/10.1016/j.neuroimage.2013.05.079..

[4] J. Cao et al., "Brain functional and effective connectivity based on electroencephalography recordings: A review," Human Brain Mapping, vol. 43, no. 2, pp. 860–879, Oct. 2021, doi: https://doi.org/10.1002/hbm.25683.

[5] E. L. Newman, K. Gupta, J. R. Climer, C. K. Monaghan, and M. E. Hasselmo, "Cholinergic modulation of cognitive processing: insights drawn from computational models," Frontiers in Behavioral Neuroscience, vol. 6, 2012, doi: https://doi.org/10.3389/fnbeh.2012.00024.

[6] M. G. Orgel, "Epineurial versus perineurial repair of peripheral nerves," Clinics in Plastic Surgery, vol. 11, no. 1, pp. 101–104, Jan. 1984, Available: https://pubmed.ncbi.nlm.nih.gov/6368091/

[7] J. Cannon et al., "Neurosystems: brain rhythms and cognitive processing," European Journal of Neuroscience, vol. 39, no. 5, pp. 705–719, Dec. 2013, doi: https://doi.org/10.1111/ejn.12453.

[8] S. Vanhatalo, P. Tallgren, S. Andersson, K. Sainio, J. Voipio, and K. Kaila, "DC-EEG discloses prominent, very slow activity patterns during sleep in preterm infants," Clinical Neurophysiology, vol. 113, no. 11, pp. 1822–1825, Nov. 2002, doi: https://doi.org/10.1016/s1388-2457(02)00292-4.

[9] A. Ikeda et al., "Subdural recording of ictal DC shifts in neocortical seizures in humans," Epilepsia, vol. 37, no. 7, pp. 662–674, Jul. 1996, doi: https://doi.org/10.1111/j.1528-1157.1996.tb00631.x.

[10] M. A. Dalby, "Epilepsy and 3 per second spike and wave rhythms. A clinical, electroencephalographic and prognostic analysis of 346 patients," Acta Neurologica Scandinavica, p. Suppl 40:3+, 1969, Available: https://pubmed.ncbi.nlm.nih.gov/4979890/

[11] J. Reiher, M. Beaudry, and C. P. Leduc, "Temporal intermittent rhythmic delta activity (TIRDA) in the diagnosis of complex partial epilepsy: sensitivity, specificity and predictive value," The Canadian Journal of Neurological Sciences. Le Journal Canadien Des Sciences Neurologiques, vol. 16, no. 4, pp. 398–401, Nov. 1989, doi: https://doi.org/10.1017/s0317167100029450.

[12] AIRD RB, GASTAUT Y. Occipital and posterior electroencephalographic rhythms. Electroencephalogr Clin Neurophysiol. 1959 Nov;11:637-56. [PubMed]

[13] G. E. Chatrian, M. C. Petersen, and J. A. Lazarte, "The blocking of the rolandic wicket rhythm and some central changes related to movement," Electroencephalography and Clinical Neurophysiology, vol. 11, no. 3, pp. 497–510, Aug. 1959, doi: https://doi.org/10.1016/0013-4694(59)90048-3.

[14] A. R. Gondeck and J. R. Smith, "Dynamics of human sleep sigma spindles," Electroencephalography and Clinical Neurophysiology, vol. 37, no. 3, pp. 293–297, Sep. 1974, doi: https://doi.org/10.1016/0013-4694(74)90033-9.

[15] J. D. Frost, J. R. Carrie, R. P. Borda, and P. Kellaway, "The effects of dalmane (flurazepam hydrochloride) on human EEG characteristics," Electroencephalography and Clinical Neurophysiology, vol. 34, no. 2, pp. 171–175, Feb. 1973, doi: https://doi.org/10.1016/0013-4694(73)90044-8.

[16] I. Ahonen et al., "Multichannel SQUID systems for brain research," in IEEE Transactions on Magnetics, vol. 27, no. 2, pp. 2786-2792, March 1991, doi: 10.1109/20.133789. keywords: {SQUIDs;Magneticfieldmeasurement;Magnetoencephalography;Humans;Electroencephalography;Neurons;Instruments;Magneticheads;Laboratories;Biochemistry}

[17] J. A. Kim and K. D. Davis, "Magnetoencephalography: physics, techniques, and applications in the basic and clinical neurosciences," Journal of Neurophysiology, vol. 125, no. 3, pp. 938–956, Mar. 2021, doi: https://doi.org/10.1152/jn.00530.2020.

[18] A. M. Beres, "Time is of the Essence: A Review of Electroencephalography (EEG) and Event-Related Brain Potentials (ERPs) in Language Research," Applied Psychophysiology and Biofeedback, vol. 42, no. 4, pp. 247–255, Jul. 2017, doi: https://doi.org/10.1007/s10484-017-9371-3.

[19] R. Abreu, A. Leal, and P. Figueiredo, "EEG-Informed fMRI: A Review of Data Analysis Methods," Frontiers in Human Neuroscience, vol. 12, Feb. 2018, doi: https://doi.org/10.3389/fnhum.2018.00029.

[20] M. J. Aminoff, "Electroencephalography: general principles and clinical applications," in Electrodiagnosis in Clinical Neurology, 5th ed., M. J. Aminoff, Ed. London: Churchill Livingstone, 2005, pp. 37–84. doi: 10.1016/B978-0-443-06647-4.50008-2.

[21] "NeuroWave-Net: Enhancing epileptic seizure detection from EEG brain signals via advanced convolutional and long short-term memory networks." Accessed: Jun. 15, 2024. [Online]. Available: https://www.aimspress.com/article/doi/10.3934/bioeng.2024006

[22] F. Gaillard, "BOLD imaging | Radiology Reference Article | Radiopaedia.org," Radiopaedia. Accessed: Jun. 15, 2024. [Online]. Available: https://radiopaedia.org/articles/bold-imaging

[23] "NeuroNet19: an explainable deep neural network model for the classification of brain tumors using magnetic resonance imaging data." *Scientific Reports*. Accessed: Jun. 15, 2024. [Online]. Available: https://www.nature.com/articles/s41598-024-51867-1

[24] K.-H. Nenning and G. Langs, "Machine learning in neuroimaging: from research to clinical practice," Radiol., vol. 62, no. 1, pp. 1–10, Dec. 2022, doi: 10.1007/s00117-022-01051-1.

[25] "Machine learning in neuroimaging: Progress and challenges - PMC." Accessed: Jun. 15, 2024. [Online]. Available: https://www.ncbi.nlm.nih.gov/pmc/articles/PMC6499712/

8

Decoding Brain Signals: Perspectives from AI Powered Examination

Avula Mahathi[1], Kishor Kumar Reddy C.[1], Nastassia Thandiwe Sithole[2], and Kari Lippert[3]

[1]Department of Computer Science and Engineering, Stanley College of Engineering and Technology for Women, India
[2]Department of Chemical Engineering Technology, University of Johannesburg, South Africa
[3]Department of Systems Engineering, University of South Alabama, USA
E-mail: avula.mahathi@gmail.com; drckkreddy@gmail.com; nastassias@uj.ac.za; klippert@southalabama.edu

Abstract

One of the main tenets of neuroscience is the interpretation of brain signals, which provides information about how the mind functions and has potential uses in technology, cognitive psychology, and medicine. While traditional brain signal analysis techniques like fMRI and EEG have provided the foundation for our knowledge of neural function, they are not always suitable for processing big and complicated datasets. This field has completely changed with the introduction of artificial intelligence (AI), which offers strong tools for quicker, more thorough, and accurate analysis of brain signals. This chapter examines how AI techniques are incorporated into the interpretation of brain signals. It covers conventional procedures as well as the introduction of machine learning and deep learning methodologies and their uses. We cover the drawbacks of AI, such as biases, data privacy, and the need for transparency, as well as its advantages, such as increased speed, accuracy, and the capacity to handle massive amounts of data. This chapter illustrates how AI-driven analysis is transforming our knowledge of the brain and opening the door for future developments in neuroscience and other fields through case studies and real-world examples.

Keywords: Brain signals, AI, neuroscience, psychology, interpretation, deep learning, AI-driven, decoding, AI-powered.

8.1 Introduction

Artificial intelligence (AI) is a quickly evolving branch of computer science that focuses on constructing systems and automated machines that can accomplish things that typically involve human intelligence [3]. Artificial intelligence systems are capable of analyzing and processing enormous amounts of data thanks to the usage of neural networks, which are designed to resemble the structure of the human brain and are used to represent complicated correlations in data [3]. One of the biggest scientific challenges of our day is trying to understand the complex functions of the human brain. Understanding brain signals – the electrical and chemical impulses that indicate cerebral activity – is essential for developing neuroscience as well as for real-world uses in technology, cognitive science, and medicine [1]. The discipline has undergone a revolution with the introduction of artificial intelligence (AI), which has unparalleled prospects for deciphering the intricate language of the brain.

8.1.1 Context and significance of interpreting brain signals

Analyzing the electrical and chemical activity in the brain is a basic component of neuroscience, which includes understanding brain signals. This area of study has been essential to our knowledge of the brain's operation, its response to diverse stimuli, and the ways in which different neurological diseases may affect brain function [2]. Through the analysis of brain signals, scientists and medical professionals may identify anomalies, learn more about cognitive processes, and create treatments for neurological conditions.

Several neuroimaging methods are often used to assess brain signals, each having unique benefits as well as drawbacks:

EEG, fMRI, and MEG are essential for monitoring and analyzing the rate of brain activity, and PET is also a very significant technique used in this process [4][5][6]. In EEG, small disks termed electrodes are attached to the scalp to produce an electric brain map. It is prized for the short time scale that can be obtained, making it effective for detecting quick changes in brain activity. However, one major issue in EEG is that its resolution is relatively poor as compared to the size and complexity of the brain and thus the exact source of electrical activity cannot be determined with high precision. On the other hand, functional magnetic resonance imaging of the brain takes functional information from the variations in flow of blood which again has high spatial resolution which helps in detailed brain mapping. There are, however, certain limitations to the use of fMRI: the measure has comparatively lower temporal resolution in relation to EEG and is relatively expensive and therefore not easily accessible.

MEG works by using magnetic fields to map the electrical activity in the brain, and it can provide both temporal and spatial resolution. However, MEG cannot be performed in any center since it depends on the expensive technology that is required for the purpose [6]. PET with the ability to take images of the metabolism of radioactive tracers used in depicting the brain's functions has excellent spatial resolution but is somewhat limited in its temporal resolution [6]. It also raises issues on exposure and safety due to the application and incorporation of radioactive materials into the filament strips.

The consequential applicability of Artificial intelligence (AI) in the elucidation of brain signals offers a spectrum of opportunities for enhanced precision and intricate understanding of brain signification. Machine learning and deep learning (DL) are thus developed to improve on conventional brain imaging techniques in an endeavor to gain deeper information about the brain activity profiles as well as to escalate the efficiency of detecting and managing neurological disorders [9]. The future lies in the amodal computing paradigm, where AI algorithms can govern the processing of a large volume of brain signal data in a short span of time and provide information that might not be otherwise observable by relying on conventional data processing paradigms only. It is this integration of AI into brain signal analysis that holds potential for a better future for neuroscience research and clinical utilization to help solve existing and emerging healthcare problems and in turn develop a greater understanding of the human brain.

8.1.2 AI's potential to improve brain signal interpretation

Effectively interpreting brain signals has significant consequences for a number of fields, such as cognitive neuroscience, interfaces between brains and computers (BCIs), and medical diagnostics. However, conventional analysis techniques face considerable difficulties due to the amount and complexity of brain signal data. Artificial intelligence (AI), in particular machine learning (ML) and deep learning (DL), has become an effective tool for tackling these issues [7]. How AI is changing how we perceive brain signals is discussed below.

Incorporation of AI and ML into the medical field has greatly enhanced the diagnosis and management of neurological diseases and mental health disorders. The pattern that is obtained from the analysis of a large volume of collective EEG and fMRI data is complex, to be noted by AI algorithms in identifying diseases like epilepsy and Alzheimer's disease [24]. It makes for the early detection and accurate diagnosis of disorders which helps to improve a lot of patients. Within the mental health domain, biomarkers can be detected by the AI models from signals generated by the human brains and hence assess the best-suited treatment options which also helps in the effective management of disorders like anxiety and depression.

They are instrumental in studying large amounts of data in cognitive neuroscience where the dynamics of connectivity of different parts of the brain during cognitive tasks can be investigated [10]. What is more, it expands our knowledge of the vital cognitive processes such as memory, attention, and decision-making. In the same way, AI can replicate the intricate processes of multilateral correspondence between the brain functions and behavior, which help to understand the neural foundations of different cognitive functions.

Brain–computer interfaces (BCI) have also received a boost from the integration of AI, a major area of interaction being the use of signals from the human brain. BCIs teach individuals with motor disabilities to control devices through their brain and AI improves the accuracy and speed of the BCI [15], [16]. In addition, AI in BCIs can change according to the user's brain data which enhances rehabilitation therapy for individuals who have received brain injuries and cognitive enrichment programs.

One of the ways through which the implementation of ML algorithms in brain signal analysis is significant is due to its capability in handling large-scale data processing and analysis. This capability gives insights hitherto unseen with conventional approaches due to the fact that it reveals patterns and predictions that would have been elusive. Some of the models that are most relevant to new discoveries in their function include deep learning models that reveal hitherto unseen correlations within the brain and make substantial contributions to the understanding of how diseases progress in the brain. These advancements in AI and ML do not only broaden our knowledge about the brain but also introduce a future-based method for diagnosing and treating neurological disorders [16].

Table 8.1 highlights the advantages and disadvantages of using AI-enhanced approaches vs. conventional methods (fMRI, MEG, EEG) for brain signal analysis. In addition to highlighting how AI advancements can improve specific areas like pattern identification and accuracy, this comparison also

Table 8.1 Comparing conventional and AI-enhanced techniques for brain signal analysis.

Method	Strengths	Weaknesses
Traditional EEG	High temporal resolution, non-invasive	Low spatial resolution
Traditional fMRI	High spatial resolution, non-invasive	Low temporal resolution, expensive
Traditional MEG	High temporal and spatial resolution	Expensive, requires specialized setup
AI-enhanced EEG	Improved pattern recognition, early diagnosis capabilities	Requires large datasets for training
AI-enhanced fMRI	Enhanced spatial and temporal analysis	Computationally intensive
AI in BCIs	Improved accuracy and adaptability	Potential data privacy concerns

emphasizes the trade-offs between various brain signal analysis approaches. It also highlights new issues that arise from AI advancements, such as the need for big datasets and greater computational requirements.

8.1.3 Overview of brain signal types and challenges in interpretation

The importance of decoding signals and patterns of the cerebral cortex: goals and perspectives of neuroscience, the role of interdisciplinary cooperation in enhancing the progress of neuroscience and in the creation of novel applications in the sphere of medicine and in the sphere of technology. This subdivision serves to explain the main types of brain signals, including EEG, fMRI, MEG, and identify the main problems which appear during their analysis [18]. We will also discuss how to turn to AI to overcome these challenges and improve the analysis of brain signals.

- **Electroencephalography (EEG):** Electrodes are applied to the scalp to monitor electrical activity in the brain by electroencephalography (EEG) [4]. Because of its excellent temporal resolution, EEG is a good tool for recording rapidly fluctuating brain activity. Nevertheless, its limited spatial sensitivity makes it difficult to determine the precise site of brain activity. EEG records the brain's normal voltage and pathological voltage using metal disks which are attached on the scalp. It is popular with researchers because of its fine temporal resolution which allows the identification of rapid fluctuations in brain signals [19].
- **Advantages:**
  High temporal resolution (milliseconds).

Non-invasive and relatively inexpensive [20].
Simple to use and transport to different scenes.

- **Disadvantages:**
  Low spatial resolution [21].
  Sensitive to interference not only from continuously produced noise but from muscle activity as well as other interfering signals.
- **Functional magnetic resonance imaging (fMRI):** Using variations in blood flow, functional magnetic resonance imaging (fMRI) detects brain activity [5]. Because of its great spatial resolution, thorough brain mapping is made possible. In comparison with EEG, fMRI has a poorer temporal resolution and is more expensive and less widely available. The fMRI works by taking pictures of these changes that happen with the circulation of blood in the brain. Another source offers descriptions of the brain structures and their roles in detail.
- **Advantages:**
  High spatial resolution (millimeters).
  Salient feature to capture and identify active networks and measurability.
  Safe techniques as it does not involve invasive processes or use of radiation.
- **Disadvantages:** Low temporal resolution (seconds). Expensive and less accessible. Can be restrictive as the participant has to keep very Still, it works well in a population that cannot do so.
- **Magnetoencephalography (MEG):** MEG concerns involve the measurement of magnetic fields that originate from the activity of the brain. Like fMRI, it has high temporal resolution and equally high spatial resolution; thus, it is ideal for studying brain functions. The magnetic fields generated by brain activity are measured via magnetoencephalography (MEG) [6]. Although expensive and specific technology is needed, it provides excellent timing and spatial resolutions.
- **Advantages:**
  High temporal resolution (milliseconds).
  Good spatial resolution (millimeters).
  Non-invasive and silent operation.
- **Disadvantages:** It is very costly and the chemicals used need special instruments for their preparation. This method is used sparingly and calls for a specially shielded chamber away from magnetic interference.

The EEG, fMRI, and MEG brain signal types are compared in Table 8.2 according to important parameters including cost, mobility, primary use cases, temporal and spatial resolution, and resolution [5]. It is evident from

Table 8.2 Features of various brain signal types.

Signal type	Temporal resolution	Spatial resolution	Cost	Portability	Major use cases
EEG	High (milliseconds)	Low	Low	High	Epilepsy diagnosis, sleep studies, BCI
fMRI	Low (seconds)	High	High	Low	Brain mapping, cognitive neuroscience
MEG	High (milliseconds)	High	Very high	Low	Brain function studies, epilepsy localization

this comparison that each type of brain signal has benefits and drawbacks in terms of its functioning, meaning that it is possible to gain valuable information about its use as a research and clinical tool. For example, EEG has fairly good temporal characteristics, is relatively cheap and portable, fMRI has very good spatial characteristics, while MEG has both good temporal and final spatial characteristics although the price and portability are considerably lower.

8.1.3.1 Challenges in brain signal interpretation

Observing brain signals is difficult because it comes with certain difficulties in interpreting the data because they are complex and may not be the same at different moments [8]. Below are some common challenges and how AI can address them:

- **Common challenges**
- **Data complexity and volume:** EEG and fMRI signal data are big and diverse, which means they are a massive challenge to handle and analyze. We will show that analyzing such datasets is not feasible using conventional statistical techniques and instead, necessitates sophisticated numerical approaches.
- **Noise and artifacts:** In practice, brain signals are embedded in noise and contain artifacts, which originate from activity in voluntary muscles, electric currents that are unrelated to brain signals, and fluctuations in the environment.
- **Individual variability:** Similar to patterns of inter-transcranial electrical stimulation, brain signals may be highly idiosyncratic because of structural and functional differences across individuals as well as fluctuations in levels of consciousness [22].
- **Temporal and spatial resolution trade-offs:** Thus, distinct types of signals give different time-space resolutions and accumulate data from multiple sources represents several challenges.

- **Interpretation and validation:** It can be very challenging to create reliable models to decode such activity, and the models which are built generally have to undergo extensive testing and cross-validation before they can be considered to be accurate.

8.1.3.2 Addressing challenges with AI

- **Advanced data processing:** Artificial intelligence can accrete enormous and intricate data from the brain signals, and complex deep learning models can identify meaningful patterns and features from the data [23].
- **Noise reduction and artifact removal:** The noise and artifacts can also be found by using machine learning techniques and be trained to minimize the noise and artifacts effects in the data of the brain signals.
- **Personalized analysis:** There exists another way to induce models capable of accounting for individual variability which as it is clear is capable of producing higher accuracy as the models are trained to learn the idiosyncrasies of data from the sample population.
- **Enhanced resolution integration:** AI comes in the ability to process different types of brain signal data at the same time for a better understanding of the brain activity with increased temporal resolution that comes with EEG data, and high spatial resolution that is associated with fMRI data [25].
- **Robust model validation:** These are techniques like cross-validation used by the AI frameworks to ensure that it is sufficiently reliable and validated through cross-validation.

Many of the conventional difficulties in interpreting brain signals can be solved by researchers and physicians by utilizing AI, which will result in more accurate and perceptive studies. This comprehensive method advances both the diagnosis and treatment of neurological disorders while deepening our understanding of how the brain functions.

8.1.4 Aim and scope of the chapter

8.1.4.1 Aim

In this chapter, the focus will be made towards identifying how the use of artificial intelligence has brought about changes in the brain signal analysis field. This chapter is going to discuss different types of AI methodologies that are currently used for understanding the multifaceted signals of the human brain and the superiority of these methodologies. Hoping to empower the readers with a thorough understanding of how AI is changing neuroscience and its related applications, this chapter explores AI and its usefulness in increasing the precision, rapidity, and productivity of interpreting brain signals.

8.1.4.2 Scope

The following are the main topics this chapter covers:

- Introduction to brain signal types and challenges.
- AI techniques in brain signal analysis.
- Background and importance of brain signal interpretation.
- The role of AI in brain signal analysis.
- Overview of brain signal types and challenges in brain signal interpretation.
- Traditional and AI methods for brain signal analysis and AI applications in brain signal analysis.
- Benefits of AI-driven analysis and future directions in AI-driven brain signal analysis.

This chapter covers these topics in order to give readers a comprehensive grasp of how artificial intelligence (AI) is transforming brain signal analysis, providing fresh perspectives, and getting past conventional constraints. The transformational potential of AI in improving brain signal interpretation and its implications for future research and clinical practice will be highlighted in this chapter.

8.2 Traditional and AI Methods for Brain Signal Analysis

Electrical and magnetic signals that form intricate patterns in the brain remain essential data that contribute to neuroscience and its application to clinical evaluation and treatment. Conquering possibilities has introduced initial findings and easier interpretations but on the other hand, AI sophisticated approaches are redesigning by explaining the simplified innate difficulties and drawbacks of conventional approaches. This section aims to illustrate the current and older approaches of analyzing the brain signals and demonstrate a vast improvement in analysis through the integration of AI.

8.2.1 Traditional methods

Conventional approaches to brain signal analysis rely on established statistical and mathematical methods. Though they frequently fail to handle the great dimensionality and complexity of brain signals, these techniques have been crucial in uncovering basic features of neural activity and functioning. The analysis of principal components (PCA), independent component analysis (ICA), wavelet transforms, and Fourier analysis are some of the primary classical methodologies.

8.2.1.1 Fourier analysis

Description: Fourier analysis breaks down signals into their periodic components by converting time-domain brain signal data into the frequency domain. For the purpose of locating and examining periodic components in brain signals, this method is especially helpful.

- **Strengths:** Excellent in locating oscillatory activity and rhythmic patterns in EEG data, including beta, gamma, and alpha rhythms.
- **Drawbacks:** Makes the assumption that the signals are stationary, which is not usually the case when brain activity is dynamic and sporadic. Additionally, its ability to handle non-linear and non-periodic events is limited.

8.2.1.2 Statistical methods

8.2.1.2.1 Principal component analysis (PCA)

Description: PCA breaks down big datasets into smaller, principal component-sized uncorrelated variables by converting the original variables into fewer, uncorrelated variables. This aids in keeping the majority of the variance while streamlining the data.

- **Strengths:** Aids in noise reduction and data display. Helpful in emphasizing the key components of brain messages.
- **Drawbacks:** Makes the assumption that variables have linear relationships with one another, possibly excluding more intricate nonlinear interactions.

8.2.1.2.2 Independent component analysis (ICA)

Description: To divide mixed signals into independent sources, ICA is utilized. This technique is frequently used to eliminate artifacts from EEG data, like those brought on by muscular contractions or eye blinks.

- **Strengths:** Excellent in separating independent signal sources and removing artifacts.
- **Drawbacks:** Depends on the potentially erroneous premise that the sources are statistically independent.

8.2.1.3 Regression techniques

Description: Relationships between brain signals and other variables are modeled and analyzed using a variety of regression approaches (linear, logistic, etc.).

- **Strengths:** Good for creating predictive models and testing hypotheses.
- **Weaknesses:** Frequently make assumptions about linear correlations, which might not fully represent the intricacy of interactions between brain signals.

8.2.1.4 Transforms of wavelets

Description: By breaking down signals into wavelets, wavelet transformations enable a multi-resolution analysis of the input data. Wavelets, in contrast to Fourier transformations, provide both frequency and time localization, which makes them appropriate for non-stationary signal analysis.

- **Strengths:** Able to analyze non-stationary and transient signals. efficient at recording frequency characteristics that change over time.
- **Weaknesses:** Needs careful wavelet function selection and is computationally demanding. Comparing interpretation to Fourier analysis, one can find it more complex.

The conventional signal processing and analysis techniques utilized in brain signal analysis are contrasted in Table 8.3. Every approach is assessed according to its benefits and drawbacks. The trade-offs associated with applying conventional methods for brain signal processing are highlighted by this comparison. While both wavelet transformations and Fourier analysis are useful for understanding the frequency domain, they are not equally adept at handling non-stationary signals. Both PCA and ICA are effective in removing

Table 8.3 Analyzing and comparing traditional techniques.

Method	Strengths	Weaknesses
Fourier analysis	Identifies periodic signals	Limited with non-stationary signals
PCA	Data simplification, noise reduction	Assumes linear relationships
ICA	Artifact removal	Assumes source independence
Regression methods	Hypothesis testing, prediction	Assumes linearity
Wavelet transforms	Analyzes non-stationary signals	Computationally intensive

artifacts from data and simplifying it, but they depend on assumptions that might not always be true. While useful for testing hypotheses and making predictions, regression techniques may not work well with nonlinear data. Depending on the particular application and properties of the brain signals being studied, each methodology has advantages and disadvantages of its own, making the choice of technique essential. In conclusion, conventional approaches to brain signal analysis have paved the way for our current comprehension of brain activity and function. More sophisticated methods must be developed because they frequently fail to handle the high dimensionality and dynamic nature of brain signals. These conventional techniques offer a fundamental framework on which increasingly complicated AI-driven techniques can be built, as the complexity of brain signal data keeps rising.

8.2.2 Introduction to AI techniques

Brain signal analysis has undergone a revolutionary change thanks to artificial intelligence (AI) techniques that utilize advanced machine learning (ML) and deep learning (DL) models [7]. These developments provide new insights and capabilities that older methods cannot match, allowing researchers and doctors to address the intrinsic complexity and volume of brain signal data. Artificial intelligence (AI) approaches are very effective in finding patterns, forecasts, and relationships in big, complicated datasets – all of which are frequent in neuroscience.

8.2.2.1 Machine learning (ML)

Description: Computers can learn from data and make judgments or predictions because of a variety of models and algorithms known as machine learning (ML) [7]. The three main types of machine learning are reinforcement learning, unsupervised learning, and supervised learning. Without the need for human intervention, machine learning approaches may classify various brain activity states, forecast outcomes, and group comparable patterns in the context of brain signal analysis.

Strengths: Machine learning models may be trained to identify intricate patterns in brain signals and are proficient at managing high-dimensional data. They are especially helpful for problems involving classification, regression, and clustering since they can adjust to new data and gradually improve their performance.

Applications: ML approaches are used to categorize various brain activity types, identify abnormalities in brain signals, and forecast neurological disorders based on EEG, fMRI, or MEG data.

8.2.2.2 Supervised learning

Description: In the context of supervised learning, models are trained using labeled datasets, which contain known input data and matching output labels. When it comes to brain signal analysis, this method works especially well for tasks like regression and classification.

Strengths: When given enough labeled data, supervised learning models can spot patterns and make predictions with a high degree of accuracy. They are frequently employed for tasks like classifying cognitive states based on brain activity or identifying epileptic episodes in EEG data.

Challenges: If the training data is not typical of real-world scenarios, performance may deteriorate and requires a significant amount of labeled data for effective training.

8.2.2.3 Unsupervised learning

Description: Training models on unlabeled data with the goal of finding underlying structures or hidden patterns is known as unsupervised learning. This group includes methods like dimensionality reduction and grouping.

Strengths: Unsupervised machine learning can be used to reduce the complexity of brain signal data, find unknown patterns, and do exploratory data analysis. It can provide information that trained approaches are unable to disclose.

Applications: Used to uncover novel biomarkers for neurological diseases, minimize noise in neuroimaging data, and group comparable brain activity patterns.

8.2.2.4 Reinforcement learning (RL)

Description: Reinforcement learning is a technique used to teach models to make a series of choices by punishing bad behavior and praising good conduct. Although less frequently used in brain signal analysis, this method may find use in neurofeedback and adaptive BCI systems.

Strengths: RL models are well-suited for dynamic and interactive applications due to their capacity to learn complicated behaviors and adjust to changing contexts.

Applications: May be helpful in creating BCIs that are adaptable and capable of learning and optimizing control techniques in response to user input.

8.2.2.5 Hybrid methods

Description: The integration of AI techniques with conventional methods can effectively capitalize on the advantages of each methodology. For instance, conventional feature extraction techniques can be combined with ML models to increase the precision of categorization.

Strengths: By adding domain-specific information, hybrid techniques can increase the interpretability and performance of AI models.

Applications: Employed in multimodal brain signal analysis, which combines several data sources (such as fMRI and EEG) to create a more thorough picture of brain activity.

Researchers can overcome the shortcomings of conventional approaches and obtain greater comprehension of how the brain works by combining AI tools with standard approaches. AI-enhanced brain signal analysis is very promising for building novel brain–computer interfaces, enhancing clinical diagnoses, and furthering neuroscience research. AI is a vital instrument in the current investigation of brain signals because of its transformational potential.

8.2.2.6 Comparative evaluation

Artificial intelligence (AI) techniques offer significant improvements over classical approaches by facilitating more intricate and thorough examinations of brain signals. The principal benefits comprise:

- **Accuracy:** When it comes to pattern recognition and classification tasks, AI models – particularly deep learning – offer better accuracy than conventional techniques.
- **Efficiency:** AI can process and analyze big datasets faster, which cuts down on the amount of time needed for analysis.
- **Scalability:** AI methods can handle the growing volumes of brain signal data produced by contemporary neuroimaging technology because they are very scalable.

Artificial Intelligence (AI) tools facilitate the recognition of intricate patterns, allow for flexibility in response to novel input, and offer insights that may result in improved comprehension of cognitive processes, earlier and more precise medical diagnosis, and more efficient brain–computer interfaces (BCIs). Researchers and practitioners can take use of the advantages of both approaches by combining conventional methods with AI-enhanced techniques. In the end, this integrated strategy improves clinical results and advances our understanding of brain function by enabling more thorough and accurate analysis of brain signals. The groundwork for more in-depth talks on particular AI applications, advantages, and future possibilities in later sections is laid by this thorough examination of conventional and AI-enhanced methods.

8.3 AI Applications and Case Studies in Brain Signal Analysis

8.3.1 AI methods for analyzing brain signals

In particular, deep learning (DL) and machine learning (ML) have been applied, and this has greatly enhanced the field of brain signal analysis [7]. These models provide new insights and capabilities that go beyond conventional methods, and they perform exceptionally well in handling the complexity and volume of brain signal data. An overview of the many ML and DL models [12] used in brain signal analysis is given in this part, together with information on their designs, effects, and performance measures.

Using ordinary statistical techniques to analyze brain signal data is challenging because these data contain high dimensionality and nonlinearity; however, machine learning models are useful in analyzing such data based on pattern recognition, data classification and predictive analysis on large datasets. Some of the most important machine learning methods that have been used in this domain are as follows: the support vector machines which is suitable for classification and regression as it is based on the identification of hyperplane which differentiates between different classes of data; the random forest which enhances the classification accuracy and reliability by a combination of decision trees; and the k-nearest neighbors type of algorithms which is an easy mechanism for the data instance that is learned based on its Furthermore, other techniques such as principal component analysis (PCA) and independent component analysis (ICA) help in the data visualization improvement, noise reduction, and feature extraction from the data by projecting them to the principal components or by isolating the mixed signals and separating them into independent source signals [13]. Altogether, these models collectively improve the performance of incongruity identification, mental state categorization, and the interpretation of intricate neurosignaling, which severely boosts the sphere of brain signal processing.

The advantages, disadvantages, and uses of several machine learning models that are frequently applied to brain signal analysis are shown in Table 8.4. The salient features of every machine learning model are emphasized in this synopsis, along with their applicability to particular brain signal analysis tasks and any potential drawbacks.

Neural networks are one of the key technologies underlying the modern deep learning models that have emerged as highly successful for the brain signal interpretation because of their ability to capture nonlinear relationships in data. Field images, neuroimaging from fMRI for example, are best processed by convolutional neural networks (CNNs), whereas sequential-structured information such as EEG signals [11] which is a time series data is best interpreted by recurrent neural networks (RNNs) or long short term memory (LSTM) networks. Also, autoencoders are involved in the unsupervised learning activities like noise removal and feature extraction that helps to improve the quality of the data coming from a brain signal. Thanks to such deep learning models, breakthroughs in the identification of early symptoms, prognosis, and analysis of brain activity have been made due to the enhancement of the signals' analysis process.

AI techniques paved the way for breakthroughs not only in the field of brain signal analysis but also in the understanding of the function of the human brain. AI related to application of machines that have complex

Table 8.4 Metrics of machine learning model performance.

Model	Strengths	Weaknesses	Applications
Support vector machine (SVM)	Effective in high-dimensional spaces, versatile	Computationally intensive, less effective with large datasets	Classification of mental states, anomaly detection
Random forests	Robust to noise and overfitting, handles high-dimensional data	Interpretability can be challenging	Feature selection, classification in neuroimaging
k-nearest neighbors (k-NN)	Simple, effective for small datasets	Inefficient with large datasets, sensitive to noise	Real-time classification in BCI
PCA and ICA	Data visualization, noise reduction, feature extraction	May oversimplify data, assumptions may not always hold	Preprocessing, artifact removal in EEG

algorithms using the ML and DL techniques has made great advancement amplifying precise, fast and efficient analysis of the brain signals. Such classification methods include support vector machines (SVMs), random forests, k-nearest neighbors (k-NN) allowing for classification of different states, detection of deviations or predictions of neurological disorders from achieved EEG/fMRI/MEG [5], [6], [11]. Furthermore, DL application on DL models such as CNN, RNN, and LSTM has been employed for accurately identifying pattern and temporal structure of the brain signals thereby improving the analysis of neuroimages, time series prediction, and understanding of temporal patterns of brain activity [17]. In addition, more enhanced methods have been made possible through artificial intelligence of neuro-BCIs [14] and tailored treatment plans for neurological diseases and injuries, leading to enhanced health of patients and their quality of life. In general, AI enhances the analysis of brain signals, which significantly contributes to new neurological studies and the development of diagnostic tools and treatments in neuromedicine.

8.3.2 Case studies and practical applications

In this section, we will examine several valuable examples and practical intentions in which AI has contributed to the progress of brain signal analysis. These examples indicate how AI-based approaches work practically in different fields such as medical diagnosis, psychological disorder, cognitive scientist and brain–computer interface.

Case study 1: This is why it is important to conduct early diagnosis on patients suffering from Alzheimer's disease.

Overview: It revealed that the use of AI algorithms, especially deep learning models, can be applied in fMRI and EEG data to diagnose Alzheimer's disease in its early stages. Diagnostic approaches that humans use lack the capability to detect onset signs of the disease, while AI will analyze signs that would be considered insignificant.

Key points:

Data source: Data sets containing images of the brains of AD patients with different severity and healthy elder control subjects and sentinel fMRI and EEG signals.

Methodology: Convolutional layers for handling the imaging features while Recurrent Neural Networks for the temporal features in the EEG data.

Results: AI models exhibited heightened accuracy in detecting the ailments at their early stages and thereby recommending interventional measures.

Impact: First, diagnoses increase the quality of patient management and, second, open possibilities for new therapies.

Case study 2: For epilepsy patients, predicting seizures The KNN model can predict seizures by learning the similarities and differences between new data and a database of previous epileptic seizures.

Overview: The current AI models used in epilepsy consist of machine learning, that aims at predicting seizures by analyzing EEG data. Forecasting is highly beneficial when it comes to the patient's quality of life, especially in cases where the patient may be at risk of developing further complications.

Key points:

Data source: Entire dyadic instances taken using electrical headgear on epilepsy patients.

Methodology: Use of patterns which go with machine learning algorithms for support vector machines (SVMs) and deep learning of LSTMs (long short-term memory networks) to find pre-seizure patterns. Employing deep learning approaches where mainly CNN and RNNs to distinguish signals from the user's brain to infer his intentions.

Results: Enhanced performance and reliability of BCI systems and that is why it will allow people with impaired limbs more precise control of prosthetics, computers, as well as other technical gadgets and appliances.

Impact: Higher levels of employment, education, unprompted accessibility, and social participation among people with disabilities.

Practical application 1: AI for early detection and management of mental disorders

Overview: Brain function signs are used by AI-lead models to continuously track and evaluate mental health states. Of particular concern in this application is the identification and treatment of mental disorders like depression and anxiety.

Key points:

Data source: Collect medical records and data from patients with mental health disorders including the EEG/fMRI results.

Methodology: Use of machine learning algorithms for discovering biomarkers and studying the progression of the modified brain state.

Results: It reported that artificial intelligence models ensured appropriate observation and risk of a patient's mental state decline.

Impact: While the fact remains that patients with mental health conditions have chances of relapse, the following benefits may be considered:

Practical application 2: The following is a list of cognitive enhancement and neurofeedback

Overview: Neurofeedback is another application of AI in which individuals have the freedom to learn through enabling attention span, memory, and learning. These systems offer implied reactions to the users in real time depending on the brain activity.

Key points:

Data source: EEG signals recorded while trying to perform a certain task.

Table 8.5 Summary of key case studies and practical applications.

Case study/application	AI technique used	Data source	Key outcomes	Impact
Early diagnosis of Alzheimer's	CNNs, RNNs	fMRI, EEG	High accuracy in early diagnosis	Timely intervention and new treatments
Predicting seizures in epilepsy	SVMs, LSTMs	EEG	High predictive accuracy, reduced false alarms	Enhanced patient safety and management
Enhancing BCIs	CNNs, RNNs	EEG	Improved accuracy and speed of BCI systems	Increased independence for disabled users
AI in mental health monitoring	Various ML techniques	EEG, fMRI	Continuous monitoring and early detection of changes	Better mental health management
Cognitive enhancement and neurofeedback	Machine learning algorithms	Real-time EEG	Improved cognitive performance through neurofeedback	Applications in education and rehabilitation

Methodology: Incorporation of machine learning algorithms that is to interpret signals from people's brains and respond accordingly.

Results: Enhanced cognitive processes and learning interventions through individualized stimulation of brain function.

Impact: Use in schools, cognitive retraining and attention training, and sports performance.

A synopsis of notable case studies and real-world applications that use AI to evaluate brain signals is included in Table 8.5. It contains details about the important results, data sources, AI approaches applied, and the overall impact of these applications. This overview shows the numerous uses of AI approaches to neurological illnesses and disorders, highlighting notable advancements in patient safety, diagnostic precision, and overall efficacy of brain-computer interaction systems. These developments have an impact on mental health monitoring, patient management, healthcare, and cognitive improvement.

8.4 Implications of AI-Powered Analysis

Artificial intelligence (AI) in brain signal analysis has been a game changer in that it provides a myriad of advantages that equally make a huge impact on analyses afloat in research and clinical practices. Perhaps the most significant benefit of embracing AI is the high level of precision in the results as compared to the results attained with conventional techniques. AI models can also analyze more detailed features in the signals and appreciate more precise structural aspects of the data by utilizing high-end algorithms and deep learning architectures which enhances the diagnostic and predictive accuracy of the outcomes.

Figure 8.1 offers a graphical representation of the said equation displaying the accuracy comparison which is learnt through AI analysis.

Furthermore, AI techniques present a distinguishable addition to the rate of analysis concurrently. Collecting data using traditional procedures may involve using or partial automation of traditional approaches that is both time-consuming and costly. While the human brain takes a considerable amount of time to analyze large quantities of data, extract features and make decisions, AI algorithms are much faster in these tasks. Table 8.6 prescribes the velocity rates of analyses of different methodologies showing how the AI approaches enable efficiency.

In addition, AI is advantageous in analyzing and collecting brain signal datasets, a feature valuable in the advent of big data. Given that the number

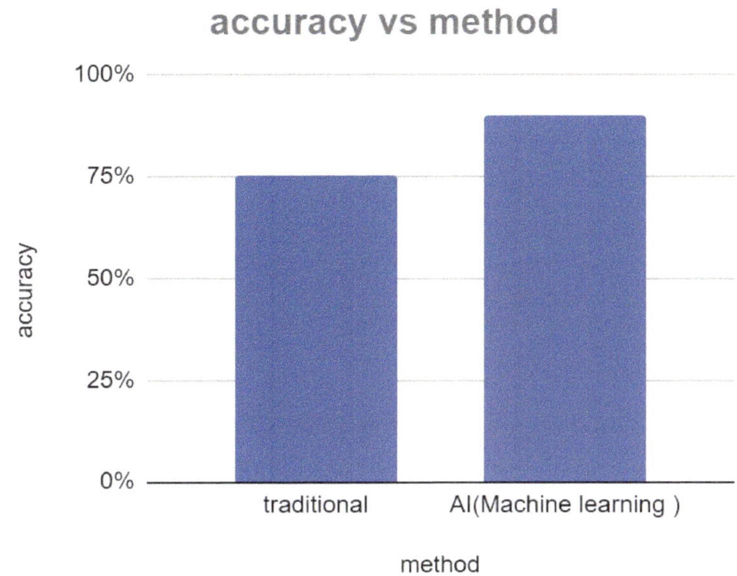

Figure 8.1 Accuracy comparison between traditional and AI methods.

of publications in neuroscience and applied practices has increased dramatically, traditional ways of processing, analysis, and storing very well might not be sufficient to handle the vast amounts of data available. As they are optimized to contain scalable architectures and integrate parallel computation capacities, AI-powered systems showcase outstanding performance in handling expansive data sets. To sum the outline, the benefits of AI-driven analysis are unprecedented boosts in precision, velocity, and expansibility, making it a staple in an advanced exploration of brain signals and diagnostic capabilities. Ultimately, it overhauls our understanding of how brain signals can be perceived and analyzed, offering new possibilities for knowledge and

Table 8.6 Rapid measures of different analysis methods.

Analysis technique	Processing speed	Resource utilization	Scalability
Traditional methods	Moderate	High	Limited
Machine learning	High	Moderate	Moderate
Deep learning	Very high	Moderate	High
Parallel processing	Extremely high	High	Very high
Cloud computing	High	Variable	High
Distributed computing	Very high	High	Very high

developments in the field of neuroscience due to the scope of abilities that AI possesses, including the capability to reveal subtleties of patterns, work faster, and process big data.

8.5 Challenges and Ethical Considerations

In adopting AI techniques in decoding brain signals, several issues arise which are briefly described below. Two examples are data sharing and misuse, and secondly the bias of AI in its functioning.

8.5.1 Data privacy issues

The primary issue with AI-driven analysis of brain signals and their correlation is that they highly depend on the gathering and use of personal data, which is rather sensitive in most cases, hence leading to privacy concerns. The large amount of neuroimaging data, including the EEGs data and fMRI data, consists of such particulars of individuals as their cognition state, mental health conditions, and neurological diseases. Sacrilegious use of this kind of information may lead to numerous privacy invasions that may harm the affected individuals in one way or another including; discrimination, and stigmatization.

Figure 8.2 provides a decoupling of the complexity of the privacy issues relating to the use of AI in the analysis of datasets obtained from brain signal analysis, as well as the need to protect the rights and privacy of individuals

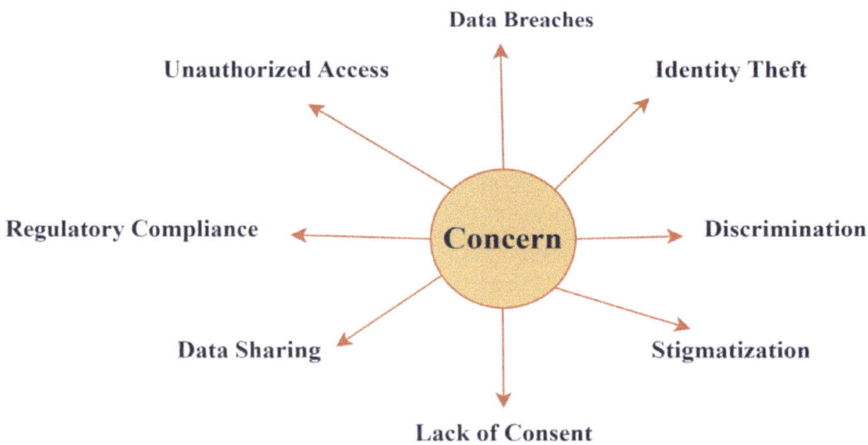

Figure 8.2 Data privacy concerns in AI.

through the incorporation of strong privacy protection policies and ethical principles.

8.5.2 Biases in AI

Another important risk factor, which is still beyond the expertise of developers, lies in the possibility of biases existing within AI systems to support and enhance discrimination and injustice in the study of human brain signals. They can be attributed to limited training datasets, the algorithms used in the model, or intrinsic human bias appearing within decision-making processes. Perverse effects in the context of brain signal analysis can result in misleading patterns, misinformation, wrong diagnoses, risk of inequalities in outcomes, and wrong treatment suggestions, which will further existing healthcare inequities and intensify social inequalities. Table 8.7 gives elaborate examples of how bias manifests with AI systems to show how the cases of bias mitigation strategies, algorithmic fairness and transparency are relevant to apply in the case of using AI to analyze brain signals.

In overcoming these issues, there is a need to employ effective privacy practices, identify, and report bias in AI-driven brain signal analysis, and follow ethical guidelines that promote fairness and accountability for AI's applications. Maintaining privacy-preserving, controlling for bias, and ensuring algorithmic equity will help ensure that stakeholder values for AI-based services of signal analysis on the brains are ethical. Table 8.8 lists the various kinds of biases that AI systems may have, along with an explanation of each type and its effects, demonstrates the several ways in which bias may appear in AI systems and the serious consequences that these biases may have for both people and society.

Table 8.7 Examples of biases in AI systems.

Bias type	Description	Impact
Gender bias	Algorithms exhibit differential performance based on gender	Gender disparities in diagnosis and treatment
Racial bias	Discriminatory outcomes based on racial characteristics	Reinforcement of racial stereotypes and disparities
Confirmation bias	Algorithms reinforce existing beliefs or stereotypes	Confirmation of biases, hindering diversity and inclusion
Sampling bias	Unequal representation of certain groups in training data	Inaccurate predictions and generalizations

Table 8.8 Current developments in technology and their effects.

Technological advancement	**Description**	**Impact**
Real-time neurofeedback systems	Systems enabling individuals to modulate brain activity in real time	Cognitive enhancement, neurorehabilitation
Wearable neuroimaging devices	Portable devices facilitating brain monitoring in diverse settings	Expanded access to brain monitoring
EEG-based brain–computer interfaces	Interfaces allowing direct communication between the brain and computers	Assistive technology, neuroprosthetics

8.6 Future Directions in AI-driven Brain Signal Analysis

The dependence on the future of brain signal analysis is set high due to the potential trends in its development and technologies that are already connected to Artificial intelligence (AI). These advancements offer the potential of transforming previous conceptions of how nerves function and organizing neurological diseases as well as improving diagnostic capability and therapeutic outcomes for a variety of ailments originating in the central nervous system.

8.6.1 Emerging trends

New trends in approaches of AI driven brain signal processing analytics encompass a vast range of approaches and solutions [27]. One such emerging trend is the use of multimodal data fusion techniques and methods that employ data from not only EEG, fMRI, and MEG but a combination of all the three. Alas, the fairly recent introduction of methods pertaining to explainable AI (XAI) allows the researchers to understand the reasons for certain AI-based performance analyses and increase the credibility of the diagnostics and other predictive frameworks based on such approaches. In addition the day has not been far off when we start using neuromorphic computing and cognitive-inspired AI architecture that will lead to more efficient and scalable algorithms to replicate the cognition process of the human brain.

8.6.2 Technological advancements

The relatively contemporary technological improvement has brightened a new future of the application of AI in brain signal analysis with an emphasis on better diagnosis and treatment in medicine. Major achievements are those that have led to developing real-time neurofeedback systems where subjects gain a level of control over the brain activity that can adapt to external cues, enhancing cognition, and rehabilitation. In addition, convenient neuro-technology in wearable portable EEG headsets as well as monitoring devices have provided greater opportunities within medical as well as other contexts to fulfill different experiments and treatments. Although at the moment, it has not been significantly developed yet, blockchain technology presents the potential to be categorized as an emerging technology more so when combined with AI in neuroscience. In particular, blockchain as a distributed database could improve the reliability of brain signal data that are considered sensitive. This has been proven in the use of this technology in areas like education, health, and financial and resource management among others. Specific to application in analyzing brain signals, the technology of blockchain can be utilized for safeguarding data and with its feature of guaranteeing the identity of records, ideal for patient privacy and data integrity in the use of AI in diagnosis and treatment. In this way, with the help of blockchain, both researchers and clinicians are able to share and analyze vast amounts of data on neurological disorders' manifestations and progress, which plays an important role in improving the situation in neuroscience [26]. Table 8.8 provides an overview of modern advancements in technology as well as its importance in multiple areas, specifically emphasizing the progressive nature of technology for studying brain signals.

The future of AI in brain signal analysis is moving towards combining different types of data and making AI decisions easier to understand. New technologies are also making it possible to monitor and respond in real-time. Using the impact of AI and combined with technology in research and clinical practice, the researchers will be able to seek new horizons of groundbreaking in the human brain processes and in brain disorders resulting in better results for the patients and better quality of lives.

8.7 Conclusion

Therefore, it can be stated that the notion of AI is a revolutionary innovation in the area of neuroscience and clinical practice regarding signal evaluation. This is due to the use of AI with enhanced capabilities of machine learning (ML) algorithms and deep learning (DL) in analyzing the brain signal data accurately at a faster rate and on a large scale. Machine learning methods, support vector machines, random forests, convolutional neural networks, and recurrent neural networks have proved to be powerful in classification, anomaly detection, decision-making, and other analysis works. However, with the prospects of integrating AI in the analysis of the brain signals, there are a number of issues and concerns that arise. The general data protection regulation can become a real issue when it comes to the use of personal data to train the algorithms due to privacy concerns and the potential issue of reinforcement learning by the AI algorithms. Specifically, to achieve responsible and fair use of AI in the context of brain signal analysis, there is a need to develop good privacy controls, eliminate biases, and follow ethical algorithms.

In the future, scientific and technological future targets, including multimodal data fusion, explainability of AI, and technological innovations in the real-time monitoring of data, are likely to provide a solid ground for further important discoveries in mapping human brain function and the development of targeted diagnostic and treatment approaches. Through enhancing the proactive role of AI in the neurosciences and technology, researchers and clinicians can advance the boundaries of neuroscience leading to enhanced healthcare for improved human use of the brain.

References

[1] Pradesh, A. SCIENTIFIC DISCOVERIES IN THE AGE OF ARTIFICIAL INTELLIGENCE.

[2] Chataut, R., Nankya, M., & Akl, R. (2024). 6G networks and the AI revolution—Exploring technologies, applications, and emerging challenges. Sensors, 24(6), 1888.

[3] Hossain, M. T., Afrin, R., & Biswas, M. A. A. A Review of the Prospects and Constraints for Using Artificial Intelligence for the Interpretation of Remote Sensing Data.

[4] Amin, U., Nascimento, F. A., Karakis, I., Schomer, D., & Benbadis, S. R. (2023). Normal variants and artifacts: importance in EEG interpretation. Epileptic Disorders, 25(5), 591-648.

[5] Beldzik, E., & Ullsperger, M. (2024). A thin line between conflict and reaction time effects on EEG and fMRI brain signals. Imaging Neuroscience, 2, 1-17.

[6] Fred, A. L., Kumar, S. N., Kumar Haridhas, A., Ghosh, S., Purushothaman Bhuvana, H., Sim, W. K. J., ... & Gulyás, B. (2022).

A brief introduction to magnetoencephalography (MEG) and its clinical applications. Brain sciences, 12(6), 788.

[7] Raghavendra, U., Acharya, U. R., & Adeli, H. (2020). Artificial intelligence techniques for automated diagnosis of neurological disorders. European neurology, 82(1-3), 41-64.

[8] Bijsterbosch, J., Harrison, S. J., Jbabdi, S., Woolrich, M., Beckmann, C., Smith, S., & Duff, E. P. (2020). Challenges and future directions for representations of functional brain organization. Nature neuroscience, 23(12), 1484-1495.

[9] Surianarayanan, C., Lawrence, J. J., Chelliah, P. R., Prakash, E., & Hewage, C. (2023). Convergence of Artificial Intelligence and Neuroscience towards the Diagnosis of Neurological Disorders—A Scoping Review. Sensors, 23(6), 3062.

[10] Hussain, I., & Nazir, M. B. (2024). Mind Matters: Exploring AI, Machine Learning, and Deep Learning in Neurological Health. International Journal of Advanced Engineering Technologies and Innovations, 1(4), 209-230.

[11] Nazir, M. B., & Hussain, I. (2024). Charting New Frontiers: AI, Machine Learning, and Deep Learning in Brain and Heart Health. Revista Espanola de Documentacion Cientifica, 18(02), 209-237.

[12] Oblong, L. M., Soheili-Nezhad, S., Trevisan, N., Shi, Y., Beckmann, C. F., & Sprooten, E. (2024). Principal and independent genomic components of brain structure and function. Genes, Brain and Behavior, 23(1), e12876.

[13] Aburakhia, S., Shami, A., & Karagiannidis, G. K. (2024). On the Intersection of Signal Processing and Machine Learning: A Use Case-Driven Analysis Approach. arXiv preprint arXiv:2403.17181.

[14] Marzetti, L., Makkinayeri, S., Pieramico, G., Guidotti, R., D'Andrea, A., Roine, T., ... & Pizzella, V. (2024). Towards real-time identification of large-scale brain states for improved brain state-dependent stimulation. Clin. Neurophysiol, 158, 196-203.

[15] Yousefi, F., & Kolivand, H. (2024). Brain signals as a new biometric authentication method using brain-computer interface. In Encyclopedia of Computer Graphics and Games (pp. 228-241). Cham: Springer International Publishing.

[16] Albahri, A. S., Al-Qaysi, Z. T., Alzubaidi, L., Alnoor, A., Albahri, O. S., Alamoodi, A. H., & Bakar, A. A. (2023). A Systematic Review of

Using Deep Learning Technology in the Steady-State Visually Evoked Potential-Based Brain-Computer Interface Applications: Current Trends and Future Trust Methodology. International Journal of Telemedicine and Applications, 2023(1), 7741735.

[17] Sadegh-Zadeh, S. A., Fakhri, E., Bahrami, M., Bagheri, E., Khamsehashari, R., Noroozian, M., & Hajiyavand, A. M. (2023). An approach toward artificial intelligence Alzheimer's disease diagnosis using brain signals. diagnostics, 13(3), 477.

[18] Majumdar, S., Al-Habaibeh, A., Omurtag, A., Shakmak, B., & Asrar, M. (2023). A Novel Approach for Communicating with Patients Suffering from Completely Locked-in-Syndrome (CLIS) via Thoughts: Brain Computer Interface System using EEG Signals and Artificial Intelligence. Neuroscience Informatics, 100126.

[19] Zhou, Z., Asghar, M. A., Nazir, D., Siddique, K., Shorfuzzaman, M., & Mehmood, R. M. (2023). An AI-empowered affect recognition model for healthcare and emotional well-being using physiological signals. Cluster Computing, 26(2), 1253-1266.

[20] Gong, C., Jing, C., Chen, X., Pun, C. M., Huang, G., Saha, A., ... & Wang, S. (2023). Generative AI for brain image computing and brain network computing: a review. Frontiers in Neuroscience, 17, 1203104.

[21] Saminu, S., Xu, G., Zhang, S., Ab El Kader, I., Aliyu, H. A., Jabire, A. H., ... & Adamu, M. J. (2023). Applications of artificial intelligence in automatic detection of epileptic seizures using EEG signals: A review. In Artificial Intelligence and Applications (Vol. 1, No. 1, pp. 11-25).

[22] Hossain, K. M., Islam, M. A., Hossain, S., Nijholt, A., & Ahad, M. A. R. (2023). Status of deep learning for EEG-based brain–computer interface applications. Frontiers in computational neuroscience, 16, 1006763.

[23] Tian, Z., Wu, Z., & Ying, S. (2024). Brain functional analysis and brain-like intelligence. Frontiers in Neuroscience, 18, 1383481.

[24] Viswan, V., Shaffi, N., Mahmud, M., Subramanian, K., & Hajamohideen, F. (2024). Explainable artificial intelligence in Alzheimer's disease classification: A systematic review. Cognitive Computation, 16(1), 1-44.

[25] Allugunti, V. R., & Reddy, K. K. C., Elango, NM, Anisha, PR (2021). Prediction of Diabetes Using Internet of Things (IoT) and Decision Trees: SLDPS. Intelligent Data Engineering and Analytics. Advances in Intelligent Systems and Computing, 1177.

[26] Anisha, P. R., Reddy, C. K. K., & Nguyen, N. G. (2022). Blockchain technology: a boon at the pandemic times–a solution for global economy upliftment with AI and IoT. Blockchain Security in Cloud Computing, 227-252.

[27] Reddy, C. K. K., Anisha, P. R., Khan, S., Hanafiah, M. M., Pamulaparty, L., & Mohana, R. M. (Eds.). (2024). Sustainability in Industry 5.0: Theory and Applications. CRC Press.

9

Neuroimaging Techniques: Innovations and Applications

Md. Mahadi Hassan[1], Farhana Yasmin[2], Mahade Hasan[3], and Chetna Sharma[4]

[1]Department of Computer Science and Engineering, Bangladesh University of Business and Technology, Bangladesh
[2]School of Computer and Software, Nanjing University of Information Science and Technology, China
[3]School of Software, Nanjing University of Information Science and Technology, China
[4]Chitkara University Institute of Engineering and Technology, Chitkara University, India
E-mail: mahedi.syn@gmail.com; farabi.eu.bd@gmail.com; farhana005.cse@gmail.com; xueyu@nuist.edu.cn; chetnakaushal3558@gmail.com

Abstract

Recent breakthroughs in neuroimaging have shed light on the nervous system's anatomy and physiology in ways never before seen. Recent developments in neuroimaging are significantly enhancing our knowledge of neurological illnesses, which in turn allows for more accurate diagnosis and therapy. Applications demonstrate its efficacy in detecting, diagnosing, and treating a range of neurological conditions, including neurodegenerative disorders. Artificial intelligence (AI) has opened up new possibilities for neuroimaging by solving current challenges. The purpose of this research is to illuminate the key neuroimaging methods and how their continued development will enhance neuroscience by illuminating their usefulness in therapeutic and medical contexts.

Keywords: EEG, fMRI, PET, disorder, BCIs, neurodegenerative, AI.

9.1 Introduction

Neuroimaging is an innovative subject of neuroscience that provides many non-invasive ways to study the brain's complicated architecture [1]. This rapidly developing area of study primarily aims to detect lesions in the brain, such as those caused by vascular disease, strokes, tumors, inflammatory illnesses, and others. In this context, neuroimaging techniques are crucial for studying the nervous system, the brain, and medical, psychological, and neurological conditions.

Structural neuroimaging and functional neuroimaging are the two main types of neuroimaging methods. Functional neuroimaging measures brain functions, while structural neuroimaging primarily visualizes and quantifies brain anatomy. There are a number of popular structural neuroimaging methods, including computed tomography (CT), magnetic resonance imaging (MRI), and diffusion tensor imaging (DTI). Such methods are very helpful in the study of brain health and the diagnosis of a wide range of diseases. Among the functional neuroimaging techniques that are frequently employed for the diagnosis of neurological disorders, we find functional near-infrared spectroscopy (fNIRS), electroencephalography (EEG), positron emission tomography (PET), and functional magnetic resonance imaging (fMRI).

Neuroimaging techniques provide critical insights into the health and function of the brain by generating images of the brain or other components of the nervous system, as well as neurological illnesses [2]. The methods have greatly improved our overall comprehension of brain illnesses by offering valuable insights into changes in cognitive systems and enhancing the diagnosis of neurological conditions. More precise information on the brain's anatomy and functions is now available due to innovative neuroimaging methods, including fMRI, DTI, MRS, fNIRS, and others. These advancements have made way for more accurate diagnoses, more effective treatment options, and a more thorough understanding of neurological health. Furthermore, emerging and specialized methods provide promising avenues for research into brain illness, tailored treatment, and function. Neuroimaging-based biomedical devices have a significant impact on brain health processes. Due to its versatile features, artificial intelligence (AI) is surely improving neuroimaging methods. AI is revolutionizing clinical translational imaging, particularly neuroimaging, for early diagnosis, prediction, and treatment of chronic pain conditions, enhancing healthcare systems and addressing various health issues.

In this study, we focused on the most recent advances in neuroimaging and their applications in prestigious medical disciplines, emphasizing their importance in recognizing and diagnosing neurological diseases. The role of AI in revolutionizing clinical neuroimaging, as well as how it can enhance brain healthcare systems, has also been discussed. Furthermore, the research looked at future opportunities, problems, and prospective advances in neuroimaging and its applications.

9.2 Structural Imaging Techniques

9.2.1 Magnetic resonance imaging (MRI)

Magnetic resonance imaging (MRI) is a type of medical imaging that uses radio waves generated by computers to produce pictures of the human body's organs and tissues. Non-invasively examining the body's organs, tissues, and skeletal systems, it creates high-resolution pictures that aid in the diagnosis of many ailments. When an organism's water molecules undergo a rotation, MRI can detect this change in the rotational axis because it is dependent on the magnetic properties of atomic nuclei [3]. Among the most versatile and powerful imaging technologies, magnetic resonance imaging (MRI) has a wide range of therapeutic uses today, including in the fields of neurology,

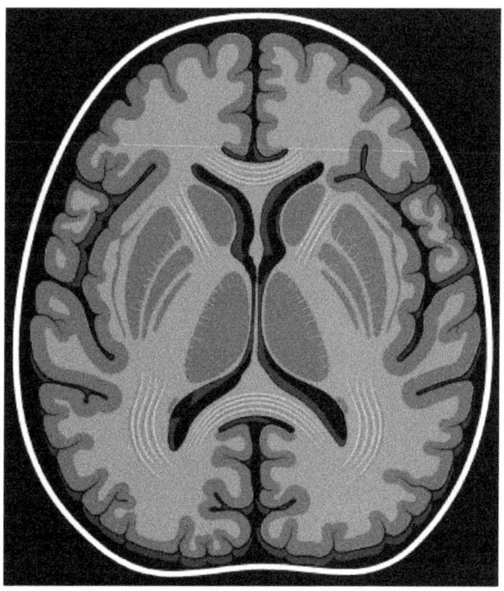

Figure 9.1 MRI scan of the human brain.

psychology, cardiology, abdominal imaging, musculoskeletal disorders, and vascular healthcare [4]. MRI's greater tissue contrast over CT and X-rays allows it to distinguish between various soft tissues such as fat, water, and muscle. Figure 9.1 shows a picture of a human brain that was scanned using magnetic resonance imaging (MRI).

Sclocco et al. [5] showed the difficulties and potential benefits of using ultrahigh-field MRI for brainstem neuroimaging. The study delved into the current and future clinical uses of ultra-high-field magnetic resonance imaging (UHF MRI) to learn more about conditions like Parkinson's disease and chronic pain. It also looked at potential ways to improve brainstem imaging with UHF MRI and how it can reveal more about neuroanatomy and neurophysiology.

9.2.2 Computed tomography (CT)

Computerized tomography (CT) scans employ X-rays to produce images of anatomical cross-sections. The area of three-dimensional (3D) brain imaging has made extensive use of this technique [6]. CT has been increasingly

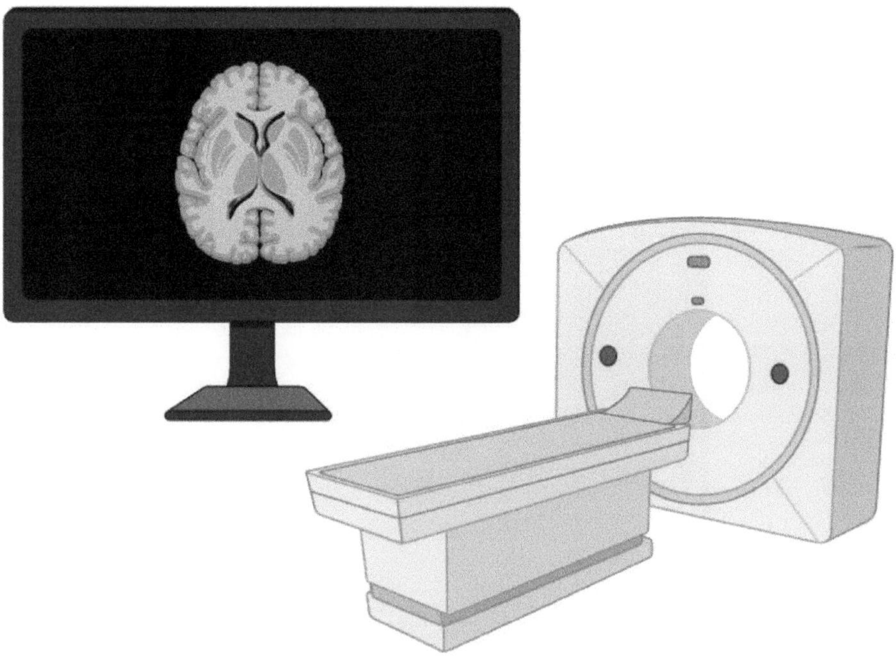

Figure 9.3 CT scan mechanism.

popular as a useful imaging tool for assessing a broad range of pathologic diseases in ED admissions since the 1990s, and its use has only grown in the last few decades [7]. For diagnostic, treatment planning, interventional, or screening reasons, CT scans may be performed on any part of the body. If it comes to diagnosing, planning therapy, and evaluating a wide range of illnesses in both adults and children, CT scans are invaluable tools. As seen in Figure 9.3, a CT scan involves a gantry mounted with a number of detectors that measure X-ray attenuation in conjunction with a spinning X-ray tube.

Kampe et al. [8] demonstrated the diagnostic utility of CT using a hybrid imaging method known as FDG-PET/CT in order to assess neurological patients in critical care. The study looked at the efficacy of FDG-PET/CT as a diagnostic tool for patients receiving care in a neurological/neurosurgical intensive care unit or a stroke unit.

9.2.3 Diffusion tensor imaging (DTI)

Diffusion tensor imaging (DTI), a non-invasive MRI method, can find out how much anisotropy and limited water diffusion there is in living tissues [9]. This novel MRI method primarily assesses microstructural changes in the brain by monitoring the tissue-level mobility of water molecules [10].

Figure 9.4 DTI of the human brain.

Applying magnetic field gradients can create a picture sensitive to diffusion in a specific direction, enabling the investigation of water diffusion. It is useful for evaluating brain injuries because it creates color pictures of the white matter tracts of nerve fibers and evaluates the flow of water molecules along the fibers. With this effective method, we can easily access the vast majority of data on the orientation of nerve fiber tracts and may utilize it to study crucial areas of neurophysiology linked to central nervous system illnesses. Figure 9.4 displays the DTI of the human brain.

Meoded et al. [11] covered the concepts and uses of DTI in pediatric neurological research. In order to navigate and explore brain activities in a unique way, the study focused on the development and innovation of neuroimaging methodologies.

9.3 Functional Imaging Techniques

9.3.1 Functional magnetic resonance imaging (fMRI)

Functional magnetic resonance imaging (fMRI) is one way to monitor brain activity based on changes in cerebral blood oxygenation [12]. It is the most advanced kind of magnetic resonance imaging (MRI), records metabolic activity in real time, provides information on brain activity, and detects subtle changes in blood flow over time induced by brain activity. During fMRI,

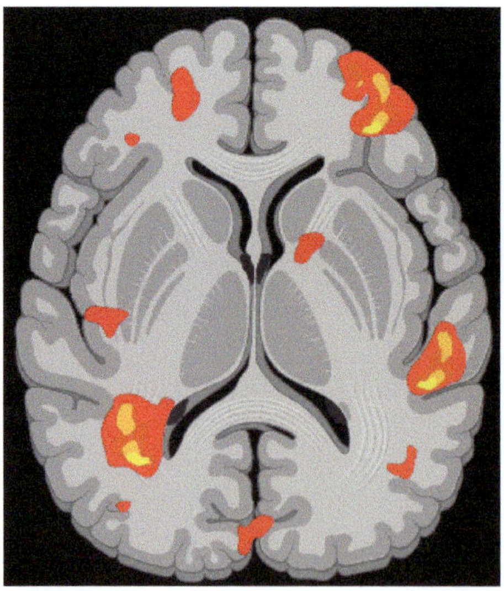

Figure 9.5 fMRI scan of the human brain.

the blood-oxygen-level-dependent (BOLD) response serves as a roundabout method for measuring electricity in neurons. It provides scientists with extraordinary tools for mapping the human brain, which allows them to investigate the bases of human thought, action, and emotion [13]. Figure 9.5 displays the fMRI scan of a human brain.

Filippi et al. [14] examined neurodegenerative conditions by evaluating the dynamic functional connectivity (dFC) in the resting-state brain using magnetic resonance imaging. In order to learn more about neurodegenerative processes and find new biomarkers for diagnosing and predicting disease outcomes, this research evaluated dFC.

9.3.2 Positron emission tomography (PET)

Positron emission tomography (PET) is a kind of radionuclide scanning and medical imaging procedure that involves administering a radioactive substance into a patient's veins or arteries and then detecting the resulting radiation. Having features like excellent sensitivity and tracer quantification

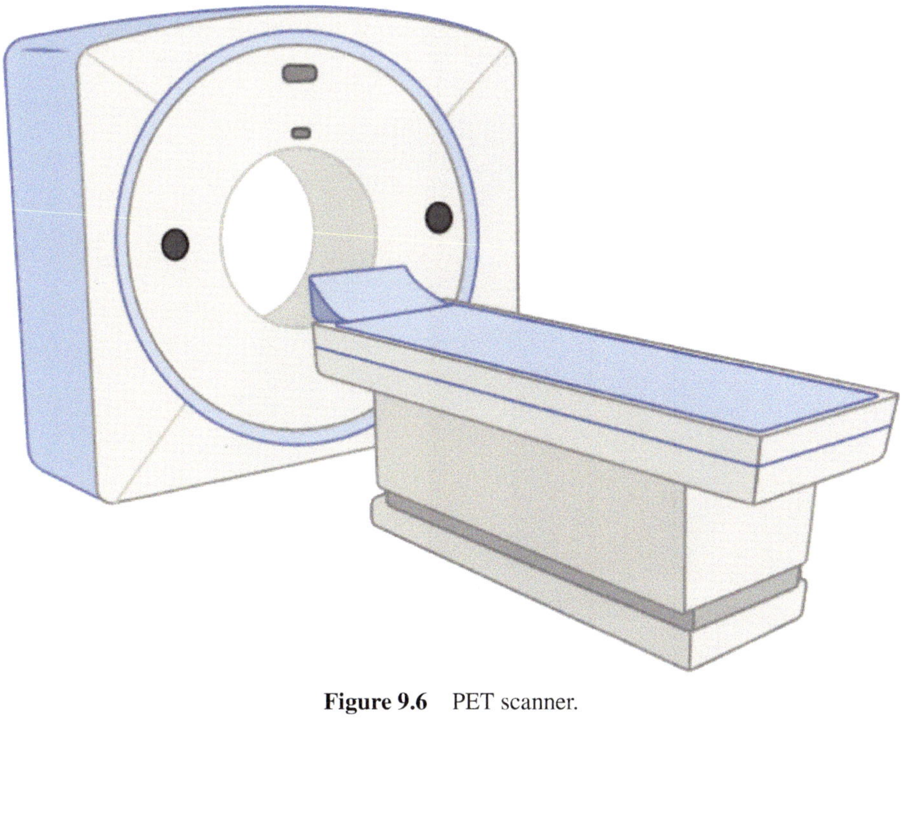

Figure 9.6 PET scanner.

capabilities, it is one of the most commonly used diagnostic imaging modalities in clinical settings [15]. Using a PET scanner and a harmless injectable radioactive substance called a radiotracer, PET imaging examinations may detect abnormal cells that take up a lot of the radiotracer, suggesting a possible health problem. PET imaging may assist in the identification of certain neurological disorders by tracking metabolic rate, blood flow, and oxygen consumption by examining different brain tissues. Figure 9.6 depicts the neuroimaging PET scanning machine.

Tan et al. [16] reviewed a list of all the radioligands used in PET neuroimaging for important biomarkers in ASD. The discovery of ASD biomarkers and the creation of appropriate PET probes lead to the benefits of ASD PET research in clinical diagnosis, medication development, and effectiveness assessment.

9.3.3 Single photon emission computed tomography (SPECT)

Single photon emission computed tomography (SPECT), known as a nuclear imaging method, may reveal molecular and functional activities in three dimensions [17]. There have been a lot of technical advancements in the field of organ-specific dedicated SPECT systems that have recently gained

Figure 9.7 SPECT image of the human brain

popularity [18], while the revolving gamma camera still has the greatest clinical value. It aids in the diagnosis of strokes, stress fractures, infections, convulsions, and malignancies by depicting the blood flow to different organs and tissues. Figure 9.7 displays the SPECT picture of the human brain.

Kalyoncu et al. [19] addressed the use of SPECT functional neuroimaging in the treatment of schizophrenia and depression. Psychiatric diseases, including schizophrenia and major depressive disorder (MDD), were the primary focus of the research, which centered on SPECT-based neuroimaging.

9.4 Emerging and Specialized Techniques

9.4.1 Magnetoencephalography (MEG)

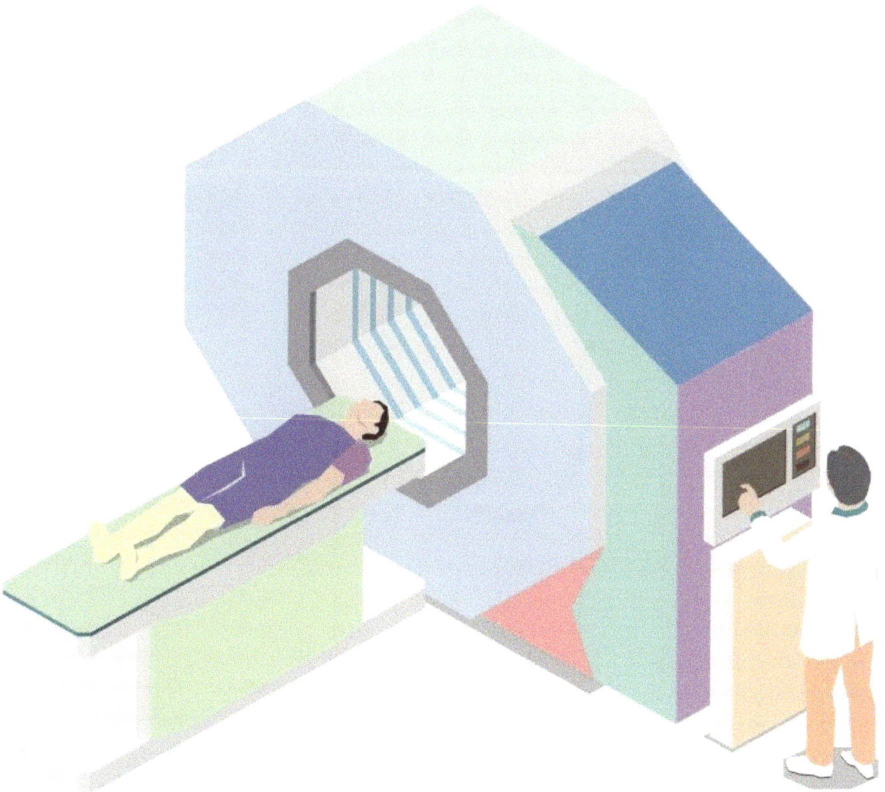

Figure 9.8 MEG system.

Magnetic resonance imaging (MRI) is a non-invasive method of studying the brain that monitors the magnetic fields generated by neurons. Researchers can measure neural activity on a millisecond basis, enabling them to analyze the temporal features of brain activity [20]. The non-invasive and radiation-free nature of MEG makes it a safe option for usage in a variety of scientific and therapeutic contexts. Figure 9.8 illustrates the MEG system.

Baillet et al. [21] discussed the features that distinguish MEG from other methods for investigating and treating brain dysfunction and function. The study demonstrated the ease of integrating MEG with various systems and techniques, thereby enabling their simultaneous performance. These systems and methods include electrophysiology, blood flow and oxygen metabolism, and brain stimulation.

9.4.2 Electroencephalography (EEG)

An electroencephalogram (EEG)'s purpose is to detect abnormalities in brain waves by measuring electrical activity in the brain. Tiny sensors mounted to

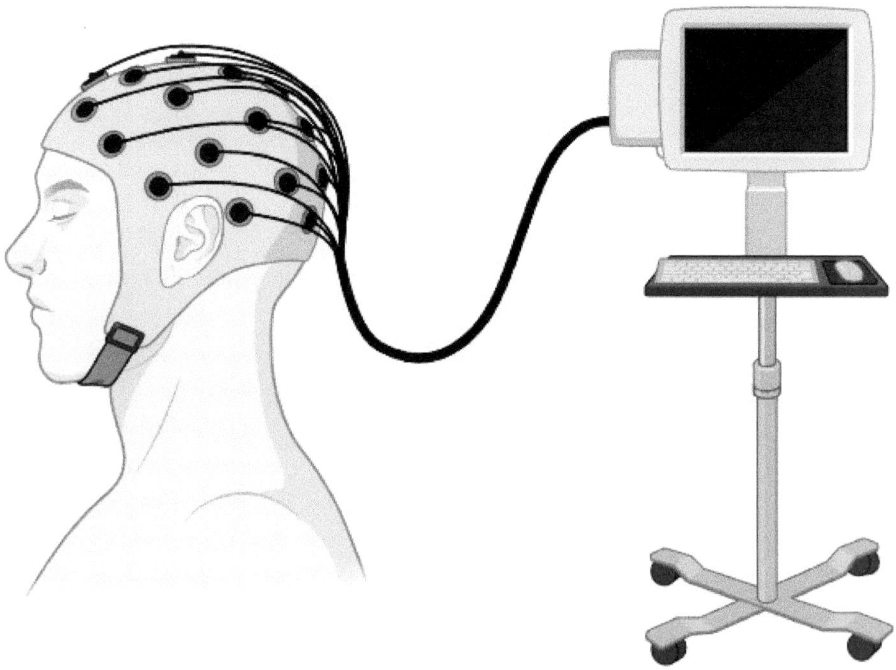

Figure 9.9 EEG machine.

the scalp capture the electrical impulses generated by the brain, making EEG devices useful for diagnosing and monitoring a wide range of brain-related diseases. Its use has improved the diagnosis of brain death, sleep disorders, and epilepsy, among other clinical concerns [22]. Figure 9.9 illustrates the attachment of the EEG machine to the human scalp.

Zhang et al. [23] conducted a comprehensive study that covered the various types of EEG signals, their various analysis methods, and the applications of these methods in the diagnosis of neurological disorders and conditions. The evaluation also categorized typical research methodologies according to the EEG signal's unique qualities.

9.4.3 Near-infrared spectroscopy (NIRS)

Near-infrared spectroscopy (NIRS) is an optical method that provides vital physiological data to the healthcare provider in real time by measuring hemodynamics and tissue oxygenation in the patient's brain [24]. Monitoring the hemodynamic responses linked to neuronal activity allows for the quantification of brain activity. NIRS objectively measures people's attention, memory, planning, and problem-solving abilities when they engage in cognitive activities. It can also track their cognitive states while they are out in the real world. Figure 9.10 shows the mounted NIRS system on a human scalp.

Sonkaya et al. [25] investigated NIRS in neurology after reviewing it as a neuroimaging tool. In addition to reviewing the characteristics, benefits, and limits of fNIRS, the study examined its principles in detail.

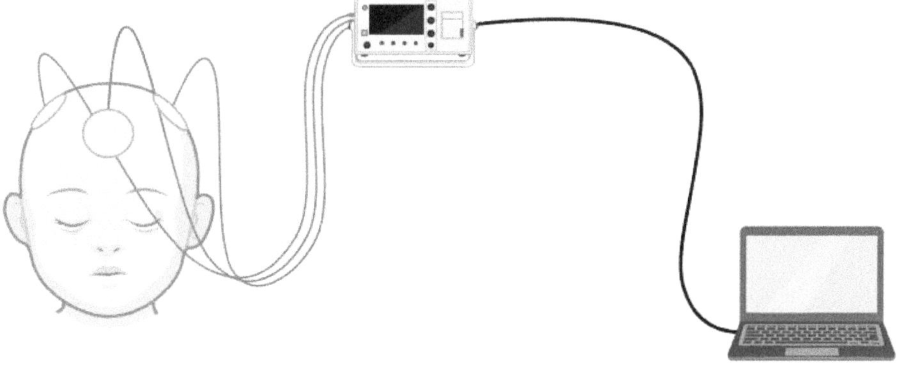

Figure 9.10 Typical NIRS system.

9.5 Advanced Applications and Innovations

9.5.1 Multimodal neuroimaging

Neuroscience research benefits from multimodal neuroimaging because it combines results from many imaging modalities, which helps to overcome the limits of any one imaging method [26]. It has recently undergone an upgrade as an imaging tool by incorporating data from other sources, including MRI, CT, DTI, EEG, and MEG. Neurology, psychiatry, neurophysiology, neurosurgery, and many more branches of neuroscience have benefited from multimodal neuroimaging methods. It is more effective than any traditional method because it assesses large-scale brain activity, connectivity, and abnormalities while overcoming the limitations of individual techniques.

9.5.2 Neuroimaging in neurodegenerative diseases

Researchers are investigating, diagnosing, and treating neurodegenerative illnesses such as Alzheimer's, Parkinson's, and progressive supranuclear palsy (PSP) using a variety of neuroimaging methods. Early detection of neurodegenerative disorders has become possible with those evolved methods, which have become the most powerful biomarker tools for these conditions. In the field of neuroimaging, MRI and PET are the imaging modalities most often used to diagnose neurodegenerative diseases [27]. Another step toward better prediction for neurodegenerative illnesses has been the broad use of neuroimaging methods.

9.5.3 Neuroimaging in psychiatric disorders

Neuroimaging plays a crucial role in psychiatry for the detection of mental diseases and the creation of novel medications [28]. Neuroimaging methods have the potential to revolutionize psychiatry by providing answers to some of the most pressing clinical problems in the profession. It is essential for diagnosing and treating mental diseases, finding structural brain lesions that cause psychosis, and predicting how a patient may respond to medical treatment. It also provides valuable information on brain processes and abnormal neural circuitry linked to a variety of mental illnesses. Major mental diseases, including trauma, stress, and related illnesses, rely on it for analysis and diagnosis.

9.5.4 Neuroimaging in cognitive and developmental neuroscience

Brain imaging is one of the domains that developmental cognitive neuroscience integrates [29]. In order to comprehend how the human brain works for cognition, neuroimaging has been an invaluable tool in revealing previously unknown details about brain anatomy, physiology, and disease. The study of cognitive and structural changes in the brain, as well as their role in tracking psychiatric characteristics via task-dependent brain changes, relies on this. When it comes to studying cognitive and developmental neuroscience, neuroimaging is crucial since it can tackle complex problems.

9.6 Applications of Neuroimaging in Biomedical Devices

Neuroimaging techniques implement biomedical devices with significant implications in the neurological health system. We discuss below the wide-ranging applications of a few of these smart devices.

9.6.1 Brain–computer interfaces (BCIs)

A brain–computer interface (BCI) is a device that can receive signals from the brain, process them, and then send those signals on to other devices to carry out certain tasks [30]. Electrophysiological and hemodynamic brain activities are both monitored by BCIs [31]. The development of BCIs holds enormous potential for improving neuropsychiatric care, particularly in situations where treatments have not been effective. Through the analysis of data and decoding of brain activity, BCIs provide customized stimulation treatment in an enclosed system. In order to keep tabs on brain activity, BCIs use neuroimaging methods, including EEG, fMRI, fNIRS, MEG, and similar methods.

9.6.2 Neurofeedback systems

Neurofeedback is a method in which a person receives information about their brain activity in order to alter and enhance it. Standard metrics for neurofeedback efficacy include changes in EEG and fMRI data. By addressing neurological disorders, neurofeedback systems improve cognitive capacities. It records and evaluates brain activity using visual or other inputs, and users develop self-modulation techniques based on this input. Individuals with neuropsychiatric problems have suggested it as a supplementary therapy option, viewing it as a tool to teach neural self-regulation to be better [32]. The most popular neurofeedback method now in use is brain function mapping based on EEG.

9.6.3 Diagnostic imaging devices

Diagnostic imaging, which involves taking pictures of the inside of the body, is used to establish a medical diagnosis. By illuminating diagnostic processes and directing less intrusive treatments, it has transformed healthcare [33]. Diagnostic imaging systems include common neuroimaging methods, including X-ray, MRI, CT, and DTI, as well as related methods for observing and analyzing the central nervous system, which aids in illness diagnosis and brain health assessment. The use of imaging devices in conjunction with these methods is crucial for the non-invasive creation of pictures of the interior structures of the body.

9.6.4 Monitoring and prognosis devices

Public health studies and consumer health technology can learn a lot from digital health monitoring [34]. Monitoring and prognosis equipment utilizes neuroimaging methods to track and predict the onset of neurological disorders in such situations. The use of wearable devices that offer dependable signal quality for neuroimaging is essential for the monitoring and prognosis of a wide range of health disorders. These technologies are improving our capacity to track brain health and forecast outcomes, and they are finding growing use in the diagnosis and prognosis of cognitive neurodegenerative illnesses. Its aid allows for the monitoring of important brain activity that contributes to neurological wellness.

9.6.5 Personalized medicine

Neuroimaging personalized medicine integrates imaging profiling, molecular diagnostics, and clinical-pathological indices to provide individualized approaches to diagnosis, prognosis, and treatment. It adapts to the specific needs of each patient, and focused investigations maximize the effectiveness of medical imaging treatments. In order to enhance brain health, neuroimaging might be part of a customized treatment strategy [35]. Through careful analysis of each patient, it is possible to provide personalized diagnostic approaches that are designed to meet their individual requirements. Over the years, tailored treatment strategies for brain health have steadily advanced through the use of neuroimaging.

9.6.6 Image-guided therapy (IGT)

Image-guided therapy (IGT) utilizes imaging data to improve the accuracy of disease localization and boost the effectiveness of treatments. Personalized treatment planning, precise targeting, therapeutic effectiveness, reduced side effects, shorter procedure durations, and responsive follow-up are the primary aims of image-guided therapy (IGT) [36]. Imaging is used by IGT for treatment planning, execution, and evaluation. Patients' quality of life is greatly enhanced by the use of image-guided therapy in the treatment of neurological conditions, including stroke and other neurological diseases. It is a successful therapy procedure with applications in many medical domains, including neuroimaging.

9.7 Artificial Intelligence in Neuroimaging

Artificial intelligence (AI) is enhancing clinical translational imaging, in particular neuroimaging, to enable early diagnosis, prediction, and treatment of chronic pain conditions that affect brain health [37]. AI systems using machine learning algorithms enable quick and accurate analysis of medical images, aiding in the detection of early-stage illnesses that may be challenging to identify using traditional techniques. Deep learning and machine learning algorithms have revolutionized the area by overcoming older approaches to performance [38]. A growing number of applications in neuroimaging are using artificial intelligence to improve imaging and analytics. These applications include clinical image prediction, neuroimage analysis, neurological diagnosis, and the development of novel features to improve healthcare systems. It is also impacting a wide range of disciplines, with a particular emphasis on human health issues including cancer, chronic sickness, and mental disorders. Yasmin et al. [39] highlighted the significance of AI in the early detection of digestive diseases using the GastroNet model. Hassan et al. [40] presented an AI model that employs a variety of machine learning algorithms to predict early CKD. Haque et al. [41] proposed a NeuroNet19 AI model that utilizes stocktickerMRI data to classify brain tumors. Using AI to improve clinical imaging and uncover new insights into it has been a huge advantage for this kind of clinical prediction.

9.8 Future Directions and Challenges

The field of neuroimaging is seeing tremendous growth and may significantly impact the years to come. It has the potential to be further developed with the help of AI in the future, resulting in cutting-edge neuroimaging technology that can tackle both current and future problems. AI has the potential to completely transform the field of neuroimaging by optimizing the process, assisting in diagnosing conditions, and boosting the effectiveness of treatments. It has the capability to automate data processing, provide quantitative conclusions, and assist in the interpretation of complex research. By successfully using artificial intelligence, we can effectively tackle the difficulties faced in neuroimaging and enhance the accuracy of diagnosis and therapy. Prioritizing the resolution of these issues and doing meticulous research should be the top priority in order to improve the field of neuroimaging. Nevertheless, there will be obstacles to overcome in the existing setting in order to do this. Complexity based on real-time applications, poor statistical power, system invasiveness, and ethical considerations are all factors that will complicate matters.

9.9 Conclusion

In the field of brain healthcare, neuroimaging technology has been an invaluable resource. This tool's extensive usage and adaptability have greatly increased its impact on the field of neuroscience. The study aimed to present important concepts and applications of neuroimaging technology, as well as how these tools might improve mental health. This study further demonstrates the role of AI in neuroimaging and its applications in the healthcare system. While neuroimaging technology has undoubtedly advanced and become an important part of the fight against neurodegenerative disorders, the study has also brought attention to the factors and obstacles that may limit further advancements in this area.

Acknowledgments

Thanks to all authors for completing this chapter successfully.

References

[1] N. Jones, "Neuroimaging: Unravelling the Brain's Mysteries," Neuroscience and Psychiatry: Open Access, vol. 6, no. 4, pp. 95–97, Aug. 2023. doi:10.37532/npoa.2023.6(4).95-97

[2] C. Yen, C.-L. Lin, and M.-C. Chiang, "Exploring the frontiers of neuroimaging: A review of recent advances in understanding brain functioning and disorders," Life, vol. 13, no. 7, p. 1472, Jun. 2023. doi:10.3390/life13071472

[3] S. Majumdar, R. Roy, M. Sen, and M. Chakraborty, "Review article on Magnetic Resonance Imaging," Advances in Intelligent Systems and Computing, pp. 731–739, Jul. 2019. doi:10.1007/978-981-13-7403-6_64

[4] T. Yousaf, G. Dervenoulas, and M. Politis, "Advances in MRI methodology," *International Review of Neurobiology*, pp. 31–76, 2018. doi:10.1016/bs.irn.2018.08.008

[5] R. Sclocco, F. Beissner, M. Bianciardi, J. R. Polimeni, and V. Napadow, "Challenges and opportunities for brainstem neuroimaging with Ultra-high Field MRI," *NeuroImage*, vol. 168, pp. 412–426, Mar. 2018. doi:10.1016/j.neuroimage.2017.02.052

[6] D. Spronk *et al.*, "Evaluation of Carbon Nanotube X-Ray Source Array for stationary head computed tomography," *Medical Physics*, vol. 48, no. 3, pp. 1089–1099, Feb. 2021. doi:10.1002/mp.14696

[7] L. C. Pescatori *et al.*, "Clinical impact of computed tomography in the emergency department in nontraumatic chest and abdominal conditions," *Emergency Radiology*, vol. 25, no. 4, pp. 393–398, Mar. 2018. doi:10.1007/s10140-018-1592-0

[8] K. K. Kampe *et al.*, "Diagnostic value of positron emission tomography combined with computed tomography for evaluating critically ill neurological patients," *Frontiers in Neurology*, vol. 8, Feb. 2017. doi:10.3389/fneur.2017.00033

[9] T. Jeon, M. M. Fung, K. M. Koch, E. T. Tan, and D. B. Sneag, "Peripheral Nerve Diffusion Tensor Imaging: Overview, pitfalls, and Future Directions," *Journal of Magnetic Resonance Imaging*, vol. 47, no. 5, pp. 1171–1189, Oct. 2017. doi:10.1002/jmri.25876

[10] W.-S. Tae, B.-J. Ham, S.-B. Pyun, S.-H. Kang, and B.-J. Kim, "Current clinical applications of diffusion-tensor imaging in neurological disorders," *Journal of Clinical Neurology*, vol. 14, no. 2, p. 129, 2018. doi:10.3988/jcn.2018.14.2.129

[11] A. Meoded, G. Orman, and T. A. G. M. Huisman, "Diffusion weighted and diffusion tensor MRI in pediatric neuroimaging including Connectomics: Principles and applications," *Seminars in Pediatric Neurology*, vol. 33, p. 100797, Apr. 2020. doi:10.1016/j.spen.2020.100797

[12] L. Hay, A. H. B. Duffy, S. J. Gilbert, and M. A. Grealy, "Functional magnetic resonance imaging (fmri) in design studies: Methodological considerations, Challenges, and recommendations," *Design Studies*, vol. 78, p. 101078, Jan. 2022. doi:10.1016/j.destud.2021.101078

[13] A. Bowring, C. Maumet, and T. E. Nichols, "Exploring the impact of analysis software on task fmri results," *Human Brain Mapping*, vol. 40, no. 11, pp. 3362–3384, May 2019. doi:10.1002/hbm.24603

[14] M. Filippi, E. G. Spinelli, C. Cividini, and F. Agosta, "Resting state dynamic functional connectivity in neurodegenerative conditions: A review of Magnetic Resonance Imaging Findings," *Frontiers in Neuroscience*, vol. 13, Jun. 2019. doi:10.3389/fnins.2019.00657

[15] M. M. Khalil, "Positron Emission Tomography (PET): Characteristics and performance," *Basic Sciences of Nuclear Medicine*, pp. 319–341, 2021. doi:10.1007/978-3-030-65245-6_13

[16] Z. Tan *et al.*, "Positron emission tomography in the neuroimaging of autism spectrum disorder: A Review," *Frontiers in Neuroscience*, vol. 16, Apr. 2022. doi:10.3389/fnins.2022.806876

[17] R. Massari and G. S. Mok, "Editorial: New trends in single photon emission computed tomography (SPECT)," *Frontiers in Medicine*, vol. 10, Dec. 2023. doi:10.3389/fmed.2023.1349877

[18] B. F. Hutton, "The origins of SPECT and SPECT/CT," *European Journal of Nuclear Medicine and Molecular Imaging*, vol. 41, no. S1, pp. 3–16, Nov. 2013. doi:10.1007/s00259-013-2606-5

[19] A. Kalyoncu and A. S. Gonul, "The emerging role of SPECT functional neuroimaging in schizophrenia and depression," *Frontiers in Psychiatry*, vol. 12, Dec. 2021. doi:10.3389/fpsyt.2021.716600

[20] F. M. Hammond and S. Katta-Charles, "Magnetoencephalography," *Encyclopedia of Clinical Neuropsychology*, pp. 2065–2066, 2018. doi:10.1007/978-3-319-57111-9_50

[21] S. Baillet, "Magnetoencephalography for brain electrophysiology and imaging," *Nature Neuroscience*, vol. 20, no. 3, pp. 327–339, Feb. 2017. doi:10.1038/nn.4504

[22] A. Weintraub and J. Whyte, "Electroencephalography," *Encyclopedia of Clinical Neuropsychology*, pp. 1–3, 2016. doi:10.1007/978-3-319-56782-2_24-3

[23] H. Zhang *et al.*, "The applied principles of EEG analysis methods in neuroscience and clinical neurology," *Military Medical Research*, vol. 10, no. 1, Dec. 2023. doi:10.1186/s40779-023-00502-7

[24] F. Lange and I. Tachtsidis, "Clinical Brain Monitoring with TIME DOMAIN NIRS: A review and future perspectives," *Applied Sciences*, vol. 9, no. 8, p. 1612, Apr. 2019. doi:10.3390/app9081612
[25] A. R. Sonkaya, "The use of functional near infrared spectroscopy technique in neurology," *NeuroQuantology*, vol. 16, no. 7, Jun. 2018. doi:10.14704/nq.2018.16.7.1688
[26] S. Liu *et al.*, "Multimodal Neuroimaging Computing: The workflows, methods, and platforms," *Brain Informatics*, vol. 2, no. 3, pp. 181–195, Sep. 2015. doi:10.1007/s40708-015-0020-4
[27] C. G. Schwarz, "Uses of human MR and PET imaging in research of Neurodegenerative Brain Diseases," *Neurotherapeutics*, vol. 18, no. 2, pp. 661–672, Apr. 2021. doi:10.1007/s13311-021-01030-9
[28] J. C. Masdeu, "Neuroimaging in psychiatric disorders," *Neurotherapeutics*, vol. 8, no. 1, pp. 93–102, Jan. 2011. doi:10.1007/s13311-010-000 6-0
[29] A. Matejko and D. Ansari, "Developmental cognitive neuroscience and learning," *Encyclopedia of the Sciences of Learning*, pp. 961–966, 2012. doi:10.1007/978-1-4419-1428-6_605
[30] J. J. Shih, D. J. Krusienski, and J. R. Wolpaw, "Brain-computer interfaces in medicine," *Mayo Clinic Proceedings*, vol. 87, no. 3, pp. 268–279, Mar. 2012. doi:10.1016/j.mayocp.2011.12.008
[31] L. F. Nicolas-Alonso and J. Gomez-Gil, "Brain Computer Interfaces, a review," *Sensors*, vol. 12, no. 2, pp. 1211–1279, Jan. 2012. doi:10.3390/s120201211
[32] L. A. Weber, T. Ethofer, and A.-C. Ehlis, "Predictors of neurofeedback training outcome: A systematic review," *NeuroImage: Clinical*, vol. 27, p. 102301, 2020. doi:10.1016/j.nicl.2020.102301
[33] M. Friebe and A. Illanes, "Advancements in medical imaging and image-guided procedures: A potential—or rather likely—paradigm shift in diagnosis and therapy: Understand disruption and take advantage of it!," *Applied Sciences*, vol. 13, no. 16, p. 9218, Aug. 2023. doi:10.3390/app13169218
[34] M. Emish and S. D. Young, "Remote wearable neuroimaging devices for health monitoring and Neurophenotyping: A scoping review," *Biomimetics*, vol. 9, no. 4, p. 237, Apr. 2024. doi:10.3390/biomimetics9040237
[35] J. H. Cole, "Neuroimaging studies illustrate the commonalities between ageing and brain diseases," *BioEssays*, vol. 40, no. 7, Jun. 2018. doi:10.1002/bies.201700221

[36] J. Yanof and M. Kuhn, "New technology for image-guided therapy," *Philips Research*, vol. 6, pp. 139–146. doi:10.1007/1-4020-4384-8_9

[37] U. K. Patel et al., "Artificial Intelligence as an emerging technology in the current care of neurological disorders," Journal of Neurology, vol. 268, no. 5, pp. 1623–1642, Aug. 2019. doi:10.1007/s00415-019-09518-3

[38] E. J. Bacon et al., "Neuroimage analysis using Artificial Intelligence Approaches: A systematic review," Medical & Biological Engineering & Computing, Apr. 2024. doi:10.1007/s11517-024-03097-w

[39] F. Yasmin et al., "Gastronet: Gastrointestinal polyp and abnormal feature detection and classification with Deep Learning Approach," IEEE Access, vol. 11, pp. 97605–97624, 2023. doi:10.1109/access.2023.3312729

[40] Md. M. Hassan et al., "A comparative study, prediction and development of chronic kidney disease using machine learning on Patients Clinical Records," Human-Centric Intelligent Systems, vol. 3, no. 2, pp. 92–104, Feb. 2023. doi:10.1007/s44230-023-00017-3

[41] R. Haque, Md. M. Hassan, A. K. Bairagi, and S. M. Shariful Islam, "Neuronet19: An explainable deep neural network model for the classification of brain tumors using magnetic resonance imaging data," Scientific Reports, vol. 14, no. 1, Jan. 2024. doi:10.1038/s41598-024-51867-1

10

Intelligent Movement-Controlled Brain–Computer Interface System Based on EEG Signal

Khin Eaindray Htun[1], May Thu Kyaw[2], Htun Htun[3], Kyaw Kyaw Oo[4], Si Thu Aung[5], and Md. Mehedi Hassan[6]

[1]School of Information Technology, King's Mongkut University of Technology Thonburi, Thailand
[2]Department of Electronic Engineering, West Yangon Technological University, Myanmar
[3]Department of Solution, Partner Associates Int'l Co. Ltd., Myanmar
[4]Department of Data & Insights, Capgemini Singapore Pte. Ltd., Singapore
[5]Department of Mathematics, The State University of New York, USA
[6]Computer Science and Engineering Discipline, Khulna University, Bangladesh
E-mail: mehedihassan@ieee.org

Abstract

Brain–computer interfacing (BCI) system allows neurologically disordered patients to communicate with the outside world and add an extra control modality to their abilities. This chapter primarily analyzes the electroencephalography (EEG)-based signals for four types of movements, that is, left, right, top, and bottom, and uses a machine learning model to classify them to use as a movement-controlled BCI system. EEG non-invasive signals acquired from the brain were processed and analyzed, employing the discrete wavelet transform (DWT) for feature extraction, and utilizing the artificial neural network (ANN), convolutional neural network (CNN), and K-nearest neighbors (KNN) for classification. We presented the precision, recall, and F1 scores of the different algorithms. ANN achieved an accuracy of 85.27% in combination with a feature extraction method of discrete wavelet transform. Other types of classification algorithms such as CNN and KNN were also measured to be inferior to using ANN. The proposed combination also yielded promising results with precision, recall, and F1 score of 0.85, 0.85, and 0.85, respectively. These metrics demonstrate the method's ability to accurately classify positive instances while maintaining a good balance between precision and recall. Our approach of combining DWT filter and ANN classification relatively reduced the time of feature extraction and achieved higher accuracy than current complex classification methods.

Keywords: Brain Computer Interface, Electroencephalography, Discrete Wavelet Transform, Artificial Neural Network, Movement Classification.

10.1 Introduction

Brain–computer interfacing (BCI) is a system that controls the computer application directly through the information from brain signals. BCI is widely used in the medical domain, for prosthesis control or for the treatment of neurological disorders [1]. Due to this progress, BCI is gradually getting introduced outside the medical domain. Gaming and virtual reality have been potentially growing fields and in the near future, the world will heavily rely on them. BCI in a game fashion, allows the players to control the game without relying on their bodies and the physical controllers [2]. Today's EEG-based BCI technology allows neurological disordered patients to communicate with the outside world. It also becomes a helping hand for the "situational disabled" users, who are able to use only some parts of their body, to add an extra control modality to their abilities.

The underlying objective of brain–computer interfaces is to replace, extend, augment, or enhance human functions by detecting brain activity, extracting information from that activity, and then transforming those data into outputs [3]. Users can control machines mentally on account of BCIs. The potential is limitless, despite the fact that most technology is currently at an experimental stage. BCIs have the potential to replace lost abilities in speech or movement [4]. By using the nerves or muscles that move the hand, people may regain control of their bodies. BCIs have also been used to support individuals in improving the residual functionality of damaged pathways, where physical movement is necessary. BCIs can also boost the overall performance of humans, such as by instructing a drowsy driver to completely wake up. Additionally, the body's natural outputs can be strengthened by a BCI, possibly with the assistance of an additional hand [5].

Several methods have been used to assess brain activity for BCIs. BCIs often link to the brain using wearable or implantable devices. Surgically affixing a BCI to brain tissue is a common method of implanting them. Individuals who have had serious physical traumas or neuromuscular abnormalities could be more suitable with it. The interference from surrounding tissue is reduced when BCIs are implanted since they monitor signals directly from the brain. Infection and rejection are surgical concerns, nevertheless [6]. In order to monitor brain activity, wearable BCIs on the scalp sometimes require a cap with conductors. Electroencephalography (EEG) is the primary method used by the majority of wearable BCIs to assess brain activity [7, 8]. There are significant BCI evaluations that focus on certain applications, for instance, wheelchair control, BCI mobile robot, and virtual reality and gaming [9].

Most researchers have been interested in EEG modeling and classification. For instance, Monesi and Sardouie [10] offer extended common spatial pattern (ECSP) analysis, a new technique for special extraction that makes use of preexisting data knowledge to provide a wider variety of characteristics than classic PSC analysis. Kumar [11] offers a single-band CSP framework that makes extensive use of tangent spatial mapping (TSM) ideas. The band-pass filter used in the combined technique, known as CSP-TSM, is applied to MI-EEG signals.

To increase the precision of the EEG MI classification, Amin et al. [12] suggested a multi-class technique that integrates convolutional neural networks (CNN) with various techniques and structures. The spatial and temporal characteristics of the raw EEG data are extracted using this method using a variety of sophisticated characteristics. In using MCNN and CCNN, it is said to be novel demonstration. Fusion approaches developed by Amin et al. [12] exceed all cutting-edge machine learning and deep learning strategies for EEG categorization. According to the BCI dataset "Competition IV 2a," the presented MCNN approach has an accuracy of 75.7%. The self-coding-based proposed CCNN approach enhances EEG topic classification by more than 10%.

This chapter makes a significant contribution to the field of EEG-based brain–computer interface (BCI) systems by conducting an in-depth investigation and evaluation of diverse datasets comprising brain control signals from various sources. The primary focus of this study revolves around the analysis of feature extraction techniques utilizing the discrete wavelet transform (DWT). The overall contributions of this paper are as follows:

- This chapter contributes to the field of EEG-based bracomputer interface (BCI) systems by investigating and evaluating diverse datasets of brain control signals from various sources.

- The study focuses on analyzing the outcomes of feature extraction using the discrete wavelet transform (DWT) technique.
- It compares the performance of classification methods, including artificial neural network (ANN), convolutional neural network (CNN), and K-nearest neighbors (KNN), in interpreting EEG signals for BCI applications.
- The findings provide insights into the strengths and weaknesses of the investigated techniques, offering valuable knowledge for the development and implementation of BCI technologies.
- The paper's contribution lies in its systematic approach, utilizing diverse datasets, feature extraction techniques, and classification methods to provide a comprehensive understanding of EEG-based BCI systems.

10.2 Related Works

BCI technology will be integrated and will take the gaming industry. It is the next level of enthusiasm. Controlling a computer game with your own brain is an exciting application. The main purpose of BCI system is rehabilitation for serious injuries. Recently, innovating it to use entertainment has become an inspiration.

According to reports, the method that is commonly applied involves the use of classification algorithms [12]. One or more classification algorithms are selected from a wide range of options while designing the classification stage. Multiple classification methods, including support vector machines (SVM), neural networks (NN), linear discriminant analysis (LDA), Bayesian classifiers, K-nearest neighbor (KNN), deep learning (DL), and various iterations of deep learning, are explored within the research of EEG-based BCI that is already published.

LDA identifies the optimal linear combinations which include feature vectors and they best characterize the properties of a given input wave. LDA separates a variety of object classes. Hyperplanes are used to complete this quest. By identifying the projection that maximizes the distance between class means and minimizes the interclass variance, the isolation hyperplane can be obtained [13].

Enhancing the practicality of EEG utilization in various applications and reducing reliance on qualified specialists requires the capacity to achieve a reliable automated classification system for EEG signals [14]. It is important to note that despite the enormous advancements made in traditional BCI systems over the past several decades, significant obstacles persist in the field of

EEG categorization. Difficulties come from a variety of genetic or situational EEG abnormalities, a low SNR, and reliance on knowledge of experts for feature extraction. Additionally, the vast majority of current machine learning studies focus on static data, making it impossible to effectively classify rapidly varying brain signals [15].

An advanced type of neural network architecture with a focus on exploring spatial information is known as a convolutional neural network (CNN). In CNN, there is a minimum of one layer that uses a convolution operator to map the input to output [16, 17]. According to research, CNN is capable of capturing the distinct relationships between features and patterns linked to various brain signals [15].

A variety of discriminative filter bank common spatial patterns were used to extract the features for [18], which investigated the classification of multiple MI using CNN (DFBCSP). A 92.75% classification accuracy was achieved in [19] where, in contrast, introduced a time-frequency image obtained through wavelet transform together with a CNN-focused technique for identifying EEG motor movement. Tayeb et al. successfully used this model within a real-time control system for robotic arm operation to accurately classify unprocessed EEG MI signals, attaining an accuracy of 84%. For the classification of MI EEG, a CNN-driven multilayered feature integrated model was developed in [21]. CNN was used for EEG MI classification in three further trials, with reported classification accuracy ranges of 80, 93, and 86% [22–24]. A further point to be made is that SSVEP-based BCI systems have also used the CNN model.

In [25] focus BCI game controller using Emotiv EPOC headset and API. The research takes into account two cognitive and expressive approaches/methods. The training and later use of the input signals in the C# game as stimuli with the assistance of external references, files, and libraries is represented by the data processing algorithm section. Training a subject on the cognitive suite of the Emotiv EPOC Control Panel is the fundamental logic used in our system. Users are instructed to focus their attention on a certain activity, such as pushing or pulling, for a predetermined amount of time. Signals recorded during this time are kept in a training file specific to the subject. The common ones are thinking push, pull, left, right, rotate, and move, among other activities that a subject may train for. The training file for the specific user is loaded into the application while the game is being played. To find a match between stored thinking and current thoughts, the training file is compared to the brain signals collected during gameplay. The action in the game that corresponds to the relevant thinking is carried out after the signal received during the game matches one of the pre-stored thought signals.

The authors of [26] show how BCI can be used to control mobile games. The created BCI system collects brain impulses initially, filters out noise from unprocessed signals, and then extracts a list of features. Following classification, the extracted features are used to control the game. In this investigation, three distinct reaction types have been taken into account. These include motor control, SSVEP in short for steady-state visually evoked potential and P300 ERP, event-related potential. Three healthy individuals between the ages of 20 and 23 were used to test the system. The test individuals had no prior BCI experience. Following a training session, each participant was instructed to concentrate on the smartphone screen while wearing an EEG headset. This selection of the appropriate neuron for the operation or task the user wishes to carry out was then made.

10.3 Methodology

We analyzed the outcomes of feature extraction using the discrete wavelet transform (DWT) and fast Fourier transform (FFT). The classification methods artificial neural network (ANN), convolutional neural network (CNN), K-nearest neighbors (KNN), and support vector machine (SVM) were used in research, and assessment metrics (precision, recall, and F1 score) will be demonstrated in comparison to other algorithms. The system diagram of the brain–computer interface system is shown below in Figure 10.1. During the process, the EEG non-invasive signals are acquired from the brain via BCI2000 system. Then, the signals are processed by segmenting them into the same amount of time to unify the data dimension. The feature extraction and selection methods are used to minimize unnecessary data and time

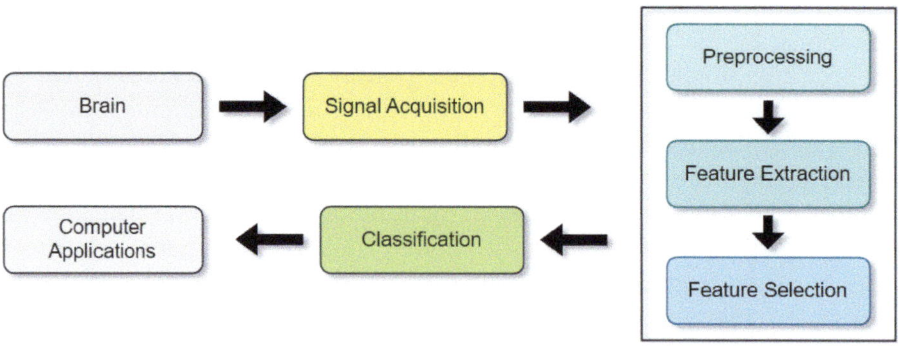

Figure 10.1 Diagram of brain–computer interface system.

complexity. The most used feature extraction methods are time frequency distributions, fast Fourier transforms and wavelet transforms. After selecting features from the raw data, the processed data are put into the machine learning model to distinguish the desired outputs, which are again used as a command to a computer system.

10.3.1 Data acquisition: Acquiring EEG signals from human brain

The EEGs were read at the sample rate of 160Hz from 64 electrodes according to the international 10-10 system in which the electrodes are excluded as in Figure 10.2. The digits below the electrodes specify the order from which they arise in the records [27]. The data are in EDF format in the dataset.

10.3.2 Dataset

The dataset that is used in this paper is selected from the Physionet, named EEG motor movement imagery dataset, containing over 1500 recordings from 109 subjects [27].

During the experiment, the subjects are to perform the motor/imagery tasks and the EEG signals of 64 channels are recorded using BCI2000 system. Each of the subjects completes these experimental runs: baseline runs that last for one minute twice and runs for the following tasks, each last for two minutes thrice:

- The subject opens and closes the left or the right fist when a target emerges on the corresponding screen side.
- The subject images opening and closing the left or the right fist when a target emerges on the corresponding screen side.

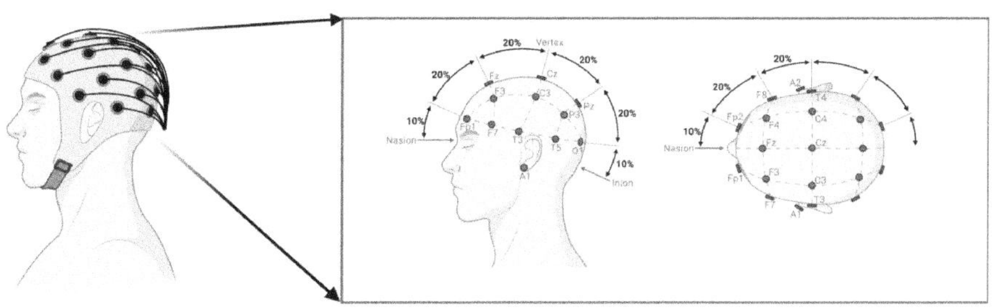

Figure 10.2 64 electrodes according to the international 10-10 system (excluding electrodes Nz, F9, F10, FT9, FT10, A1, A2, TP9, TP10, P9, and P10) that the input signals are taken from.

- The subject opens and closes both fists or both feet when a target emerges on the top or the bottom side of the screen respectively.
- The subject images opening and closing both fists or both feet when a target emerges on the top or the bottom side of the screen respectively [27].

The EEG recordings are then processed and segmented only for four imagery movement tasks, left and right, top, and bottom movements [28].

10.3.3 Data preprocessing

The dataset from physionet contains motor/imagery tasks other than the four MI tasks. The dataset consists of 1526 EEG recordings, with each recording containing 14 runs, 84 trials per each run. In our work, only the four MI

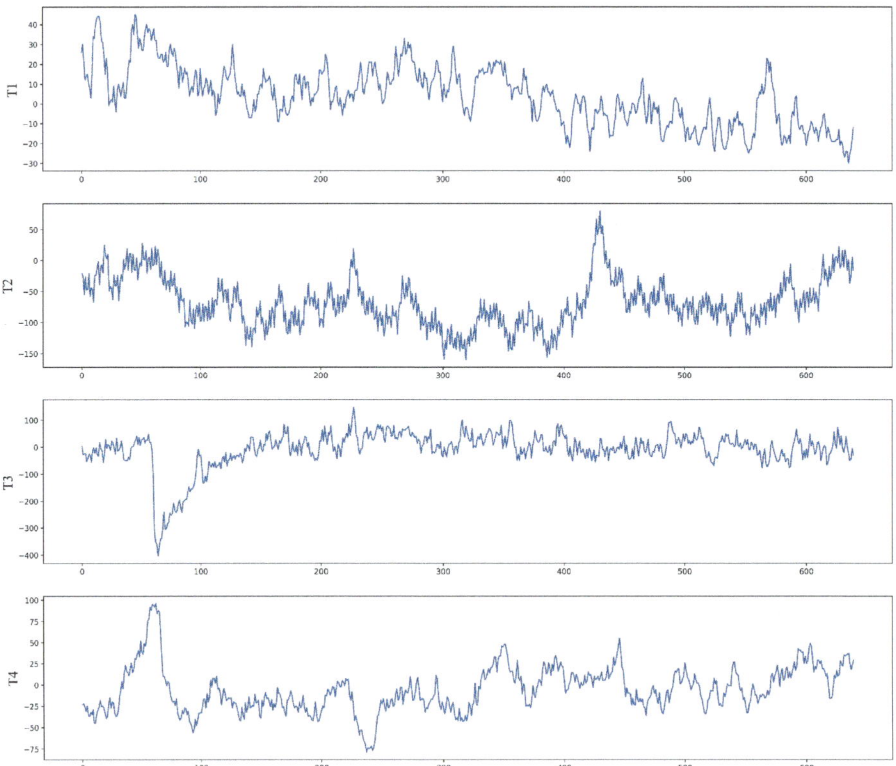

Figure 10.3 Random EEG signals for four MI tasks after segmentation from the original recordings.

tasks are chosen from them. Among all the trials, the desired four MI tasks are included in 6 runs and 21 trials per run for each subject. From all the 64 channels, the EEG signals are segmented four seconds for each task to unify the data dimension since all the tasks are performed around four seconds. Then, the four MI tasks of the signals are separated and labeled as T1, T2, T3, and T4 [28]. Figure 10.3 shows the random signals of each task after segmentation from the raw dataset.

After segmentation, the signals are visualized to find the relationship between the four tasks as in Figures 10.4 and 10.5. From the figures, some of the similarities between the same task and differences in different tasks are apparent. For instance, in the case of task 1 activity, the signals typically commence at a low value and gradually intensify over time. A majority of the task 1 signals display an increase in value between the 600th and 640th cycles. In contrast, the task 2 signals start at a higher amplitude and tend to decrease towards the end. Similarly, there is a notable disparity between the task 3 and 4 signals at approximately 100 and 200 cycles from the beginning. Specifically, the task 3 signals experience a steep ascent during these cycles, whereas the task 4 signals undergo a significant decline, presenting a completely opposite pattern compared to the task 3 signals. These corresponding differences between different tasks are important to take into consideration while performing feature extraction.

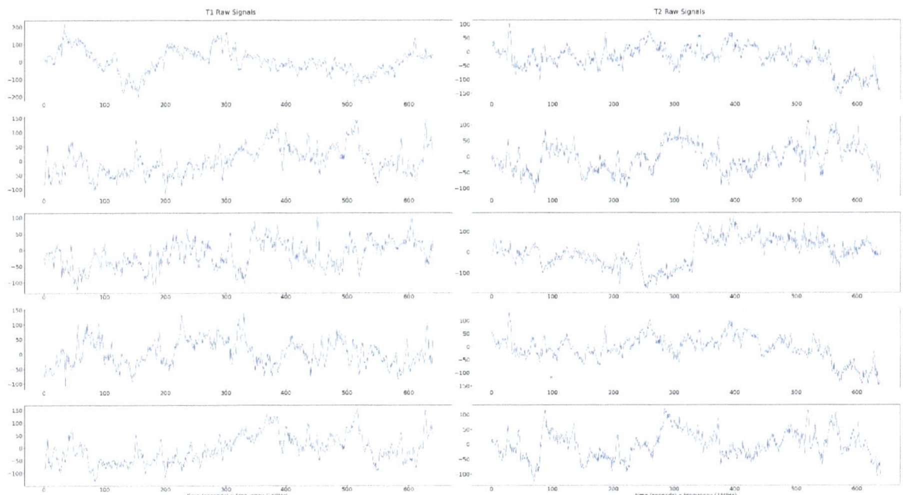

Figure 10.4 Selected raw EEG signals of task 1 and task 2 activities for visualization. Five signal samples for task 1 and task 2 from left to right.

Figure 10.5 Selected raw EEG signals of task 3 and task 4 activities for visualization. Five signal samples for task 3 and task 4 from left to right.

10.3.4 Feature selection

From the above observation, the differences of signals of each task are mostly changes in time and they are differed reciprocally. They can be said to have a negative correlation between T1 and T2, T3, and T4. However, as the signals are not on point value, plus the noises from the signals are also needed to be filtered before the classification. Therefore, in this work, features are selected from the signals series by applying discrete wavelet transform for the purpose of using the DWT filter banks which are calculated as in eqn (10.1) and (10.2). The EEG signals are passed through the low-frequency-pass filter and high-frequency-pass filter as in Figure 10.6. The output contains the detail coefficient from the high-frequency-pass filter and the approximation coefficient from the low-frequency-pass filter as expressed in the below equations [27, 28].

$$y_{\text{low}}[n] = \sum_{k=-\infty}^{\infty} x[k]g[2n-k] \tag{10.1}$$

$$y_{\text{high}}[n] = \sum_{k=-\infty}^{\infty} x[k]\,h[2n-k]. \tag{10.2}$$

As a result of this decomposition, the time series signals have been reduced by half while doubling the frequency detail. Therefore, the input data dimension is also reduced to half of the original data dimension.

The detail coefficients are chosen to be used as inputs for classification models here because the high-pass filter reduces the noise and indicates where important details and prominent changes are located in the signals. The approximation coefficients are almost the same as the original signals and they are not necessarily changed due to the low-pass filter [29]. Only the first order filtering is used for the accelerated pace and simplicity of computing.

Figure 10.7 demonstrates one set of samples for each task which includes original signal and the detail coefficients of the signal after applying the DWT high-frequency-pass. The results after applying the filters, as can be seen in the figures, are emphasized in the changes in amplitude of the signals. Moreover, from Figures 10.8 and 10.9 where the color maps of original signals and detail coefficients are presented respectively, the contrast in

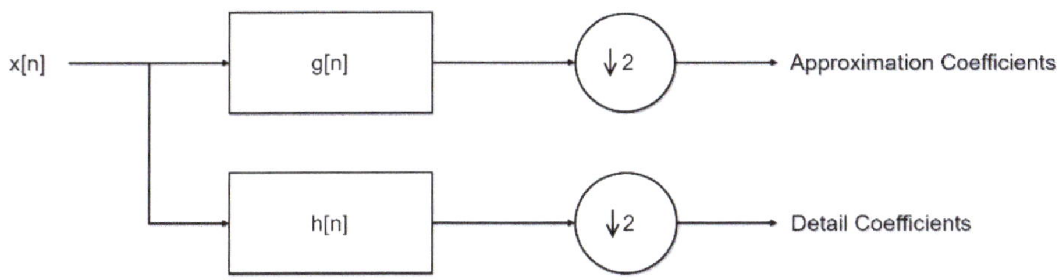

Figure 10.6 Block diagram of discrete wavelet transform (level 1) filtering analysis.

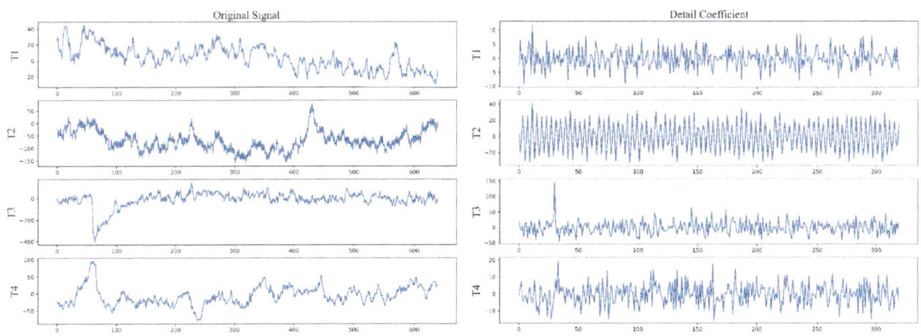

Figure 10.7 Original signals after normalization vs. detail coefficients of the original signals after applying DWT filter.

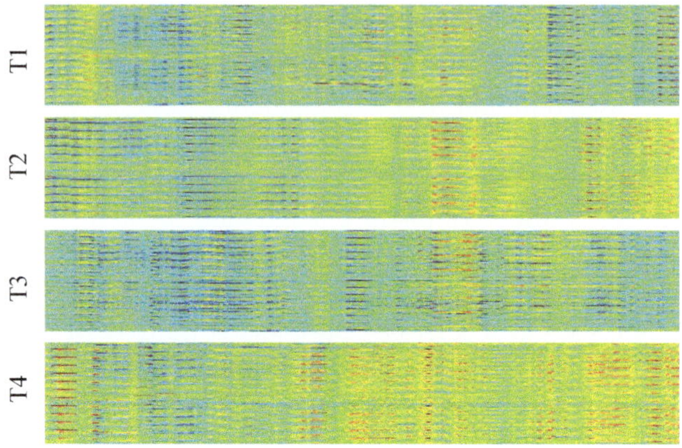

Figure 10.8 Original signals color maps of samples that are randomly selected. Each sample (640 × 1) corresponds to each task.

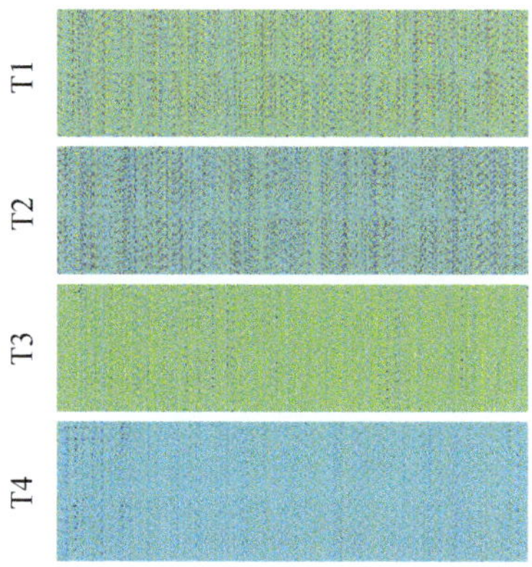

Figure 10.9 Feature color maps of samples that are randomly selected after applying discrete wavelet transform. Each sample (320 × 1) corresponds to each task.

amplitude of signals between the original and detail coefficients is expressed clearly. Although some differences can be noticed in the original signals, the amplitudes are similar and not readily apparent. However, in the detail coefficients, the values are more distinguishable within each class (see Figure 10.9 below).

10.3.5 Classification method: Artificial neural networks

Artificial neural networks (ANNs) are inspired by the sophisticated system operations of a human brain where billions of neurons work together connectedly. TensorFlow ANNs are used in our work, with the input layer of dimension 320 × 1, three hidden layers with 512 neurons, 256 neurons and 128 neutrons respectively. Then, a dropout layer with a rate of 50% is added to prevent overfitting the training dataset. The last layer along with the Softmax activation function contains four outputs for the classification of the four motor imagery tasks. The architecture of the ANNs model is described in Figure 10.10 and the system architecture in Figure 10.11.

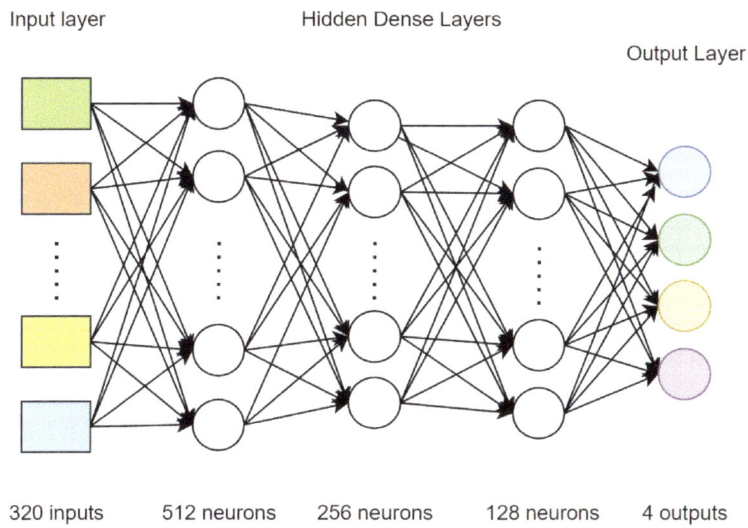

Figure 10.10 Artificial neural networks architecture diagram.

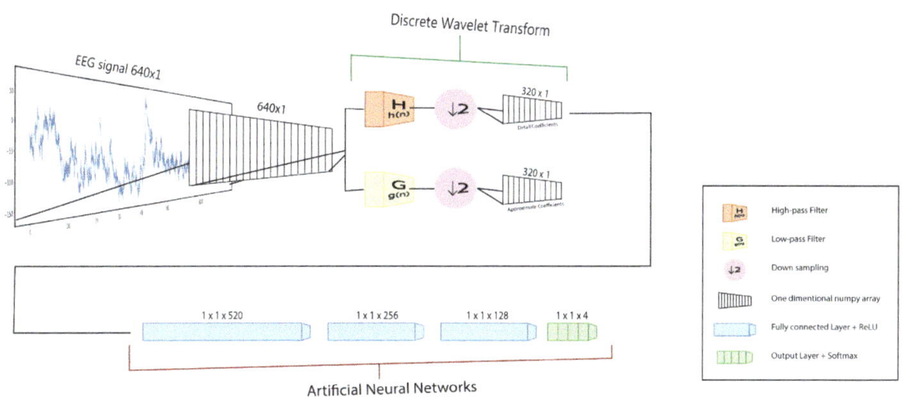

Figure 10.11 System diagram including the discrete wavelet transform filter and neural network architecture.

10.4 Results

10.4.1 Classification accuracy

The dataset was split into a ratio of 80:20, with 80% allocated for the training set and 20% for the test set. The processed dataset is then tested with multiple models including the artificial neural networks, convolutional neural networks, K-nearest neighbors, etc. Among them, using ANNs yields the best results with an accuracy of 85.27%.

The confusion matrix in Figure 10.13 displays the classification results of each respective tasks. T1, T2, T3, and T4 achieved 87%, 84.9%, 85.2%, and 84% at their corresponding peak accuracy as shown in Figure 10.12 (a). The precision, recall, and f1 score are also reliably consistent around 85% each as per Figure 10.12 (b). Figure 10.13 shows two plots of AUC-ROC curve and precision-recall curve of each MI task among the four left, right, up, and down movement controls. The AUC-ROC curve assesses the effectiveness of a binary classification model by analyzing the balance between true positive and false positive rates, whereas the precision-recall curve evaluates the precision and recall interplay, offering valuable insights into the model's performance at various classification thresholds. The 97% AUC-ROC value and above 90% precision-recall AUC value indicate that the proposed model is effective in making accurate predictions and has a high discriminatory power. Figure 10.14 depicts the variation in loss and accuracy metrics of the artificial neural network (ANN) model throughout the training and testing phases. This assists in determining the performance stability of the model

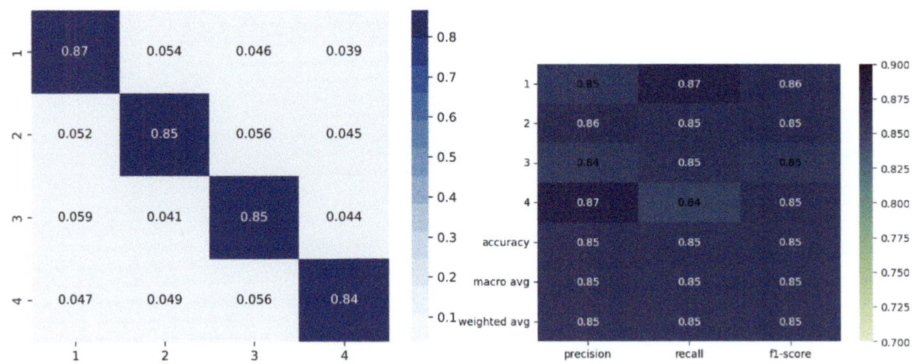

(a) (b)

Figure 10.12 (a) Confusion matrix of accuracy percentages per each motor/imagery task. (b) Precision-recall-f1 score matrix.

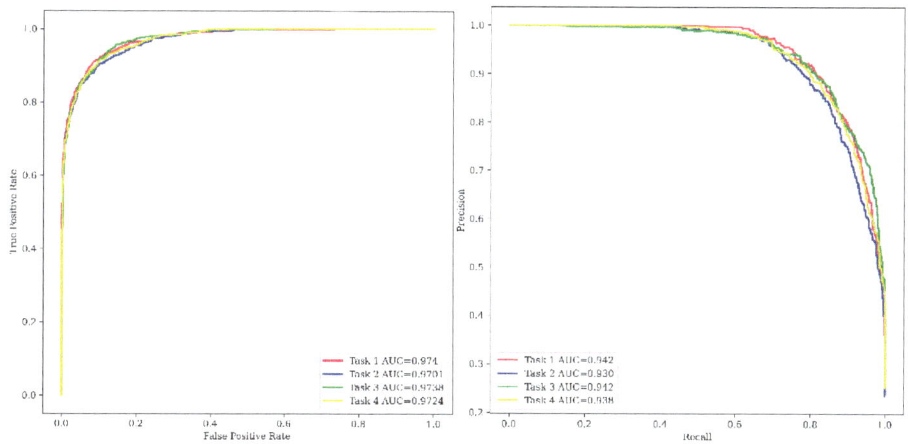

(a) (b)

Figure 10.13 (a) Area under ROC curve of individual MI tasks. (b) Precision-recall curve of individual MI task.

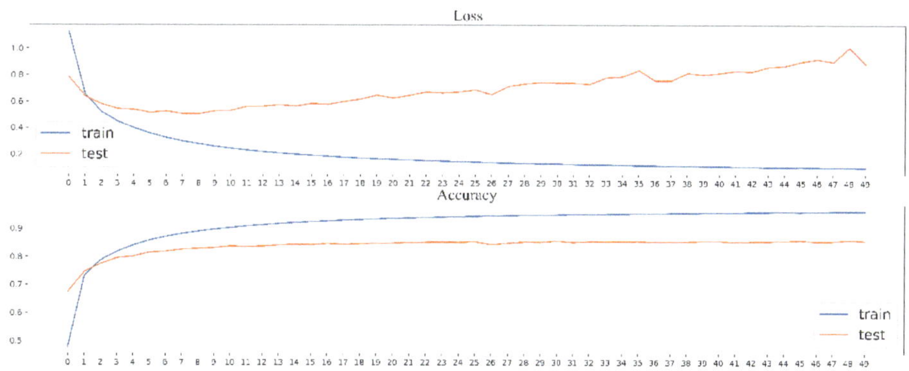

Figure 10.14 Plots depicting the loss and accuracy of the proposed model were generated for both training and testing at each epoch. Loss plot displayed above, and accuracy plot displayed below.

during training. Even though the result of the model in accuracy is consistent throughout the training and validation process, the slightly elevated loss of the model during validation may be attributed to inconsistencies in the input data, as mentioned earlier in the data preprocessing section. While most signals within the same class exhibit similarities in their features, there are minor differences present that contribute to the observed increase in loss.

10.4.2 Model comparison

The models are compared according to the techniques of dropout, batch normalization and hyperparameter tuning. Figure 10.15 illustrates the validation accuracy of various models throughout the training process. In the initial iterations, the accuracy exhibits the highest growth rate, gradually stabilizing as training progresses. However, it continues to fluctuate above 80% during the course of training. The maximum accuracy is achieved by the proposed work. The results of various classification machine learning models are also trained together with the proposed model and compared as per Figures 10.16 and 10.17, to show the effectiveness of our proposed model comparison with the other type of classification models of our research. The accuracy attained from the trained models, ANN, CNN, and KNN are 85.275%, 73.63%, and 54.325% successively. Figure 10.16 presents a comparison of the training accuracies per epoch, highlighting the observable performance differences.

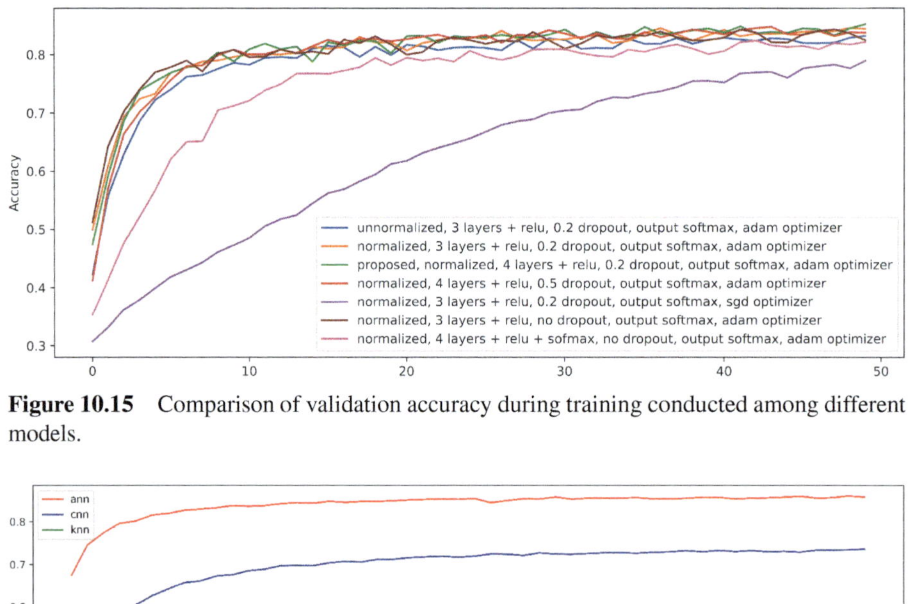

Figure 10.15 Comparison of validation accuracy during training conducted among different models.

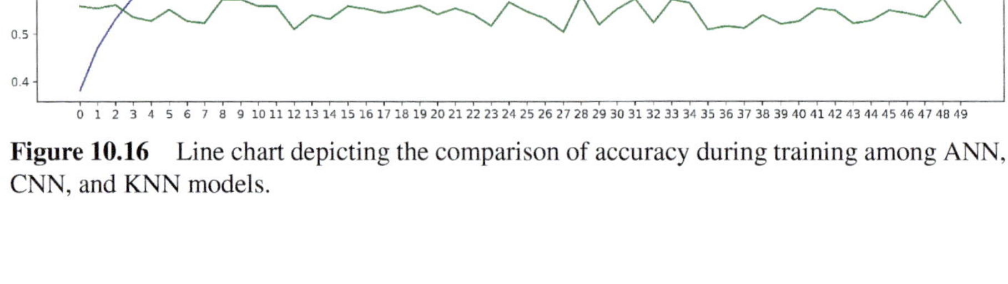

Figure 10.16 Line chart depicting the comparison of accuracy during training among ANN, CNN, and KNN models.

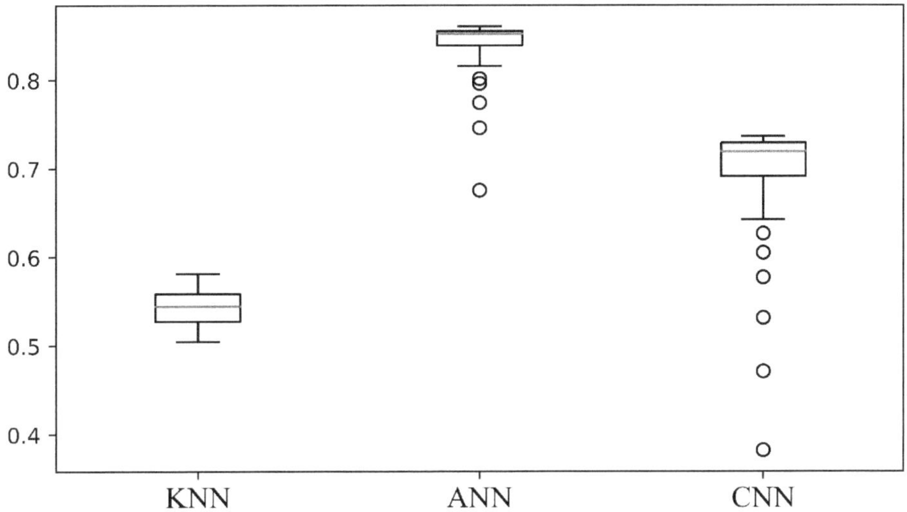

Figure 10.17 Boxplot chart depicting the comparison of accuracy among ANN, CNN, and KNN models.

Figure 10.17 displays a boxplot representing these accuracies, indicating that ANN achieved the highest mean accuracy and overall performance, followed by CNN, while KNN had the lowest performance among the three methods.

The proposed model also outperformed 90% of the previous works on motor imagery tasks using EEG signals. Table 10.1 shows the comparison of the previous works [23], [19–33] to this work by describing the methods used and the maximum accuracy that was achieved.

Table 10.1 Overview of the previous works executing the four-movement motor/imagery tasks in comparison to this work.

Work	Dataset	Method	Maximum accuracy
Hyun et al. (2013)	EEGMMI Dataset	CSP-LDA[a]	66.65%
Does et al. (2018)	EEGMMI Dataset	CNNb	80.10%
Pinheiro et al. (2018)	EEGMMI Dataset	ANN[c]	74.96%
Ma et al. (2018)	EEGMMI Dataset	RNNsd	68.20%
Alwasiti et al. (2020)	EEGMMI Dataset	DMLe	64.7%
Chen et al. (2020)	BCI IV 2a Dataset	FBSF[f] & CNNb	72.0%
Proposed method	EEGMMI Dataset	**DWT[g]-ANN[c]**	**85.27%**

[a] Common spatial pattern and linear discriminant analysis, [b] Convolutional neural network, [c] Artificial neural network, [d] Recurrent neural network, [e] Deep machine learning, [f] Filter-bank spatial filtering, [g] Discrete wavelet transform.

10.5 Discussion

We analyzed and compared EEG-based BCI systems with various datasets of brain control signals from various sources. The discrete wavelet transform is used to contrast the results of feature extraction. Artificial neural network (ANN), convolutional neural network (CNN) and K-nearest neighbors (KNN) are utilized as classification techniques and are evaluated using the cleaned-up dataset. Among these, using ANNs results in the highest accuracy. The proposed DWT-ANN has an overall classification accuracy of 85.27% on the Physionet EEG dataset. Left, right, forward, and backward achieved a maximum worldwide average accuracy of 87%, 84.9%, 85.2%, and 84%, respectively. Therefore, 85% of the time, the proposed model is able to detect the imagery movement signals from the brain correctly.

This result has been further compared with previous works, including those conducted in recent years. As depicted in Table 10.1 above, the dataset of EEG motor imagery movements comprises several studies employing different techniques. In [29], the application of Common spatial pattern and linear discriminant analysis on the given dataset led to a classification accuracy of 66.65% for motor movements. Similarly, in [23], [31], and [33], neural network architectures such as CNN and RNN were considered, resulting in classification accuracies of 80.1%, 68.2%, and 72% respectively. In contrast, [30] and [32] explored contrasting ideas, achieving accuracies of 74.96% and 64.7% respectively. A similar study in [34] practiced the same feature extraction method, DWT and classified the signals using support vector machine and received a medium accuracy. Among these studies, the proposed method exhibits a higher level of accuracy, surpassing the results achieved by the other approaches.

These findings demonstrate the efficacy of the proposed methodology in accurately classifying different motor intentions from EEG signals. The utilization of ANNs, in conjunction with the DWT feature extraction technique, showcases promising results, highlighting the potential for developing efficient and reliable BCI systems. Such advancements can have significant implications in assisting individuals to control external devices and interfaces using their mind, paving the way for enhanced communication and interaction for everyone, especially individuals with motor disabilities.

10.6 Strength and Limitations

In this paper, the target is to yield a high-accuracy movement classification without heavy features extractions and computational complexity for the speedy process of classifying the movement control during a game. The feature signals are extracted using a high-frequency-passed filter of discrete wavelet transform function. The high-frequency detail components can help capture the fast nature of event-related potentials, allowing for more accurate detection and analysis. However, by using only a high-pass filter, there is a risk of losing or significantly reducing these low-frequency components, potentially diminishing the ability to capture certain aspects of motor control related signals. On the other end, Haar wavelet (db1) is the first and fastest of discrete wavelet transform due to its simple nature of first-ordered formulation. The simple combination of DWT's simple high-pass function and neural networks is a novel approach that stands out due to its lightweight and simplicity. This will be utilized in gaming applications to enhance user experience and interaction and well-suited for real-time analysis on gaming devices with limited processing capabilities. Its simplicity allows for quick implementation and low latency, enabling seamless integration into gaming systems.

10.7 Future Direction

Future studies could go a number of possible paths to enhance the proposed strategy. The research's objective is to implement the proposed solution to a practical BCI and evaluate how well it functions. We plan to integrate the BCI technology into the gaming industry. A BCI system based on EEG signals might provide a futuristic gaming industry with an immersive and creative gaming experience. Through BCI technology, which is based on EEG, users can control and participate in video games using their brain activity. Real-time game control would be implemented using the classified commands. Users may move characters, select items, activate skills, or even alter the game's environment by using only their brain activity. Brain activity patterns of the user may be used to teach the system how to adapt and develop over time. This adaptive learning would help the BCI system understand the user's goals and game preferences while also improving accuracy and responsiveness. In spite of the fact that research into EEG-based BCI gaming systems is still in its initial stages, they have the potential to produce compelling and distinctive interactive experiences by letting users directly manipulate the activities directly with their brain. Further, this study will support facilitating the enhanced mobility of people with the situational disabilities or even complete disabilities, enabling them to actively participate in various social and physical activities with greater ease and confidence.

10.8 Conclusion

This research proposed an EEG-based imagery movements classifier with DWT and ANN to process and classify the movement tasks of brain. The focus was on examining EEG signals and conducting feature extraction by applying a high frequency filter using the discrete wavelet transform (DWT). To achieve improved outcomes, the classification techniques utilized were artificial neural network (ANN), convolutional neural network (CNN), and K-nearest neighbors (KNN). Among these methods, the classification using ANN achieved the highest accuracy rate of 85.27%. Regarding the four motor movements (left, right, upward, and downward), they achieved average accuracy rates of 87%, 84.9%, 85.2%, and 84% respectively, which are considered the highest accuracy rates worldwide. The precision, recall and f1-score are also harmonious at 85% with a high AUC percentage value of ROC and precision-recall curve, 97% and 93% overall.

The novel approach of combining the simple and fast DWT Haar high-passed filter and ANN classification not only significantly reduces the duration of feature extraction but also surpasses the accuracy achieved by existing intricate classification. Its lightweight nature and simplicity make it an attractive choice for resource-constrained gaming devices, ensuring smooth operation and enhancing user engagement in gaming environments.

References

[1] Shih, J. J., Krusienski, D. J., & Wolpaw, J. R. Brain-computer interfaces in medicine. *Mayo clinic proceedings* **87**, 268-279 (2012).

[2] Plass-Oude Bos, D. et al. Brain-computer interfacing and games. *Brain-Computer Interfaces: Applying our Minds to Human-Computer Interaction* 149-178 (2010).

[3] Schalk, G., & Allison, B. Z. Noninvasive brain–computer interfaces in *Neuromodulation* 357-377 (Elsevier, 2018).

[4] Bi, L., Fan, X. A., & Liu, Y. EEG-based brain-controlled mobile robots: a survey. *IEEE transactions on human-machine systems* **43**, 161-176 (2013).

[5] Wolpaw, J. R. Brain–computer interfaces in *Handbook of clinical neurology* **110**, 67-74 (Elsevier, 2013).

[6] Leuthardt, E. C., Moran, D. W., & Mullen, T. R. Defining surgical terminology and risk for brain computer interface technologies. *Frontiers in Neuroscience* **15**, 599549 (2021).

[7] Casson, A. J., Abdulaal, M., Dulabh, M., Kohli, S., Krachunov, S., & Trimble, E. Electroencephalogram. *Seamless healthcare monitoring: advancements in wearable, attachable, and invisible devices*, 45-81 (2018).

[8] Casson, A. J. Wearable EEG and beyond. *Biomedical engineering letters* **9**, 53-71 (2019).

[9] Leeb, R. et al. Self-paced (asynchronous) BCI control of a wheelchair in virtual environments: a case study with a tetraplegic. *Computational intelligence and neuroscience* **2007**, (2007).

[10] Monesi, M. J., & Sardouie, S. H. Extended common spatial and temporal pattern (ECSTP): A semi-blind approach to extract features in ERP detection. *Pattern Recognition* **95**, 128-135 (2019).

[11] Kumar, S., Mamun, K., & Sharma, A. CSP-TSM: Optimizing the performance of Riemannian tangent space mapping using common spatial pattern for MI-BCI. *Computers in biology and medicine* **91**, 231-242 (2017).

[12] Amin, S. U., Alsulaiman, M., Muhammad, G., Bencherif, M. A., & Hossain, M. S. Multilevel weighted feature fusion using convolutional neural networks for EEG motor imagery classification. *Ieee Access* **7**, 18940-18950 (2019).

[13] Abdulkader, S. N., Atia, A., & Mostafa, M. S. M. Brain computer interfacing: Applications and challenges. *Egyptian Informatics Journal* **16**, 213-230 (2015).

[14] Ramesh, D., & D'Souza, N. A. One-step fabrication of biomimetic PVDF-BaTiO3 nanofibrous composite using DoE. *Materials Research Express* **5**, 085308 (2018).

[15] Lotte, F. et al. A review of classification algorithms for EEG-based brain–computer interfaces: a 10 year update. *Journal of neural engineering* **15**, 031005 (2018).

[16] Wysocka, M., Wysocki, O., Zufferey, M., Landers, D., & Freitas, A. Biologically-informed deep learning models for cancer: fundamental trends for encoding and interpreting oncology data. Preprint at https://arxiv.org/abs/2207.00812 (2022).

[17] Zhang, R. et al. Using brain network features to increase the classification accuracy of MI-BCI inefficiency subject. *Ieee Access* **7**, 74490-74499 (2019).

[18] Olivas-Padilla, B. E., & Chacon-Murguia, M. I. Classification of multiple motor imagery using deep convolutional neural networks and spatial filters. *Applied Soft Computing* **75**, 461-472 (2019).

[19] Xu, B. et al. Wavelet transform time-frequency image and convolutional network-based motor imagery EEG classification. *Ieee Access* **7**, 6084-6093 (2018).

[20] Tayeb, Z., Fedjaev, J., Ghaboosi, N., Richter, C., Everding, L., Qu, X., ... & Conradt, J. Validating deep neural networks for online decoding of motor imagery movements from EEG signals. *Sensors* **19**, 210 (2019).

[21] Amin, S. U., Alsulaiman, M., Muhammad, G., Bencherif, M. A., & Hossain, M. S. Multilevel weighted feature fusion using convolutional neural networks for EEG motor imagery classification. *Ieee Access* **7**, 18940-18950 (2019).

[22] Tang, Z., Li, C., & Sun, S. Single-trial EEG classification of motor imagery using deep convolutional neural networks. *Optik* **130**, 11-18 (2017).

[23] Dose, H., Møller, J. S., Iversen, H. K., & Puthusserypady, S. An end-to-end deep learning approach to MI-EEG signal classification for BCIs. *Expert Systems with Applications* **114**, 532-542 (2018).

[24] Wang, Z. et al. Short time Fourier transformation and deep neural networks for motor imagery brain computer interface recognition. *Concurrency and Computation: Practice and Experience* **30**, e4413 (2018).

[25] Aslam, M. J., Hayat, S., Moetesum, M., & Siddiqi, I. A Brain Computer Interface (BCI) based Game Controller using Expressive and Cognitive Signals, (2016).

[26] Coulton, P., Wylie, C. G., & Bamford, W. Brain interaction for mobile games. *Envisioning Future Media Environments*, 37-44 (2011).

[27] Goldberger, A. L. et al. PhysioBank, PhysioToolkit, and PhysioNet: components of a new research resource for complex physiologic signals. *Circulation* **101**, e215-e220 (2000).

[28] Hou, Y., Zhou, L., Jia, S., & Lun, X. A novel approach of decoding EEG four-class motor imagery tasks via scout ESI and CNN. *Journal of neural engineering* **17**, 016048 (2020).

[29] Kim, H. S., Chang, M. H., Lee, H. J., & Park, K. S. A comparison of classification performance among the various combinations of motor imagery tasks for brain-computer interface. *Neural Engineering (NER)*, 435-438 (2013).

[30] Pinheiro, O. R., Alves, L. R. G., & Souza, J. R. D. EEG signals classification: Motor imagery for driving an intelligent wheelchair. *IEEE Latin America Transactions* **16**, 254-259 (2018).

[31] Ma, X., Qiu, S., Du, C., Xing, J., & He, H. Improving EEG-based motor imagery classification via spatial and temporal recurrent neural networks. *IEEE engineering in medicine and biology society*, 1903-1906 (2018).

[32] Alwasiti, H., Yusoff, M. Z., & Raza, K. Motor imagery classification for brain computer interface using deep metric learning. *IEEE Access* **8**, 109949-109963 (2020).

[33] Chen, J., Yu, Z., Gu, Z., & Li, Y. Deep temporal-spatial feature learning for motor imagery-based brain–computer interfaces. *IEEE Transactions on Neural Systems and Rehabilitation Engineering* **28**, 2356-2366 (2020).

[34] Chum, P., Park, S. M., Ko, K. E., & Sim, K. B. Optimal EEG feature extraction using DWT for classification of imagination of hands movement. *Journal of Korean Institute of Intelligent Systems* **21**, 786-791 (2011).

Acknowledgements

We appreciate the works by the EEG Motor Movement/Imagery Datasets for the source of data used in our study.

Author Contributions

K.E.H: Writing, Formal Analysis, Data Collection; M.M.H: Original Writing, Reviewing, Project Administration, M.T.K: Formal Analysis, Draft Writing H.H: Draft Writing, Data Collection, Formatting; K.K.O: Data Collection, Formatting; S.T.A: Final Writing, Formal Analysis; F.Y: Formatting, Visualization; S.M.S.I: Mentoring, Funding, Writing

Competing interests

The Authors declare no Conflicts of Interests.

Additional information

Correspondence and requests for materials should be addressed to Md. Mehedi Hassan

11

Decoding Brain Signals for Connectivity Analysis with Machine Learning: Innovations and Applications in Neuroscience

Thota Akshitha[1], Kishor Kumar Reddy C.[1], Manoj Kumar Reddy D.[2], and Srinath Doss[3]

[1]Department of Computer Science and Engineering,
Stanley College of Engineering and Technology for Women, India
[2]Department of Electrical and Electronics Engineering,
Vardhaman College of Engineering, India
[3]Faculty of Engineering and Technology, Botho University, Botswana
E-mail: thotaakshitha39@gmail.com; drckkreddy@gmail.com;
dadireddymanojkumarreddy9@gmail.com;
srinath.doss@bothouniversity.ac.bw

Abstract

Recent advances in decoding the brain signal with the help of machine learning have significantly contributed to the current paradigms of understanding neural activity and the creation of newer and superior neurotechnologies. Understanding the review of methodologies and the progress of brain connectivity analysis from the perspective of machine learning, the chapter offers a deeper insight into the corresponding field starting with equipment and procedures for acquiring the EEG signals followed by important steps in pre-processing of the data like noise elimination, filtering, and segmentation. Moving forward, the chapter continues to explain the application of WPD in decomposing EEG signals into several frequency bands including alpha, theta, gamma, beta, and delta bands and how these decomposed EEG signals serve as a base for connectivity analysis. A substantial part is devoted to the exploration of how signals in the brain are interpreted by the networks in terms of connectivity characteristics by using graph theoretical measures and explains how machine learning models can be used in analyzing the test data that classifies the brain connectivity states using obstructive sleep apnea (OSA) as a case study, and comparing the connectivity between OSA patients and normal subjects. Concluding by briefly highlighting the current limitations and future directions of EEG-based connectivity analysis, as well as the existing controversies and capacities for integration with other methodologies, that the importance of machine learning in developing the field of neuroscience was stressed effectively throughout this chapter.

Keywords: Neurotechnologies, machine learning, brain connectivity, electroencephalogram signals (EEG signals), wavelet packet decomposition (WPD), obstructive sleep apnea (OSA), neuroscience.

11.1 Introduction to Brain Signal Processing and Connectivity Analysis with Machine Learning

Neural signal acquisition refers to the capturing and recording of neural signals, neural signal analysis is the identification and quantification of signals while neural signal interpretation is used in the discovery of how the brain functions and the identification of disorders associated with the brain [2]. The most common brain signals utilized in neuroscientific studies are electrical (EEG), magnetic (MEG), and metabolic. In detail, the most common type of EEG is preferred due to its high temporal resolution and its lower costs compared to other methods. Some of the critical issues in processing the signal involving the brain include brain signal acquirement, which involves the use of electrodes placed on the scalp or implanted in the brain to capture the neural activities [1]. Next is the preprocessing stage where signals are filtered to eliminate contaminants and other distortions originating from eye movement and electrical interferences. Connectomics is the study of the connections and connections, how the different brain areas communicate with each other, which of course is important since the breakdown of the brain function is based on the cooperation of its networks. By structural connectivity we mean the links between the networks within the brain which are normally analyzed based on diffusion MRI scans for the white matter fibers. Seeded functional connectivity refers to the identification of correlations between signals originating from different brain regions for synchronized activity; the calculations are usually derived from data obtained from EEG, MEG, and fMRI. The knowledge of brain connectivity is helpful in diagnosis and prognosis of neurological and psychiatric diseases, in the neural substrates study and behavior comprehending, and in the neurorehabilitation approaches treatment planning. It encompasses this extensive analysis for purposes of diagnosing and monitoring diseases such as epilepsy, autism, and schizophrenia.

Data science powered by ML has emerged as an invaluable technology for neuroscience to develop optimal ways to process massive and intricate brain signals. This is because the neural data can be analyzed by the existing statistical methods and the ML algorithms to understand relationships that would otherwise be challenging to identify. Further, ML models can provide outcomes of neurological function and detect certain diseases from patterns of variables thus helping with timely diagnoses and unique treatment regimens. In addition, it is essential for the creation of the BCI, through which a human can directly control, for instance, a computer, using just signals originating in the brain and therefore providing new opportunities for people with a limited or no possibility of movement. A combination of state-focused artificial intelligence techniques with conventional neuroscience methodologies brings about a better understanding of the structure and relationships in the brain to provide efficient diagnostic procedures and treatment measures.

11.2 Brain Signals: Acquisition and Preprocessing

The signals detected in the brain are essentially underlain by the acquisition process together with the necessary preprocessing to make the data suitable for analysis. This section, therefore, focuses on the method of EEG signal acquisition, and then the corresponding preprocessing step employed to improve the quality of the extracted signals as well as gain valuable information from it.

EEG signal acquisition techniques

EEG or electroencephalography on the other hand is one of the important techniques applied in recording brain electrical activity. They are electrical recordings made from electrodes that are attached to the scalp that dissect the firing patterns of the neurons that summarize the function of the brain. Electrode arrangement comes in many forms like the 10–20 system where different spatial resolution and coverage area can be reciprocated. Researchers are privileged to arrange for specific brain region or complete brain picture [9]. Furthermore, the collection of raw EEG data can be ensured according to the given requirements of specific research background since it might be applied to the clinic for diagnostic purposes or in research laboratories for experimental studies. In summary, the approaches towards EEG acquisition are flexible that allow obtaining information of the brain in an intact state, and thus, EEG can offer a unique view on neural processes.

Preprocessing steps

Noise reduction, filtering, and segmentations are some of the primary methods that help to improve the result of voice activity detection in speech. Unfortunately, artifacts can be quite misleading, and thus prior to conducting any EEG data analysis, preprocessing steps are greatly needed to improve signal quality. Artifact mitigation and suppression via software-based algorithms, such as artifact rejection and independent component analysis (ICA), are used to remove unwanted sound originating from muscles, eye movement, or external disturbances. Next, the high-pass and low-pass as well as the band-pass filters are used to select the desired frequencies while excluding the other unwanted frequencies and thereby eliminate noise. There are two types of processing, namely, band-pass filtering, which filters EEG signals within a specific range, and segmentation, which partitions the continuous EEG signal into epochs or windows further assisting in the analysis of a specific state or

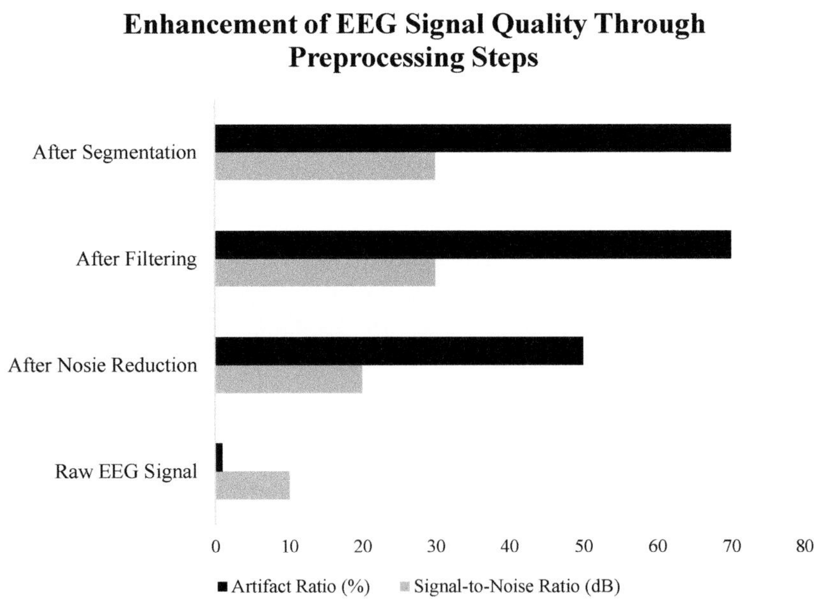

Figure 11.1 Enhancement of EEG signal quality through preprocessing steps.

event. As with these preprocessing steps, the researchers want to make sure that there is clean, reliable EEG data for analysis and interpretation [9].

As shown in Figure 11.1 the strength of connection of certain areas in the brain towards one another is depicted, thus suggesting comparative neural coupling values. Out of all the model's highlighted areas, the limbic seemed to have the most connections with a connection value of 0. 95, which means that the level of interregional communication currently is quite high. On the contrary, the connectivity measured at 0 reflects relatively distributed and weaker connection in the occipital lobe. The value of 53 implies that these cells had significantly lower connectivity to other regions of the brain. This makes it easier to compare regions with focus on form and arrangement of connectors in the neurological regions. Knowledge of these drawings in contact strength offers useful information on the functional architecture and possible functions of areas of the brain in cognition and action. These kinds of assessments are important for yielding insight into already-known structures such as neural networks and their role in neurologic functionality and pathology.

Feature extraction methods

The two types of this analysis are time and frequency domain analysis. Feature extraction is a critical step in making a diagnosis of brain activity by utilizing EEG signals to isolate important features from unimportant ones. Time domain analysis involves the calculation of simple statistics of amplitude of the EEG signal and their variance and skewness to explain temporal features of the signal. On the other hand, frequency domain analysis is focused on conversion of the EEG data into the spectral forms using Fourier or wavelet transforms and can unveil the power spectrum density function for different frequencies. From these domains, characteristics that detail the temporal progression and frequency of signal fluctuations in neural activity are extracted; these provide informative inputs for subsequent machine learning processes as they strive to decipher the function and connectivity of the brain [20].

11.3 A Review of Machine Learning Methodologies for Brain Connectivity Examination

Artificial intelligence (AI) techniques have significantly transformed the big picture of the easily usable and easily predictive systems designed with brain connectivity. The following section describes several types of ML techniques as well as broad types of architectures that are relevant to the analysis of connectome data.

- **Supervised learning approaches**
  In supervised learning one trains a model on data which has been classified or tagged in some form or other depending on the nature of the particular learning issue at hand: we are more precisely predicting outcomes or categorizing the data into classes, the classes being predefined. Within the field of analysis of connectivity of the brain, pattern recognition or supervised learning can be applied to classify the patterns linked to certain neurological disorders or certain states of mind. For instance, connectivity features derived from EEG or fMRI signals can be used as predictors for training decision support algorithms such as support vector machines (SVMs), random forest, and neural networks for the classification of patients into diagnostic categories or for prognosis of treatment response. Supervised learning can be used to build a model that helps explain the extent of disconnection and how it can relate or is connected to neurological diseases [20].
- **Unsupervised learning and clustering methods**
  Unsupervised learning and clustering methods also pertain to this category, as they involve working on raw data without being given a preexisting model as guidance. Clustering is self-organization of data in taxonomy which is unsupervised learning paradigm and does not require prior labeling. Generally, clustering algorithms – are the most used algorithms in the process of grouping identical brain connectivity patterns due to their features, including k-means clustering and hierarchical clustering. In this regard, clustering strategies segregate regions based on their functional or structural connectivity patterns and unearth information regarding the architecture of brain networks. Such insights assist in the identification of subtypes of neurological disorders, dissection of the functional modules within the brain, and linking the connectomes architecture to cognition.
- **Deep learning architectures**
  These instruments include ML which refers to the use of algorithms for making decisions based on data, and DL, a sub field of ML that uses neural networks with multiple layers to learn representations of data. An attribute of large-scale and high-dimension data, RNNs and CNNs hold potential in brain connectivity analytics owing to their suitability in processing connectivity matrices obtained from multiple neuroimaging

modes. CNNs can learn spatial features from images of the human brain and analyze temporal relationships in functional data equivalent to EEG signals using RNNs. Further, deep learning methods work on the feature extraction and pattern recognition at multiple layers of the networks and help in accomplishing the learning of end-to-end functional relationships in connectivity data to improve the classification and discovery of potential biomarkers for neurological disorders. In using these techniques in machine learning, the researchers will therefore obtain a profound better understanding of the structure and functioning of brain networks to foster breakthroughs in neuroscience studies and clinical practice [13, 16].

11.4 Introduction to Brain Networks

These are the parts that make up the graph: There are two different types of parts on a graph: Nodes: These are the points or the shapes that connect the edges, through which the flow travels. Through brain networks, neuroscience researchers are able to bring into focus essential theoretical abstractions that would help them to understand how different parts of the brain are connected and how they work in unison. This section initiates an exploration into the foundational components of brain networks: nodes and edges are additional the CPC (Canonical Polyadic Component) architecture may still accommodate more elements. In the context of a brain's topology, nodes represent entities that correspond to certain anatomical or functional parts of the brain present in the brain's complex structure. These entities refer to a structural hierarchy, beginning with single cortical fields or subcortical nuclei up to supermodel functional networks that integrate numerous areas. Every node displays certain features and performs some tasks and operations, which indicate that complex neural activity takes place in different parts of the brain at the same time due to the connections that link the nodes. On the other hand, edges perfectly illustrate the relationship or connection between two or more nodes within the same network [4].

These connections can exist in the form of structural or functional networks, delineating white matter fascicles accessible via tools like diffusion MRI or statistical associations derived from the correlation analyses of neural activities across the two networks inferred from functional MRI/ EEG data [5, 17]. These edges capture the complex temporal dependencies between the brain regions in terms of exchanging either detailed information or a simple signal, which in turn provides rather crucial information about the mechanisms that control the function of the brain. By providing researchers the ability to define and further explore the utilization of nodes and edges in brain network studies, there are key organizational constraints and constant interactions in the brain that can be better understood and addressed, enabling new development of neuroscience studies and potential treatments.

As shown in Figure 11.2, the strength of connection of certain areas in the brain towards one another is depicted, thus suggesting comparative neural coupling values. Out of all the model's highlighted areas, the limbic seemed to have the most connections with a connection value of 0. A value of 95 means that the level of interregional communication currently is quite high. On the contrary, the connectivity measured at 0 reflects relatively distributed and weaker connection in the occipital lobe. A value of 53 implies that these cells had significantly lower connectivity to other regions of the brain. This makes it easier to compare regions with focus on form and arrangement of connectors in the neurological regions. Knowledge of these drawings in contact strength offers useful information on the functional architecture

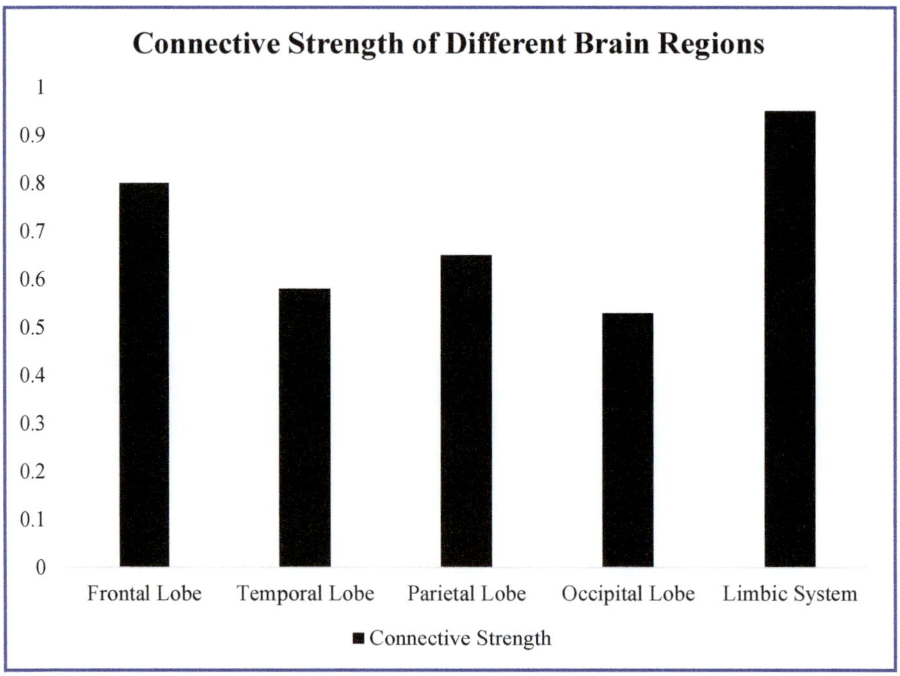

Figure 11.2 Connective strength of different brain regions.

and possible functions of areas of the brain in cognition and action. These kinds of assessments are important for yielding insight into already-known structures such as neural networks and their role in neurologic functionality and pathology.

11.4.1 Graph theoretical measures: Closeness centrality, degree centrality, path length

Energetic properties of a graph depend on the graph theoretical measures that provide quantitative understanding of the structural and functional organization of brain connectivity. In this section, we delve into three key graph theoretical measures: Distance variables that are of interest include degree, centrality, and path length [4]. Degree is assumed as an essential centrality measure that calculates the number of connections or the edges in the network attached to a specific node. The maximum values computed in degree centrality are referred to as the hubs and these nodes are significant in interconnecting information in the brain regions. Every graph possesses a degree distribution through which it is possible to know the connectivity profile of the network which may extend to the visibility of highly interconnected regions and faults [18].

Centrality measures are indices of nodes in the network that would show the extent of the centrality of the nodes. The betweenness centrality indicates how much a node is placed in the middle of the shortest paths connecting the other nodes and hence provides an indirect measurement of the potential communication bridges. Closeness centrality quantifies the effectiveness of a node in conveying information throughout a network, considering proximity to all other nodes. Offering information path length refers to the average of the shortest path connecting the nodes in the network. It measures the integration of the brain networks, as well as the speed of the communication between different areas of the brain – shorter connection lengths denote that the areas are closely connected and function smoothly. Connectivity density and connection strength refer to two key dimensions of resting state functional and structural networks, which provide complementary approaches to studying the architecture and activity of large-scale brain networks.

As shown in Figure 11.3, the graph theoretical measures of different broke areas and the fields, degree, betweenness centrality, closeness centrality path length are depicted. These measures allow us to infer essential information regarding the position and function of some of the brain regions within large, interconnected networks. For instance, the limbic system gets various

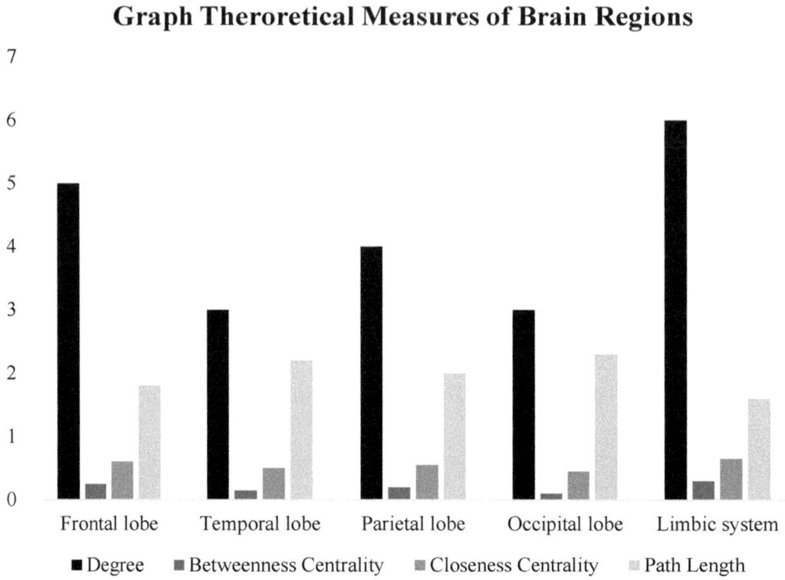

Figure 11.3 Graph theoretical measures of brain regions.

connectivity values of 6 which may suggest that this area is like a hub where it is highly connected with other areas near it. On the other hand, the connectivity map of the occipital lobe shows the least connectivity value of 3, which means less direct connection. Eccentricity focuses on how a region is situated in between the parts of the network; as such, the grandees connexons show a higher betweenness centrality, where the frontal lobe and limbic system have been located as putative routers of information. Closeness centrality defines how easily information is transmitted in the network, and the limbic system received the highest score in this regard, meaning that it will be easy for it to feel and interact with other parts of the brain. Short path length equals to the mean of minimal distance between a region and all remaining regions in the net, the minimum value of path length is 1, correspondingly is the limbic system. 6, indicating a good flow of information/communication within the social network in the organization. In general, these measures are relatively useful in shedding light on the anatomical organization and connectivity of different and larger neural networks of the brain.

Functional connectivity

Functional connectivity refers to the mathematical approach by which the relationships or synchrony of the signal can be captured across different areas in the brain. It captures the level of coupling between neural oscillations and can be used to evidence the efficiency of connective coupling of the brain regions throughout different cognitive operations and functions [3], [15]. Noninvasive neuroimaging methods, including functional magnetic resonance imaging (fMRI), electroencephalography (EEG), and magnetoencephalography (MEG), are used to investigate functional connectivity, frequently by calculating the coherence of intrinsic oscillations. Functional alterations in connectivity also occur as a result of differences in signal regulation as related to brain function and are frequently linked to neurological or psychiatric conditions [11].

Structural connectivity

On the issue, structural connectivity attests to the real or physical links existing between distinct brain regions, based on the white matter tracts that enable neural interfacing. This sort of connectivity gives an understanding on the structural basis that underlies functional connectivity networks within the brain [4]. Functional connectivity is studied using Jitter and functional Magnetic resonance imaging while structural connectivity is examined using diffusion tensor imaging (DTI) and diffusion spectrum imaging (DSI). In this case, alterations in structural organization of the brain, as white matter lesions or fiber tracts abnormally, may affect neural transmissions which lead to neurocognitive deficits or neurological conditions [3]. The combination of functional and structural connectivity provides ensemble information for analyzing the network organization and function of the brain, which can enhance understanding of brain function in a healthy state and in diseases.

- **Signal decomposition techniques**
  Another method used for decomposing a signal is wavelet packet decomposition (WPD). Wavelet packet decomposition is a signal transformation tool that is more generalized than wavelet transform, which is applied to decompose signals into frequencies. While conventional power time frequency or Fourier analysis works with a fixed frequency sinusoidal basis function, WPD has some level of adaptability in the analysis of non-stationary and transient signal like EEG signals. WPD on the other hand refines the original signal by splitting the signal into sub bands in each level of decomposition and finally the feature extraction is obtained frequency selective while maintaining the original time and frequency localization.

Wavelet packet decomposition is a complex process that can be subdivided into a few preliminary steps. First, by the use of wavelet transform, the signal is pulled apart into groups, the groups containing approximation coefficients and detail coefficients at each level of transform. Subsequently, decomposition tree is traversed to highlight specific decomposition that may be in relation to frequency bands or resolution. This kind of selective decomposition enables one to selectively perform an analysis of the kind and frequency number components considered important in any related study [6].

- **Band separation: Alpha, theta, gamma, beta, delta**
  Subsequently, it identifies the different frequencies of the decomposed signal that represent the various rhythms of EEG namely alpha, theta, gamma, beta, and delta. Like the case with neurons, every frequency band has its own waveform that corresponds to certain cognitive state or physiological activity. For instance, alpha band oscillations are usually related to relaxed wakefulness and beta band oscillations are associated with active cognitive processing and motor activity [21].
- **Feature extraction from decomposed signals**
  To feature extraction from the decomposed signals we utilize wavelet transform and Fourier transform. After the signal is separated into different bands, analyses of features are then performed depending on the specified bands. Some of these may include the mean field/mean value/average value, variance, skewness, power spectral density, entropy, or any other spectral measure. These features are able to record temporal and spectral properties of neural activity within each frequency band for the purpose of analysis.
- **Absolute value computation of EEG bands**
  In this process, we measure the absolute value of the EEG bands, so as to determine the strength of the waves in each band. To characterize the amplitude values of the eight frequency bands of recorded EEG, the signals obtained are transformed to their absolute values. This step removes phase information and keeps the envelope of the oscillatory activity, which is more suitable for reliable estimation of band-specific power.
- **Correlation matrix formation**
  Finally, correlation matrices are created to measure the functional connectivity between functional networks in the respective frequency bands of EEG. Pearson product-moment correlation coefficients reflect the strength of a linear relationship or, statistical dependence between the

two frequency bands that can help to understand the temporal structure of coordination between activities across the region. Wavelet Packet Decomposition and several subsequent analytical methods provide a versatile tool for researchers to study the intricate nature of brain signals and understand the meaningful underlying connectivity patterns which define the proper function of the brains, and their dysfunctions as well.

As shown in Table 11.1 the step-by-step procedure that involves the WPD processing of the EEG signals followed by connecting brain analysis is provided. The process starts with the help of WPD that filters all EEG signals and analyzes them in terms of frequency components giving the time-frequency analysis. This is succeeded by the iteratively conducted WPD where particularized channels are suspected of detailed exploration. The signals are subsequently partitioned into these unique bands, including alpha, theta, gamma, beta, and delta, in correlation with the dissimilar cognitive and neural conditions. From these decomposed signals, features that are significant are derived, and other comprehensive features that are inclusive

Table 11.1 Steps in brain signal decomposition and connectivity analysis.

Step	Description	Purpose/outcome
Signal decomposition Techniques: Wavelet packet decomposition (WPD)	Decomposes the EEG signal into various frequency components using wavelet functions.	Provides a detailed time-frequency representation of the signal.
Detailed process of WPD	Involves iterative splitting of the signal into high and low-frequency components.	Enables the isolation of specific frequency bands for further analysis.
Band separation: Alpha, theta, gamma, beta, delta	Separates EEG signals into distinct frequency bands (alpha, theta, gamma, beta, delta).	Facilitates the study of different cognitive and neural states associated with each frequency band [21].
Feature extraction from decomposed signals	Extracts meaningful features from each frequency band, such as power and phase information.	Captures essential characteristics of neural activity for analysis.
Absolute value computation of EEG bands	Computes the absolute values of the features within each frequency band.	Quantifies the magnitude of neural activity in each band.
Correlation matrix formation	Forms a matrix of correlations between the absolute values of different frequency bands.	Visualizes the functional connectivity and relationships between different brain regions.

of the power and phase characteristics are derived. The absolute values of these features are then calculated and can be used to measure the activity of the neurons within the band. Next, a correlation matrix is generated to show the total tolerance and shared similarities of synchronous neural activities of various brain regions with corresponding values. This detailed procedure offers good protection against the development of not very successful analysis of brain connectivity, and enables to distinguish overlooked delicate links and patterns in the brain's neural network.

11.5 Brain Connectivity Analysis Tools and Techniques

Continuing from the sections, where the text provided the general information about the methods of the brain connectivity analysis, explaining the practical use of the program tools, such as Brain Net Viewer and Brodmann Atlas. These tools truly offer significant assistance in representing and analyzing connectivity patterns within the human brain to efficiently facilitate the understanding of complex network architectures between various regions.

- **Using the readings from Brain Net Viewer and Brodmann Atlas**
  There are many software tools used to perceive and analyze the connectivity data of the brain, but Brain Net Viewer is one of the most important and effective programs in this regard. It enables the researchers to reconstruct the structural and functional networks within the brain in the third dimension effectively enabling exploration of connectivity densities in various regions of the brain. Brodmann Atlas is integrated into Brain Net Viewer which provides anatomical labels and lets the researchers to select and label section of a brain in the network.
- **Node degrees and adjacency matrices**
  This task will further seek to reveal a node degree and adjacency matrices. Node degrees and adjacency matrices have been established as a basic evaluation of the connectivity patterns underlying brain networks. The node degree refers to the number of connections or edges, which each node in the network has, meaning that it reflects the amount of connectivity of each area in the brain. Indirect graphs, in contrast, in the form of adjacency matrices contain the information about the entire network and allow one to perform quantitative analysis and visualization of network topology [18].
- **Real time graphical view and analytic of connectivity alteration**
  Connectivity dynamics and the mechanism used in depicting and analyzing alterations in connectivity density of brain networks are facilitated

by visualization techniques. Time-course analysis, which is supported by the dynamic visualization tools, can be used to assess the time evolution of the connectivity and comparison of different conditions in experiments can be achieved. This idea of combining the analysis of connectivity across baseline and a later time point is helpful in assessing dynamic changes in brain networks and identifying specific attributes of a network related to a certain cognitive process or neurological disorder [7].

Connectivity analysis methods can be segregated into two broad categories, based on the type of statistical techniques used for analysis.

- **Statistical analysis**
  Statistical analyses are also imperative when it comes to measuring and testing the hypotheses about observed connectivity in brain networks [2]. There is always use of correlation analysis, regression analysis, and hypothesis testing in the determination of the magnitude and presence of relations between two different areas of the brain. They can also be used to estimate the probability and randomness whereby valuable connectivity measures can be obtained and their reliability and reproducibility evaluated.
- **Mann–Whitney–Wilcoxon**
  The Mann–Whitney–Wilcoxon test is a statistical test that is used when comparing two conditions and the main purpose of this test is to examine the differences in connectivity values between two groups of subjects or two different experimental conditions. It might also be significant to compare connectivity indices between various groups; for instance, the patients with a certain neurological disorder and the individuals without it, so that the investigators can define the changes in the topology of the connections that are related to the occurrence of the corresponding condition [7]. Consequently, the Mann–Whitney–Wilcoxon test is therefore a suitable and effective measure that can be employed even under conditions of non-normality or small sample sizes.
- **Connectivity pattern analysis between OSA and healthy subjects**
  The results of this study are mainly based on the differences in network topology measures between OSA patients and healthy controls. As mentioned, connectivity measures, including values of functional connectivity strength or place in a network could also be different in OAS patients, and thus alterations in the organization of the brain network could be revealed by comparing them. It could help to compare the patterns and potentially identify the neural pathways responsible

for sleep disturbances and possible targeted approaches for treating cognitive and sleep impairments in OSA [22].

11.6 Case Study: Obstructive Sleep Apnea (OSA) Classification

This coursework focuses on the classification of obstructive sleep apnea (OSA) using the readily available dataset known as ISRUC Database and explores neural and physiological effects related to both OSA and normal sleep period. The analysis unfolds across two primary sections:

Section 1: Dataset description and preprocessing

The ISRUC database, featuring physiological signals recorded during sleep research, provides numerous records including EEG, ECG, and respiratory information. In continuing the-section, we take a lot of time to explain the characteristics of the dataset while stressing on it because of its importance as the foundation of the research. The following section provides details of our preprocessing framework that objectively describes the processes which we employed to maintain data authenticity. This helps in getting a clean dataset, which is essential for carrying out a wide range of analyses with minimal artifacts and noise. With the dataset primed, we then proceed to feature extraction phase where, given the physiological signals, employing state-of-the-art signal processing, we derive relevant features. This preparatory phase provides a strong framework behind which further analysis is covered, paving the way for further investigation.

Section 2: Connectivity analysis decision and classification outcome

As a critical part of our investigation, we now perform connectivity analysis in order to discover features that can distinguish OSA patients from normal subjects. On the basis of the connectivity metrics estimated from the raw EEG data, we assess the amplitude and phase coupling of the neural process within various networks of the brain. Bedside polysomnography enables us to compare topical and temporal connectivity of neurons in patients with OSA to their healthy equivalents to distinguish singular modes of neural operating inherent to OSA. After that, using more refined approaches in network classification, such as the incorporation of support vector machines and neural networks, the subjects are classified into either the OSA or healthy

Table 11.2 Signal types and preprocessing steps.

Signal type	Description	Preprocessing steps
EEG	Electroencephalography signals	1. Filtering for noise removal
		2. Artifact removal
		3. Epoching for segmentation
ECG	Electrocardiography signals	1. Filtering for noise removal
		2. R-wave detection
Respiratory	Respiratory signals	1. Filtering for noise removal
		2. Detection and removal of artifacts

category by their connectivity patterns [22]. The previous interpretation of classification outcomes not only provides the understanding of the utility of our proposed methodology in OSA identification but also offers the information regarding the neurobiology of sleep disorder.

As shown in Table 11.2 various frequency domain signals and the preprocessing that is usually involved in a signal of the type are summarized. Electroencephalography can capture the electrical brain activity of a patient. For each signal type, there are specific preprocessing steps, which consist in processes such as noise removal, artifact detection, and segmentation of the signals, and these are the preliminary processes required by the engineers before proceeding with the actual analysis of the signals. They presented guidance for researchers and practitioners engaged in acquiring and analyzing biomedical signals so that entitled data preprocessing can be properly done, before further analysis or interpretation of data.

11.7 Research on New Directions in Studying Brain Connectivity through ML

New advancements in signal processing have greatly improved the ad hoc detection of neurological connections. It must be pointed out that contemporary techniques that arise from technical user fields and adaptive filtering, empirical mode decomposition (EMD), and wavelet transforms have significantly enhanced the capability of finding valuable information in BCI recordings. Adaptive filters continuously change one or more filter parameters in accordance with signal characteristics, which presents increased and improved noise cancelation and artifact elimination, thus facilitating clean signal capture in actual field settings. There are other techniques such as empirical mode decomposition (EMD) that work decomposition of a signal into its intrinsic mode functions (IMFs), which are different modes of oscillation present in the signal. As stated earlier, this method is quite useful when we work with signals that are nonlinear and non-stationary, which we expect to encounter in the dataset originating from the human brain. EMD enables the extraction and analysis of individual frequency band information that elucidates thc functionality of the brain and connections. Wavelet transforms then include the wavelet packet decomposition (WPD) and the Hilbert-Huang transform (HHT) which produce excellent time-frequency analysis and can identify specific transient neural occurrences. WPD can provide a rather loose definition of how signals can be analyzed into sub bands, in the manner described for isolating alpha, beta, and gamma bands. On the other hand, HHT is a combination of EMD with Hilbert transform, which provides time-frequency signal representation in an adaptive manner and provides both amplitude and phase [12].

These techniques have expanded the capacity in differentiating newer and more refined connections that were otherwise not recognizable, which have increased the understanding of complex neurological networks behind neurons and cognitive processes, as well as various neurological disorders. Machine learning (ML) models have reshaped the decisions in the past few decades about brain connectivity analysis as they provide effective tools for large and complex data analysis [13]. More so, the current innovative techniques in deep learning have performed well in predicting the connectivity of the brain. For instance, convolutional neural networks (CNNs) are especially vital for identifying spatial hierarchies in brain imaging data. The application of convolutional layers in CNNs allows for the configuration of precisely identifying different patterning and relations in the structural and functionality of the brain.

The study of temporal sequences is the major strength of recurrent neural networks (RNNs); specifically, long short-term memory (LSTM) neural networks are quicker at handling temporal sequences making them suitable for analyzing time series data as is the case with EEG signals. The use of LSTMs will enable the model to identify long-term patterns in connectivity change and its dynamics over time. Many brain networks can be directly modeled with a relatively novel type of ANNs called graph neural networks (GNNs). GNNs can learn high-order relational structures to capture general connectivity patterns if brain areas are treated as the nodes and the connections between them as the edges. This approach is more natural and matches the graph theoretical model of brain connectivity allowing for a closer representation of neural interactions. Transfer learning, another revolutionary innovation, enables practical repurposing of models by fine-tuning them to specific tasks with little retraining. Specifically, this technique is extremely helpful for discovering patterns in brain connectivity, as labeled data can be difficult to obtain. In this case, transfer learning leads to improved performance and better generalization since the knowledge from the other domains is utilized.

Another class of methods that has found application recently is the ensemble methods used for combining multiple models, which can also help to counterbalance overfitting and improve computational efficacy in analysis of brain connectivity. Combination methods including bagging, boosting, and stacking allows for amalgamation of various models to improve the shortcomings of a single model, they therefore enhance the accuracy of the prediction. The use of these sophisticated advanced ML models has been attributed to the effectiveness of classification as well as predicting neurological disorders. For example, fusing connectivity data into it, ML models can distinguish healthy controls from patients with epilepsy, Alzheimer's, or schizophrenia [9].

11.7.1 Real-world applications and future directions

The significance of application of higher order signal processing and intelligent signal processing methodologies in functional connectivity research is unrivaled and has the potential to revolutionize not only the research front but also the clinical practice in the near future. These innovative approaches are bringing about changes to the diagnosis, management and treatment of neurological disorders, and add value in a number of settings. In the clinical front, the utilization of these techniques is having its impact in diagnosing and managing diverse neurological disorders that range from epilepsy, Alzheimer's disease, and schizophrenia. For example, the application of machine learning (ML) to optimize EEG connectivity patterns has emerged as a critical tool in the diagnosis of epileptic seizures. Such connectivity patterns may represent predisposing conditions that can be detected clinically, allowing for early intervention and probable seizure prevention – issues that would drastically change the patient's quality of life [9].

Similarly in the field of neurodegenerative disease, Alzheimer's in particular using EEG and fMRI data and the help of ML models we are able to diagnose patients with early signs of cognitive decline, thereby getting them treated early enough before the progress of Alzheimer's disease accelerates. Schizophrenia is one of such diseases where certain connectivity in different networks of the human brain had been disrupted; the new types of analytical methods also can help to make more accurate diagnostic and monitoring of the disease, which leads to individual approaches in treatment [19]. Among those innovative technologies, brain–computer interfaces (BCIs) are applying these developments to enable individuals with severe paralysis to regain the ability to move their limbs. BCIs identify the brain signals corresponding to ability of movement and then convert these signals into signals that can control other objects, for example, the robotic limbs or the cursor on the computer monitor. The combination of current modern techniques in signal processing and in machine learning provides increased accuracy and efficiency in these interfaces to patients [8]. For example, the motorized mobility system enables paraplegic and quadriplegic patients who have spinal cord injuries to control their environment and hence boost their self-determination and enhanced lives. For the future, there is a good potential for synergistic multimodal paradigm that utilizes EEG, fMRI or MEG signals at the same time [11].

Finally, focusing on the future perspectives, one can note a great promise for using the combinations of the various types of brain signals, for instance, EEG, fMRI, and MEG. These approaches give better understanding and wider views of the connectivity of the brain due to accounting for both the time and space aspects of the firing neurons. For instance, EEG is rich in temporal data while fMRI data is rich in spatial data. Integration of such imaging techniques results in building a broader and comprehensive framework of the understanding of brain connectivity and improved diagnosis and treatment [11].

Nevertheless, with these technologies in the future, there will be concerns of ethical concerns and data privacy not addressed. Some risks for privacy include the ability to collect an individual's neural data because signals in the brain are inherently confidential as they reveal the individual's thoughts and actions. Moreover, enhanced data protection measures coupled with ethical standards will become practically crucial in controlling the exploitation of these innovations. In addition, the informed consent and the extent to which data will be used will prove to be touch point in the terms of the external public [17].

The continuation of working on the explainable AI models will also be highly significant in the further examination of the patterns in brain connectivity. These models provide elucidation to machine learning analysis therefore they are easily understandable by clinicians or researchers [14]. XAI makes sure that whilst the rigid and intricate processes governing the learning and decision-making algorithms in a complex ML system are difficult to decipher, the final decision that is being made is easily understandable by clinicians. It is much more constructive when an explanation can be given to the patient that is understandable for them, especially in a clinical context were providing a reason behind diagnosing a certain condition or providing a treatment is so crucial. These advances will further complement the advancement of neuroscience networks in the future and provide more novel therapeutic and better patient care in the future of neurological disorders. Signal processing techniques; machine learning; and multimodal analyses shall advance in understanding of deep and complex functioning of the human brain to establish new approaches to diagnostics and treatments, and personally designed treatment regimes. This continuous advancement holds the potential of refining diagnostic and management models of various ailments and conditions which broadly falls under the neurological category, thus improving the quality of life of many patients.

11.7.2 Opportunities and concerns in customizing and employing the analysis of structural connection and integrative multimodal approaches

Further advancements in BCI as a field also bring both challenges and exciting possibilities as it grows; this includes some of the following aspects: ethical issues, data privacy, and expanded, multimodal opportunities. This call for more frequent and extensive use of advanced human brain connectivity analysis is not without ethicist's implications and data privacy issues. Neural information is very sensitive as it discloses complex aspects of the thinker's and or doer's mind, mood, action, and plans [10]. It also opens many concerns towards the potential misuse of such information that may endanger personal privacy and consume the control of autonomy [16]. For neural data the focus should remain on privacy and protection against unauthorized access. Secure protection solutions such as storage as well as transmission security and access control measures to ensure that the data cannot be accessed and manipulated by unauthorized parties are proper security measures that should be taken. In the treatment of the brain, it is necessary to set certain ethical standards to cover all the processes related to data acquisition, processing and sharing. The implementation of these guidelines should respect the principle of informed consent in regard to usage of the respective data and the data subjects' potential exposure to the various risks. Assurance of transparency in practices managing data is also crucial to build and sustain people's trust. AHNs and other researchers and clinicians reviewing/directing the project must clearly outline the reasons for data collection, how data confidentially is maintained, and how the data collected will be used/shared. Such openness can in turn tend to the security and confidence in the moral application of brain connectivity analysis innovations [1].

Overall, one can state that integrative multimodal approaches seem to hold vast potential for the future of brain connectivity analysis. These approaches relate to the integration of one or more signals from EEG, fMRI, and MEG forms in an effort to offer integrated connectivity of the brain. Interestingly, each modality also has its own set of advantages and benefits. In EEG, recording of the electrical activity is very rapid and distinguishes the rapid, fluctuating neural activity while fMRI has highly localized information on where active changes are taking place in the brain. MEG is situated in between and provides excellent temporal and relatively good spatial resolution [11]. Such coordination is effective in avoiding the shortcomings of the techniques mentioned, as it offers a more detailed look at how the brain performs. For instance, EEG-fMRI can show direct current source estimates of EEG riding on the fMRI-BOLD signal oscillation and provide information on the neural basis of cognition and cognitive dysfunction. By integrating these modalities, there are better results when modeling the brain connectivity hence improving the diagnosis and treatment methodology.

Implementing such a solution in a real-world application requires information fusion at various levels, which is a technical and methodological challenge. The dissimilarities in the way data are captured, also the spatial and temporal aspects of the data, and the noise aspect should also be taken into consideration [12]. Current work in utilizing algorithms and intensive use of machine learning and deep learning strategies is in the process of being designed with a view of enabling the appropriate fusing of multimodal data, but these call for a standard validation and standardization. As a result, getting similar solutions requires interdisciplinary cooperation of neuroscientists, engineers, data scientists, and clinicians when working on such technologies as multimodal approaches. It is also possible that new collaborations could be established to create joint methods that would also be scientifically sound and clinically applicable. By fine-tuning the approaches, the multimodal techniques as the field develops will have the possibility in the future to dramatically change the brain connectivity analysis. In the context of the experiments that can be derived from gene editing they can contribute to the enhanced understanding and differentiation between the functional and structural connectivity of the human brain: paving way for innovative diagnostic methods and individualized therapeutic techniques for neuropsychiatric disorders. Thus, the effort to integrate various techniques in the field of brain connectivity is in a position to respond to the need and to improve our capabilities to address the various problems related to chronic brain disorders in the future.

11.8 Real-World Implications of Neuroscience Research in the Treatment of Substance Abuse and Mental Disorders

Therefore, the current study is exceptionally important because its conclusions may have applications in neuroscience and thus beneficial effects on clinical practice. One of the main applications consists of the advances in diagnostic facilities: they recognize the early stages of such diseases as epilepsy, Alzheimer's disease, schizophrenia and others depending on the peculiarities of the connectivity [15, 19]. This is because early diagnosis is another approach which can warrant timely management measures to slow the rate at which the disease unfolds, and the fate of the patients involved sealed. Altogether, machine learning models for identifying different connectivity profiles that predict individual's response to treatment has the potential of ushering in algorithms for personalized treatments [14]. This method makes it easier to determine minimal treatment doses for each client and steer clear of complications due to undesired neural properties. Using signal processing techniques and Alamut codes coupled with machine learning techniques in neurorehabilitation, the quality of life for patients with motor disability can greatly be enhanced. Such interfaces allow for the brain to interact with external equipment and gadgets, reviving lost motor skills and increasing the level of functioning in daily life [8].

In addition, integrative multimodal approaches provide the direction with an idea that is more comprehensive than connectivity research alone. This kind of holistic integration can contribute to the enhancement of theoretical classifications of cognitive processes, the identification of neural substrate pathways and networks, as well as the modification of current paradigms in conceptualizing and approaching different brain diseases and disorders in the future. Though, as the discipline evolves, issues of ethical implications need to be met, and the privacy of data collected needs to be warranted [5]. Adopting comprehensive guiding principles of ethics and protection of data is essential in the ability to build public trust and avoid mishandling of brain connectivity analysis technologies. Collectively, the realization that brain connectivity analysis is poised for an evolutionary leap through the application of fresh signal processing paradigms and novel ML models augurs well for neuroscience research and clinical applications. These technologies afford far greater understanding of brain conditions, advance diagnostic and therapeutic vehicles, and create the substratum for individualized medicine and innovative neurorehabilitation. Novel applications of trackable implants and similar devices will remain contingent on teamwork between medical professionals and technologists, as well as adherence to the highest standards of medical ethics when these technologies are fully realized for the benefit of human life quality enhancement [10].

11.9 Conclusion

The discussion given on the application of brain connectivity analysis, largely via techniques like advanced signal analysis and machine learning methodologies, has shed the following insights. Firstly, there are new techniques of signal processing like wavelet packet decomposition (WPD) and Hilbert-Huang transform (HHT) that have improved to some extent the calculation of valuable features from brain signals. These techniques allow identification of connection parameters that were essentially unidentifiable earlier, which offer better understanding of neural activity. Secondly, research in the last few years has explored the efficiency of deep learning models, which include CNNs, RNNs, and GNNs in predicting brain connectivity with great effectiveness. It is about these models that could work independently to extract features from raw data and classify and or predict neurological conditions in an enhanced way with little or no human intervention on feature creation.

Recording multiple signal modalities, including fMRI-EEG, fMRI-MEG, and EEG-MEG, using integrated multimodal methods leads to a better understanding of the connect omics field. This approach avoids drawbacks and provides deeper understanding of the neural mechanisms of different processes in the brain and various disorders. Lastly, the application of more abstract and intricate analysis of brain connection topography is quite significant in detecting neurological disorders, identifying them at their early stages, and even in developing treatment plans for each and every individual patient. Moreover, these techniques are significant in neurorehabilitation, especially in operating BCIs that helps to facilitate motor movements in locked-in patients.

References

[1] Reddy, C. Kishor Kumar, Sahithi Reddy Pullannagari, Srinath Doss, and P. R. Anisha. "Dactylology Prediction Using Convolution Neural Networks", In Fostering Cross-Industry Sustainability with Intelligent Technologies, IGI Global, 2024.

[2] G. Toholu and Ş. Yol, "A Short Tutorial: From Neuroscience to Detection of Pseudo Epileptic Seizure", IEEE, Medical Technologies Congress (TIPTEKNO), 2021.

[3] H. Chateau-Laurent and F. Alexandre, "Towards a Computational Cognitive Neuroscience Model of Creativity", IEEE 20th International Conference on Cognitive Informatics & Cognitive Computing (ICCI*CC), 2021.

[4] Neudorf, J., Kress, S. & Borowsky, R., "Comparing models of information transfer in the structural brain network and their relationship to functional connectivity: diffusion versus shortest path routing", Springer, Brain Structure and Function, 2023.

[5] Anisha, P. R., C. Kishor Kumar Reddy, Nhu Gia Nguyen, Megha Bhushan, Ashok Kumar, and Marlia Mohd Hanafiah, "Intelligent Systems and Machine Learning for Industry: Advancements, Challenges, and Practices", Taylor & Francis Group, CRC Press, 2022.

[6] W. Yu, H. Kang, G. Sun, S. Liang and J. Li, "Bio-Inspired Feature Selection in Brain Disease Detection via an Improved Sparrow Search Algorithm," IEEE Transactions on Instrumentation and Measurement, 2023.

[7] Shusha Rina, Natalia, Denis Yukhnenko, Stepan Botman, Viktor Sapunov, Vladimir Savinov, Gleb Kamyshov, Dmitry Sayapin, and Igor Voznyuk., "Modern Methods of Diagnostics and Treatment of Neurodegenerative Diseases and Depression", MDPI, Diagnostics 13, 2023.

[8] S. Dixit, A. Gaikwad, V. Vyas, M. Shindikar and K. Kamble, "United Neurological study of disorders: Alzheimer's disease, Parkinson's disease detection, Anxiety detection, and Stress detection using various Machine learning Algorithms," IEEE, International Conference on Signal and Information Processing (Icon SIP), Pune, India, 2022.

[9] A. Dev et al., "Exploration of EEG-Based Depression Biomarkers Identification Techniques and Their Applications: A Systematic Review," IEEE Access, 2022.

[10] N. P., G., Singh, B.K., "Analysis of reading-task-based brain connectivity in dyslexic children using EEG signals", Springer, Medical & Biological Engineering Computing, 2024.

[11] Baghernezhad, S., Daliri, M.,"Age-related changes in human brain functional connectivity using graph theory and machine learning techniques in resting-state fMRI data", Springer, Geoscience, 2024.

[12] Fourdain, S., Provost, S., Tremblay, J., Vannasing, P., Doussau, A., Caron-Desrochers, L., Gallagher, "Functional brain connectivity after corrective cardiac surgery for critical congenital heart disease: a preliminary near-infrared spectroscopy (NIRS) report", Taylor & Francis, Child Neuropsychology, 2023.

[13] Aafjes-van Doorn, K., Kamsteeg, C., Bate, J., Aafjes, M., "A scoping review of machine learning in psychotherapy research", Taylor & Francis, Psychotherapy Research, 2020.

[14] Sun, K., Rajabtabar, M., Samadi, S., Rezaie-Balf, M., Ghaemi, A., Band, S. S., &Mosavi, A., "An integrated machine learning, noise suppression, and population-based algorithm to improve total dissolved solids prediction", Taylor & Francis, Engineering Applications of Computational Fluid Mechanics, 2021.

[15] Suvarna Alladi,Faheem Arshad Sr.,Raghavendra Kenchaih, Bapiraju Surampudi, Avanthi Paplikar, "Neural correlates of cognitive resilience differ between experiences of bilingualism and education: A cortical surface-based morphometry study in dementia", Wiley, Alzheimer's & Dementia, 2021.

[16] R. Rajput, V. Sukumar, P. Patnaik, P. Garg and M. Ranjan, "The Cognitive Analysis for an Approach to Neuroscience", IEE, International Conference on Automation and Computation (AUTOCOM), Dehradun, India, 2024.

[17] N. Hayward et al., "A novel ultrasonic haptic device induces touch sensations with potential applications in neuroscience research", 2020 IEEE International Ultrasonics Symposium (IUS), 2020.

[18] C. W. Lerche et al., "First Performance Results of a UHF-MRI Compatible BrainPET Insert for Neuroscience", IEEE Nuclear Science Symposium, Medical Imaging Conference and International Symposium on Room-Temperature Semiconductor Detectors (NSS MIC RTSD), 2023.

[19] M., Kumar, C., Ditterline, B.L., Drazin, D., Dietz, N., "Machine Learning in Neuro-Oncology, Epilepsy, Alzheimer's Disease, and

Schizophrenia", Machine Learning in Clinical Neuroscience, Springer, 2022.

[20] Mrzygłód, B., Regulski, K., Opaliński, A., "Machine Learning Studies in Materials Science", Machine Learning in Industry, Management and Industrial Engineering, Springer, 2022.

[21] Khan, Y.F., Kaushik, B., Koundal, D., "Machine Learning Models for Alzheimer's Disease Detection Using Medical Images", Data Analysis for Neurodegenerative Disorders, Cognitive Technologies, Springer, 2023.

[22] Rajeswari, J., Jagannath, M. "Brain connectivity analysis-based classification of obstructive sleep apnea using electroencephalogram signals".*Sci Rep***14**, 5561 (2024).

12

Interpreting Electroencephalogram Brain Signals: Insights from AI-driven Analysis

Hirak Mondal[1], Anindya Nag[2], Md. Mushfiqur Rahman[3], and Anupam Kumar Bairagi[2]

[1]Department of Computer Science and Engineering,
North Western University, Bangladesh
[2]Computer Science and Engineering Discipline,
Khulna University, Bangladesh
[3]Department of Statistics, University of Dhaka, Bangladesh
E-mail: hirakmondal.cse@gmail.com, anindyanag@ieee.org,
anupam@cse.ku.ac.bd, mushfiq.dustat@gmail.com

Abstract

Advancements in healthcare technologies have recently resulted in the incorporation of sophisticated artificial intelligence (AI) models for rapid disease diagnosis. Nevertheless, the absence of transparency in these models, commonly known as "black boxes," presents difficulties in comprehending and interpreting their judgments. Explainable AI (XAI) has arisen as a remedy for this problem by providing explanations for the outputs of machine learning (ML) and the contributions of features in illness prediction models. The primary objective of this chapter is to categorize emotions based on electroencephalogram (EEG) signals by employing several ML models. We evaluate the effectiveness of logistic regression (LR), support vector machine (SVM), naive Bayes (NB), decision tree (DT), and extreme gradient boosting (XGB) algorithms on the EEG brain wave dataset. The average accuracy rates obtained are as follows: LR achieved a rate of 76%, SVM achieved a rate of 60%, NB achieved 72%, DT achieved 89%, and Extreme XGB earned 97% of emotion detection. This chapter's findings emphasize the significance of XAI in enhancing transparency and fostering confidence in AI-powered healthcare solutions. This underscores the necessity for healthcare AI models that are readily comprehensible and interpretable.

Keywords: Artificial intelligence, machine learning, explainable AI, emotion detection, EEG signals, healthcare technology.

12.1 Introduction

Emotion is a multifaceted phenomenon that impacts different facets of human behavior, such as personality, mood, cognition, and motivation [1]. It includes a broad spectrum of emotions, including fear, happiness, rage, and joy, which can have both good and negative connotations. It is essential to acknowledge and comprehend these emotions, as they play a vital role not only in one's personal welfare but also in disciplines such as computer science, artificial intelligence, and life sciences. AI has been instrumental in the progress of healthcare technologies, providing effective remedies for disease detection and therapy [2]. Nevertheless, the absence of transparency in AI models, commonly known as "black boxes," has generated apprehensions regarding their dependability and credibility. XAI has arisen as a remedy for this problem, offering elucidation on the decision-making mechanism of AI models. The identification of emotions depended on subjective techniques such as self-reporting, facial expressions, and behavioral signals. Nevertheless, these approaches frequently lack reliability due to the potential challenges individuals face in effectively articulating their feelings or deliberately distorting them [3]. As a result, academics began investigating alternate methodologies that do not depend on subjective reactions. An effective method involves utilizing physiological signs, including heart rate, blood pressure, breathing signals, and EEG signals, to deduce emotional states. EEG signals have demonstrated potential in capturing the fundamental brain activity linked to various emotions [4]. Through the analysis of these signals, researchers can acquire valuable insights into the patterns of brain activity that are associated with particular emotional states. The advancement of brain–computer interface (BCI) and EEG signals has yielded a more dependable approach to discerning human emotions [5]. EEG signals provide an automatic method for recognizing emotions and collecting authentic emotional reactions that are difficult to alter [6]. This technological progress has created new opportunities in areas such as education, entertainment, and security, and can assist persons afflicted with psychiatric problems. Utilization of XAI in the healthcare field, with a particular emphasis on the task of accurately categorizing emotions based on EEG signals [7]. Different ML methods, including LR, SVM, NB, DT, and XGB, are used to effectively classify emotions based on EEG signals. [8]. In addition, utilizing the local interpretable model-agnostic explanations (LIME) technique to clarify the reasoning behind the decisions made by these models, hence improving the transparency and comprehensibility of AI-powered healthcare solutions. The following are the main contributions of this chapter.

- This chapter offers comprehensive information on EEG for neuroscience and neuroimaging, specifically focusing on the phases of EEG analysis.
- This chapter compares the efficiency of various machine learning methods in detecting emotions from EEG data.
- This chapter offers the benefits of using XAI, such as LIME, to interpret the model's predictions locally and sheds light on the contributions of different EEG features to the emotion categorization outcomes.

The chapter is divided as follows: Section 12.2 provides a background on EEG for neuroscience and neuroimaging, while Section 12.3 covers the phases of EEG analysis. Section 12.4 provides related work, Section 12.5 provides materials and methods, Section 12.6 includes experimental results and discussion, Section 12.7 provides a comparison to the state of the art, and Section 12.8 concludes this chapter.

12.2 Background on EEG for Neuroscience and Neuroimaging

12.2.1 EEG significance and diagnosed conditions

An EEG test detects abnormalities in the brain's electrical activity. Electrodes will be strategically placed on various areas of your head during the procedure. The electrodes detect the minimal electrical charges generated by your brain activity. The EEG is employed for the identification of many neurological diseases. Epilepsy manifests as acute waveforms on the EEG when seizure activity occurs. Individuals with brain lesions, which can be caused by tumor tissues or blood clots, may have significantly reduced EEG activity, which is contingent upon the magnitude and location of the injury. This examination can be used to identify factors that contribute to decreased brain activity, such as narcolepsy, certain psychoses, Alzheimer's disease, and other conditions EEG is a useful tool for monitoring the comprehensive electrical activity of the brain, particularly in situations involving drug intoxication, trauma assessment, or the evaluation of brain injury in unconscious individuals.

The EEG is an invaluable instrument for measuring cerebral blood flow during surgical procedures. The raw EEG can be classified into five frequency ranges: Gamma (γ) with frequencies above 30 Hz, Beta (β) with frequencies between 13 and 30 Hz, Alpha (α) with frequencies between 8 and 12 Hz, Theta (θ) with frequencies between 4 and 8 Hz, and Delta (δ) with frequencies below 4 Hz [9].

12.2.2 Seizure disorders

The primary purpose of EEG is to evaluate epilepsy and other seizure disorders. Epilepsy is a neurological disorder that is distinguished by the occurrence of seizures, muscle spasms, and atypical behavior or sensations. Experts have not fully comprehended the precise ethology of epilepsy. It may involve unusual connections between neurons and chemical irregularities in the brain. The primary therapeutic approach for epilepsy is the administration of anticonvulsant medication to prevent seizures [9].

12.2.3 Sleep disorders

Narcolepsy and insomnia are two distinct instances of sleep disorders. Individuals with insomnia experience difficulties in maintaining sleep. The causes can be attributed to anxiety, trauma, specific illnesses, or certain drugs. Enhancing sleep hygiene and adopting a modified lifestyle can be beneficial [10]. Symptoms of narcolepsy include excessive daytime weariness and periods of sleep. Experts suggest that a deficiency in the brain substance hypocretin can lead to narcolepsy.

12.2.4 Brain tumors

Brain tumors can manifest in several morphologies. They can either be benign (non-cancerous) or malignant (cancerous) tumors. Headaches, convulsions, sensory alterations, impaired movement, fatigue, and changes in behavior are among the manifestations of a brain tumor. Experts state that the etiology of the majority of brain tumors is unknown. Physicians employ surgical procedures, radiation therapy, and chemotherapy as therapeutic interventions for the treatment of brain tumors [10].

12.2.5 Brain injury

Severe neurological injuries can be classified as either traumatic, produced by an external force, or acquired, often resulting from a stroke or a deficiency of oxygen in the brain. The severity of the injury will dictate the symptoms, which can range from headaches and issues with balance to comas and loss of consciousness. Rest, minimizing stress, and refraining from physical activities such as reading are fundamental components of the most essential therapies. Declining cognitive function requires extended therapy [10].

12.2.6 Dementia

Dementia is characterized by symptoms such as memory loss and cognitive impairments, including difficulties with speech and comprehension. Behavioral problems may also arise. Vascular dementia and Alzheimer's disease are among the various conditions that might lead to the development of dementia. Doctors provide medication to treat dementia to enhance cognitive function and memory. They also utilize strategies for addressing behavioral difficulties.

12.2.7 Brain infections

Brain abscesses and encephalitis, which are both caused by brain infections, are examples of illnesses characterized by inflammation of the brain. Brain infections can include headaches, seizures, alterations in mental state, and behavioral or personality problems. Usually, an infection spreads from a different part of the body to the brain [11].

12.2.8 Stroke

When the blood flow to the brain is disrupted, a stroke occurs and destroys brain cells. Signs of a stroke include abrupt muscle weakness or paralysis, numbness, drooping of the face, changes in speech, or difficulty understanding words. A blood artery experiences hemorrhaging and reaches the brain, or a blockage occurs in a blood vessel, resulting in an ischemic stroke (hemorrhagic stroke) [11].

12.2.9 Attention disorders

Attention deficit disorder (ADHD) and attention deficit hyperactivity disorder ADHD are both problems characterized by a lack of attention. They exhibit challenges in sustaining concentration, providing adequate attention, and regulating their actions. Experts attribute attention difficulties to unidentified variables. Genetic factors could potentially have a role. Medical professionals employ pharmacotherapy, behavioral intervention, and social skills training to effectively manage attention deficits. An electroencephalogram (EEG) can be conducted [12].

12.2.10 Behavior disorders

Oppositional defiant disorder (ODD) and conduct disorder (CD) are behavioral abnormalities [13]. An EEG can help pinpoint the physical etiology of various disorders. Behavioral problems hinder interpersonal connections in various settings such as business, home, family, and society. Behavior modification therapy is advantageous for those experiencing these difficulties [12].

12.3 Phases of EEG Analysis

The field of biomedical signal processing can be broadly split into the following phases, with detailed explanations provided for each step

12.3.1 Signal recording

Tracking and recording bio-signals have become essential in various domains of modern medical services in recent decades. This demonstrates that the importance of physiological control systems is recognized. Electrocardiography (ECG) is widely recognized as the primary application of biosignal monitoring. The ECG has a long and well-established history in monitoring and recording biosignals, thanks to its durability and high amplitude (about 1 mV). To conduct a comprehensive diagnostic of specific heart conditions, a cardiologist may utilize either 6 or 12 leads to record the necessary data. Currently, biosignals are regularly captured as part of diagnostic procedures in the fields of neurology/neurophysiology (such as EEG and EMG) and cardiology (such as ECG and blood pressure). Some signals, like as ECG and EEG, contain an electric component and originate from electrophysiological processes. To obtain these signals with minimal artifacts, it is necessary to employ meticulous amplification and filtering techniques [12].

- Magnetic resonance imaging (MRI) utilizes radio signals and strong magnetic forces to provide precise images of the body's internal structures.
- Magnetoencephalography (MEG) is a technique that detects and records the magnetic fields produced by the electrically charged particles in the brain, allowing for the creation of a detailed map of brain activity.

- An electroencephalogram (EEG) is a diagnostic procedure that involves the use of electrodes, which are small wire chips, linked to the head to detect and measure electrical activity in the brain.
- Specialized medical facilities collect data from human intracranial electroencephalography (iEEG) by placing or implanting electrodes directly on the brain.
- Electrocorticography (ECoG) can be performed either intraoperatively, meaning during surgery or extraoperatively, meaning outside of surgery (ECoG).
- Arterial spin labeling (ASL) is a magnetic labeling technique used during MRI imaging to evaluate blood flow to the brain non-invasively, without the need for surgery. It involves magnetically labeling the blood entering the body, allowing for a precise assessment of blood flow.

12.3.2 EEG signal recording

There are two distinct methods for capturing the EEG, depending on the placement of the reference electrode [13].

- **Bipolar montage**: The bipolar montage measures the voltage difference between two electrodes placed on an electrically active area of the scalp.
- **Monopolar montage/unipolar montage**: A monopolar montage is formed by connecting a single active electrode with one or more reference electrodes. It is possible to maintain electrical neutrality for the reference electrode, and this can be accomplished in relation to brain activity. The recorded signals serve as a representation of the difference between the reference electrodes and the active regions of the brain. Common reference points for measurement include the highest point of the nose, either the left or right mastoid, the chest, and the balanced stern vertebral lead that is not located on the head.

12.4 Related Works

Recent research has achieved significant advancements in the field of emotion identification by leveraging EEG data. In the following text, we will examine a variety of recent and significant research projects.

Ahmed et al. [14] employed EEG brain wave data for the purpose of identifying and categorizing emotions, including those associated with relaxation, neutrality, and concentration. The Muse headband dataset, which consists of data from four EEG sensors, was utilized to evaluate the performance

Table 12.1 Overview of previous research.

Author	Classifier	Dataset	Accuracy (%)
Ahmed et al. [14]	RF	Brain wave	95.07
Bird et al. [15]	RF	Brain wave	87
Liu et al. [16]	DNN	SEED	84
Kumar et al. [18]	MLP	DEAP	94

of various ML models, including NB, Bayes Net, and random forest (RF). Additionally, feature selection techniques were applied during the testing process. The RF model attained the best level of accuracy, reaching 95.07%. Bird et al. [15] found EEG features and classification algorithms that may be used to accurately recognize mental states, which is essential for human–machine interaction. Based on findings from cognitive-behavioral research, we used the Muse headband which has four EEG sensors to categorize moods as either relaxed, neutral, or focused. This dataset comprises sessions from five persons, with each session lasting one minute per mental state, for the purpose of training and testing. Using five distinct signals alpha, beta, theta, delta, and gamma this study extracted features and compared many feature selection methods and classifiers, including Bayesian networks, SVB, and RF. Using 10-fold cross-validation, these models achieved an accuracy of above 87% using just 44 out of over 2100 attributes. Liu et al. [16] investigated the correlation between different cultures and emotions by analyzing EEG and eye movement data. The study obtained an accuracy rate of 84% by employing differential entropy (DE) features and a deep neural network (DNN) classifier on the SEED dataset. Chen et al. [17] proposed that micro-states represent concise scalp potential patterns used for the spatio-temporal assessment of EEG signals. This approach employed a feature extraction method known as "k-mer," inspired by genomics. Evaluation using the DEAP dataset revealed that the combination of fine and coarse parameters significantly enhances classification performance. Kumar et al. [18] was supposed to create classification models using the SEED and DEAP datasets in order to accurately differentiate between various human emotions. The researchers employed multi-layer perceptron (MLP) and convolutional neural network (CNN) methods to analyze EEG signals. They decomposed the signals into five rhythms and calculated the DE as features. The CNN method demonstrated superior performance compared to the MLP method on the SEED dataset, earning an F1 score of 93.7%. Nevertheless, both approaches yielded comparable results for the DEAP dataset, with F1 scores of 94.5% and 94% achieved for the high vs. low arousal and high vs. low valence categories, respectively. An overview of previous research is given in Table 12.1.

12.5 Materials and Methods

The proposed methodology, seen in Figure 12.1, initiates with pre-processing procedures for the data, encompassing the elimination of noise and the completion of missing values to guarantee the integrity of the data. The dataset is subsequently divided into training and testing sets, with the training set including 80% of the data and the testing set comprising 20%. Multiple ML models, including LR, SVM, NB, DT, and XGB, are utilized. Their performance is assessed by the utilization of criteria such as accuracy, precision, recall, and F1 score, which ultimately determine the selection of the most proficient model. Ultimately, the LIME approach is employed to offer clarity and openness to the model's predictions.

Algorithm 12.1. Algorithm for our proposed model.

Input: CSV file path of the dataset

Step 1. Load the dataset from the specified file path into a DataFrame (df)

Step 2. Import necessary libraries and modules

Step 3. Map labels in the 'Label' column to numerical values using label_mapping_reverse

- *label_mapping_reverse* = {*'sad': 0, 'neutral': 1, 'happy': 2*}

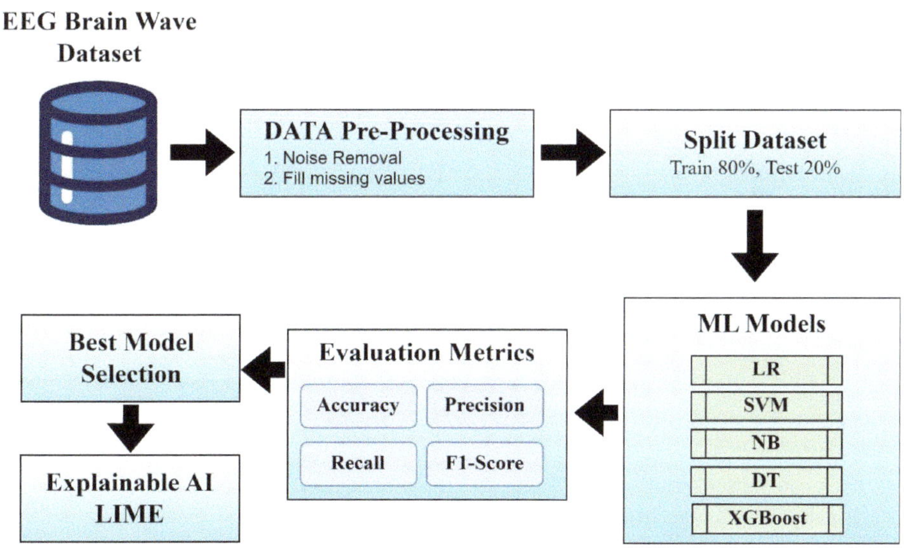

Figure 12.1 Overall workflow of the proposed emotion detection model.

- *df['Label'] = df['Label'].map(label_mapping_reverse)*

Step 4. Separate features (X) and target (y) from the DataFrame
- *X = df.drop('Label', axis=1)*
- *y = df['Label']*

Step 5. Split the data into training and testing sets
- *X_train, X_test, y_train, y_test = train_test_split(X, y, test_size=0.2, random_state=42)*

Step 6. Initialize classifiers:
- *lr = LogisticRegression()*
- *svm = SVC()*
- *nb = GaussianNB()*
- *dt = DecisionTreeClassifier()*
- *xgb_model = xgb.XGBClassifier()*

Step 7. Define a list of classifiers
- *classifiers = [('Logistic Regression', lr), ('SVM', svm), ('Naive Bayes', nb), ('Decision Tree', dt), ('XGBoost', xgb_model)]*

Step 8. Train and evaluate each classifier:
For each (clf_name, clf) in classifiers:
a. Fit the classifier on the training data
- *clf.fit(X_train, y_train)*
b. Predict the target values for the test data
- *y_pred = clf.predict(X_test)*
c. Calculate accuracy of the predictions
- *accuracy = accuracy_score(y_test, y_pred)*
d. Print accuracy
- *Print(f"Accuracy for {clf_name}: {accuracy}")*
e. Print classification report
- *Print(f"Classification Report for {clf_name}:")*
- *Print(classification_report(y_test, y_pred))*
f. Print confusion matrix
- *Print(f"Confusion Matrix for {clf_name}:")*
- *cm = confusion_matrix(y_test, y_pred)*
- *Print(cm)*
g. Plot confusion matrix
- Plot the confusion matrix using seaborn heatmap

Step 9. Initialize a LIME explainer with the training data

- *explainer=lime_tabular.LimeTabularExplainer(X_train.values,mode ='classification',feature_names=X_train.columns,class_names=y_train.unique(),discretize_continuous=True)*

Step 10. Select a random sample from the test data
- *sample_idx = np.random.randint(0, len(X_test))*
- *sample = X_test.iloc[sample_idx]*

Step 11. Explain the instance using LIME and the XGBoost model
- *explanation=explainer.explain_instance(sample.values,xgb_model.predict_proba)*

Step 12. Display the explanation in a notebook
- *explanation.show_in_notebook()*

Output: Accuracy, classification report, confusion matrix for each classifier, and LIME explanation for a sample instance

12.5.1 Data description

The dataset used in this study was sourced from Kaggle [19]. Data collection involved four participants, consisting of two males and two females. Each participant's EEG data was recorded for 60 seconds across three distinct states: 0 ('sad'), 1 ('neutral'), and 2 ('happy'). The recordings were made using a Muse EEG headband, which utilized dry electrodes to capture signals from the TP9, AF7, AF8, and TP10 EEG placements. The distribution of the data among three distinct states is seen in Figure 12.2.

12.5.2 EEG signal pre-processing

The pre-processing of EEG signals is a critical step to ensure the accuracy and reliability of subsequent analysis. Initially, the raw EEG signals were subjected to noise removal techniques to eliminate artifacts and unwanted interference [20]. To reduce common sources of noise, such as power line interference, muscle activity, and eye movements, band-pass filters and artifact rejection algorithms were employed. Following noise removal, the data underwent a thorough inspection for missing values. Missing data points were identified and handled through interpolation methods to maintain the integrity of the continuous signal. This pre-processing pipeline ensured that the EEG signals were clean and complete, providing a robust foundation for further analysis and classification tasks.

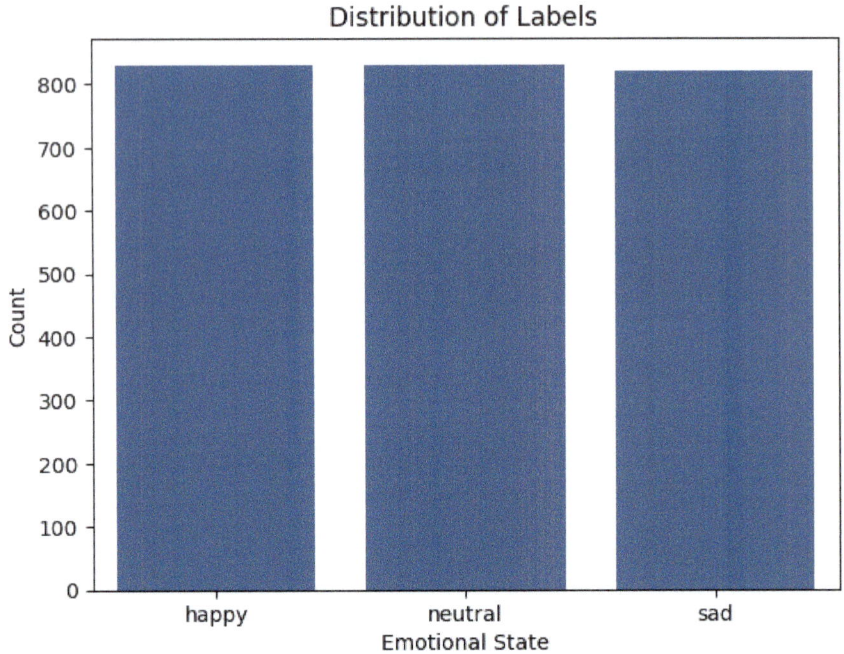

Figure 12.2 Distribution of data between three different states: Sad, neutral, and happy.

12.5.3 Machine learning classifiers

We used a thorough examination of several ML techniques to automatically diagnose emotional states. Our focus was on categorizing emotions into three distinct categories: 0 ('sad'), 1 ('neutral'), and 2 ('happy'). To do this, we divided our EEG feature data into two distinct subsets: 80% of the data was allocated to the training dataset, while the remaining 20% was assigned to the test dataset. Subsequently, we used five distinct machine learning techniques on this dataset: LR, SVM, NB, DT, and XGB. By using this method, we were able to assess and contrast the efficacy of various algorithms in precisely detecting emotional states using EEG data.

12.5.3.1 Logistic regression (LR)

LR is a simple and very efficient method for tackling problems related to binary and linear classification. The logistic regression model is a statistical approach used for binary classification, which may also be expanded to handle multiclass classification. It has similarities with the Adaline and perceptron models [20].

12.5.3.2 Support vector machine (SVM)

SVMs are robust supervised learning models used for classification and regression. The hyperplane is used to maximize the margin between classes, making SVM efficient for datasets with more dimensions than samples [21].

12.5.3.3 Naive Bayes (NB)

NB algorithm is a powerful classification approach that uses Bayes' theorem and assumes independent attributes. The phrase "naive" assumes that a class characteristic is independent of other features, which is seldom true in real-world data [22].

12.5.3.4 Decision tree (DT)

The commonly used and simply understood DT models may be utilized for classification and regression. Based on input qualities, a decision tree divides data into subgroups. DTs are readily interpretable and graphically representable, making them useful for understanding the model's decision-making process [22].

12.5.3.5 Extreme gradient boosting (XGB)

XGB is a more advanced gradient-building method. It's utilized for classification and regression. Gradient boosting sequentially teaches weak learners (typically decision trees) to correct one other's errors [23].

12.5.4 Evaluation metrics

To assess the performance of the ML algorithms in diagnosing emotional states, we employed several evaluation metrics: accuracy, precision, recall, and F1 score [30], [31]. These metrics provide a comprehensive evaluation of the model's predictive capabilities. Where TP is a true positive, TN is a true negative, FP is a false positive, and FN is a false negative [23].

12.5.4.1 Confusion matrix

A confusion matrix is necessary for classifier evaluation. It helps understand the model's particular types and levels of errors [23].

12.5.4.2 Accuracy

Accuracy is the percentage of occurrences categorized properly. It is defined as:

$$\text{Accuracy} = \frac{\text{TP} + \text{TN}}{\text{TP} + \text{TN} + \text{FP} + \text{FN}}. \tag{12.1}$$

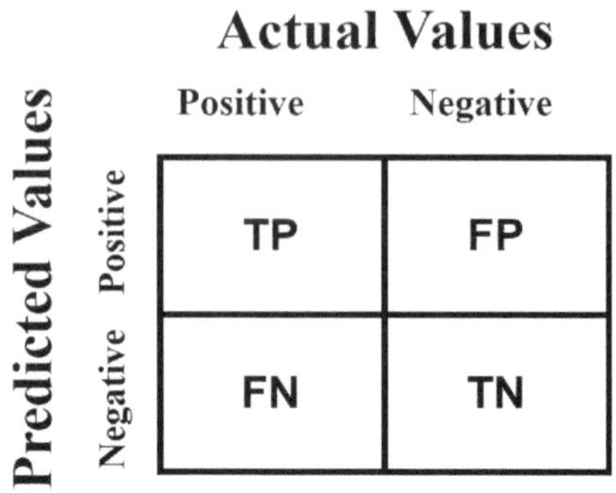

Figure 12.3 The typical structure of a confusion matrix.

12.5.4.3 Precision

Precision measures the proportion of accurate positive predictions to all positive forecasts [24]. The equation for precision is:

$$\text{Precision} = \frac{\text{TP}}{\text{TP} + \text{FP}}. \tag{12.2}$$

12.5.4.4 Recall

Recall, often called sensitivity or true positive rate, is the ratio of accurate positive forecasts to positive instances [25], [26]. The equation for the recall is:

$$\text{Recall} = \frac{\text{TP}}{\text{TP} + \text{FN}}. \tag{12.3}$$

12.5.4.5 F1 score

We can reduce feature space and improve model training speed and accuracy by identifying the primary prediction factors [27], [28]. The equation for the F1 score is:

$$\text{F1score} = 2 \times \frac{\text{Precision} \times \text{Recall}}{\text{Precision} + \text{Recall}}. \tag{12.4}$$

These evaluation metrics allowed us to rigorously compare the performance of the five ML algorithms (LR, SVM, NB, DT, and XGB) in diagnosing emotional states from EEG signals.

12.5.5 Local interpretable model-agnostic explanations (LIME)

LIME facilitates comprehension of the decision-making process of intricate models by locally approximating them with interpretable models. LIME facilitated the identification of the specific aspects of the EEG signals that had the greatest impact on the categorization of emotional states by providing reasons for individual predictions [29]. Transparency is essential for verifying the dependability of the models and understanding the fundamental patterns in the data, which eventually enhances the credibility and comprehensibility of automated emotion detection [23].

12.6 Experimental Result and Discussion

In this chapter, we employed five distinct ML models: LR, SVM, NB, DT, and XGB. A comparison of the performance of the models is provided in Table 12.2 XGB achieves a higher accuracy of 97%, while DT achieves a second height accuracy of 89%, and SVM has the lowest accuracy of 60%. Figure 12.4 depicts the confusion matrix for all applied ML models.

The dataset has three discrete classes: 'sad', 'neutral', and 'happy'. To assess the performance of the model, we calculate the accuracy, precision, recall, and F1 score for each class. Class-wise analyses of model performance are presented in Table 12.3.

An XAI framework, such as LIME, is necessary for the explanation of black box models, including the XGB ML algorithms. The model's prediction process may be examined within this framework, which also explains how specific features impact those predictions. Verifying the model's predictions for every data instance across all classes is essential for establishing trust in the ML model's ability to recognize brainwave-feeling emotions. We can decrease the feature space and enhance model training speed and accuracy by identifying the main elements that influence prediction results. This chapter aimed to build clinical confidence in emotion prediction using ML algorithms. It used the LIME model and the XGB classifier to examine the individual contribution of EEG data to identifying brainwave-feeling emotions and prediction accuracy. Figure 12.5 shows the difference between the initial XGB model's forecast of 0.97 and the current prediction level of 1.00.

Table 12.2 Comparison of model performance.

ML model	Accuracy (%)	Precision (%)	Recall (%)	F1 score (%)
LR	76	93	93	93
SVM	60	90	98	84
NB	72	92	92	92
DT	89	91	93	92
XGB	**97**	**100**	**100**	**98**

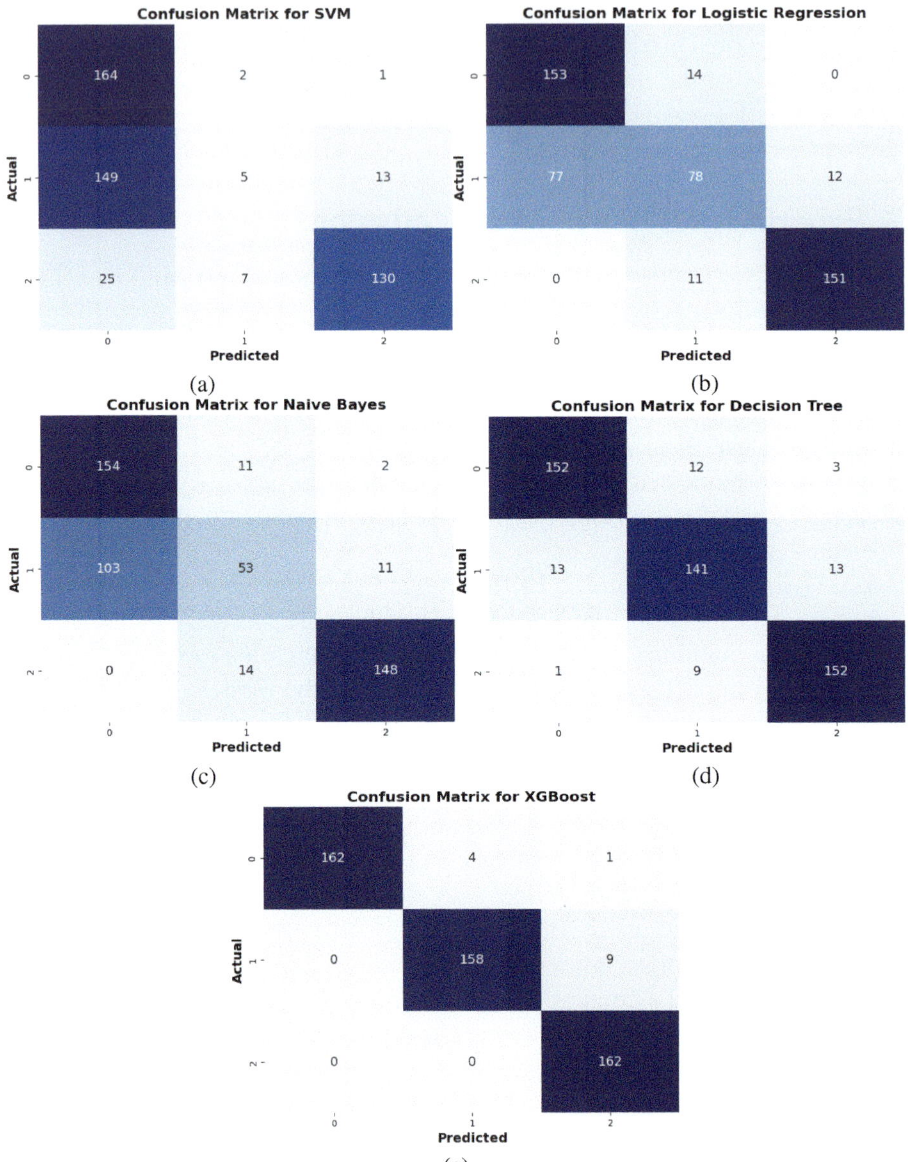

Figure 12.4 Confusion matrix of all applied ML models: (a) LR, (b) SVM, (c) NB, (d) DT, (e) XGB.

Table 12.3 Class-wise model performance.

ML model	Class	Precision (%)	Recall (%)	F1 score (%)
LR	Sad	65	93	77
	Neutral	77	44	56
	Happy	93	93	93
SVM	Sad	49	98	65
	Neutral	36	03	06
	Happy	90	80	84
NB	Sad	60	92	73
	Neutral	68	32	43
	Happy	92	91	92
DT	Sad	89	89	89
	Neutral	86	85	86
	Happy	91	93	92
XGB	Sad	**100**	**97**	**98**
	Neutral	**98**	**95**	**96**
	Happy	**94**	**100**	**97**

Figure 12.5 Explanation of XGB generated by LIME.

12.7 Comparison with State-of-the-Art Techniques

XGB model achieved an accuracy of 97% on the brain wave dataset, surpassing the performance of earlier models like RF with 95.07% and MLP with 94%. This indicates a significant enhancement in the ability to anticipate emotions from EEG data. A comparison of the performance of the model to that of past research is presented in Table 12.4 and can be found here.

Table 12.4 Comparison of the performance of the model to that of earlier research.

Author	Classifier	Class-wise analysis	Model exploitability	Accuracy (%)
Ahmed et al. [14]	RF	No	No	95.07
Bird et al. [15]	RF	No	No	87
Liu et al. [16]	DNN	No	No	84
Kumar et al. [18]	MLP	No	No	94
Proposed	**XGB**	**Yes**	**Yes (LIME)**	**97**

The earlier research we compared would not employ XAI to increase model exploitation for emotion identification from EEG data and would not analyze class-wise accuracy. However, we identify a class-wise accuracy and increase model exploitation for emotion recognition from EEG data using XAI. The use of XAI specifically LIME, has proven crucial in providing transparency and understanding of model decisions, which is essential for building trust in AI-driven healthcare solutions. While our findings are promising, further research is needed to explore the scalability of these models in real-world applications and to refine the interpretability of AI decisions. In the future, we should focus on expanding the dataset to include more diverse emotional states and populations, integrating multimodal data sources for more robust emotion recognition, and enhancing the user interface for practical deployment in healthcare settings. Additionally, investigating the application of other advanced AI techniques and their explainability could further improve the accuracy and reliability of emotion recognition systems.

12.8 Conclusion

We explored the classification of emotions from EEG signals using various ML models. The XGB model achieved the highest accuracy at 97%, demonstrating its effectiveness in emotion detection. To address the opacity of AI models, we employed XAI techniques, specifically the LIME method, to provide transparency and interpretability of the model decisions. This approach is crucial for enhancing trust and usability in AI-driven healthcare solutions. The integration of physiological signals like EEG in emotion recognition offers a more objective and reliable method compared to traditional subjective approaches. Our findings underscore the potential of EEG-based emotion recognition systems to revolutionize fields such as mental health monitoring, human–computer interaction, and education. The application of XAI further ensures that these systems are transparent and interpretable, fostering greater trust in their use. This chapter marks a significant step towards developing reliable, interpretable, and practical emotion recognition technologies that can be widely adopted in various real-world scenarios.

References

[1] Dhingra, R. C., & Jaswal, S. R. A. (2021). Emotion recognition based on eeg using deap dataset. European Journal of Molecular & Clinical Medicine, 8(3), 3509-3517.

[2] Islam, M. S., Hussain, I., Rahman, M. M., Park, S. J., & Hossain, M. A. (2022). Explainable artificial intelligence model for stroke prediction using EEG signal. Sensors, 22(24), 9859.

[3] Cheng, X. M., Cheng, P. Y., & Zhao, L. (2009, April). A study on emotional feature analysis and recognition in speech signal. In 2009 International Conference on Measuring Technology and Mechatronics Automation (Vol. 1, pp. 418-420). IEEE.

[4] Park, S. J., Hong, S., Kim, D., Yu, J. H., Hussain, I., Park, H. G., & Benjamin, H. C. M. (2019). Development of intelligent stroke monitoring system for the elderly during sleeping. Sleep Medicine, 64, S294.

[5] Hussain, I., & Park, S. J. (2021). Quantitative evaluation of task-induced neurological outcome after stroke. Brain sciences, 11(7), 900.

[6] Aminov, A., Rogers, J. M., Johnstone, S. J., Middleton, S., & Wilson, P. H. (2017). Acute single channel EEG predictors of cognitive function after stroke. PloS one, 12(10), e0185841.

[7] Hussain, I., Young, S., Kim, C. H., Benjamin, H. C. M., & Park, S. J. (2021). Quantifying physiological biomarkers of a microwave brain stimulation device. Sensors, 21(5), 1896.

[8] S. A. Hussain and A. S. A. A. Balushi, "A real time face emotion classificationand recognition using deep learning model,"Journal of Physics: ConferenceSeries, vol. 1432, p. 012 087, Jan. 2020. doi: 10. 1088 / 1742 - 6596 / 1432 / 1 /012087.

[9] Shoeibi, A., Ghassemi, N., Khodatars, M., Moridian, P., Alizadehsani, R., Zare, A., ... & Gorriz, J. M. (2022). Detection of epileptic seizures on EEG signals using ANFIS classifier, autoencoders and fuzzy entropies. Biomedical Signal Processing and Control, 73, 103417.

[10] Sharma, P. C., Raja, R., Vishwakarma, S. K., Sharma, S., Mishra, P. K., & Kushwah, V. S. (2022). Analysis of brain signal processing and real-time EEG signal enhancement. Multimedia Tools and Applications, 81(28), 41013-41033.

[11] Raghavendra, U., Gudigar, A., Paul, A., Goutham, T. S., Inamdar, M. A., Hegde, A., ... & Acharya, U. R. (2023). Brain tumor detection and screening using artificial intelligence techniques: Current trends and future perspectives. Computers in Biology and Medicine, 107063.

[12] Galgano, M., Toshkezi, G., Qiu, X., Russell, T., Chin, L., & Zhao, L. R. (2017). Traumatic brain injury: current treatment strategies and future endeavors. Cell transplantation, 26(7), 1118-1130.

[13] Khare, S. K., & Acharya, U. R. (2023). Adazd-Net: Automated adaptive and explainable Alzheimer's disease detection system using EEG signals. Knowledge-Based Systems, 278, 110858.

[14] Ahmed, Z., & Shahid, S. (2022, December). Improving emotion detection through artificial intelligence from EEG brainwave signals. In 2022 International Conference on Artificial Intelligence of Things (ICAIoT) (pp. 1-6). IEEE.

[15] Bird, J. J., Manso, L. J., Ribeiro, E. P., Ekart, A., & Faria, D. R. (2018, September). A study on mental state classification using eeg-based brain-machine interface. In 2018 international conference on intelligent systems (IS) (pp. 795-800). IEEE.

[16] Liu, W., Zheng, W. L., Li, Z., Wu, S. Y., Gan, L., & Lu, B. L. (2022). Identifying similarities and differences in emotion recognition with EEG and eye movements among Chinese, German, and French People. Journal of Neural Engineering, 19(2), 026012.

[17] Chen, J., Zhao, Z., Shu, Q., & Cai, G. (2022). Feature extraction based on microstate sequences for EEG–based emotion recognition. Frontiers in Psychology, 13, 1065196.

[18] Kumar, M., & Molinas, M. (2022, December). Human emotion recognition from EEG signals: model evaluation in DEAP and SEED datasets. In Proceedings of the first workshop on artificial intelligence for human-machine interaction (AIxHMI 2022) co-located with the 21th international conference of the Italian association for artificial intelligence (AI* IA 2022), CEUR workshop proceedings, CEUR-WS. org.

[19] Bird, J. J., Manso, L. J., Ribeiro, E. P., Ekart, A., & Faria, D. R. (2018, September). A study on mental state classification using eeg-based brain-machine interface. In 2018 international conference on intelligent systems (IS) (pp. 795-800). IEEE.

[20] Khare, S. K., Gaikwad, N. B., & Bajaj, V. (2022). VHERS: a novel variational mode decomposition and Hilbert transform-based EEG rhythm separation for automatic ADHD detection. IEEE Transactions on Instrumentation and Measurement, 71, 1-10.

[21] Khare, S. K., Gadre, V. M., & Acharya, U. R. (2023). ECGPsychNet: an optimized hybrid ensemble model for automatic detection of psychiatric disorders using ECG signals. Physiological Measurement, 44(11), 115004.

[22] Nag, A., Mondal, H., Kabir, S. R. R., Islam, M. R., Ahmed, M. E., & Jamil, S. H. (2024, May). Emotion Decoding: An Extensive Examination of Electroencephalogram Signals Using Explainable Machine

Learning. In 2024 6th International Conference on Electrical Engineering and Information & Communication Technology (ICEEICT) (pp. 604-609). IEEE.

[23] Penzel, T., Kesper, K., & Becker, H. F. (2006). Biosignal monitoring and recording. Information Technology Solutions for Healthcare, 288-301.

[24] Rasheed, K., Qayyum, A., Qadir, J., Sivathamboo, S., Kwan, P., Kuhlmann, L., ... & Razi, A. (2020). Machine learning for predicting epileptic seizures using EEG signals: A review. IEEE reviews in biomedical engineering, 14, 139-155.

[25] Nag, A., Das, A., Sil, R., Kar, A., Mandal, D., & Das, B. (2022, December). Application of artificial intelligence in mental health. In International Conference on Intelligent Systems Design and Applications (pp. 128-141). Cham: Springer Nature Switzerland.

[26] Hassan, M. M., Rony, M. A. T., Khan, M. A. R., Hassan, M. M., Yasmin, F., Nag, A., ... & El-Shafai, W. (2023). Machine learning-based rainfall prediction: Unveiling insights and forecasting for improved preparedness. IEEE Access, 11, 132196-132222.

[27] Nag, A., Das, B., Sil, R., Hameed, A. A., Bhushan, B., & Jamil, A. (2023, July). A Survey on Image-Based Cardiac Diagnosis Prediction Using Machine Learning and Deep Learning Techniques. In International Conference on Computing, Intelligence and Data Analytics (pp. 478-491). Cham: Springer Nature Switzerland.

[28] Low, H. Q., Keikhosrokiani, P., & Pourya Asl, M. (2024). Decoding violence against women: analysing harassment in middle eastern literature with machine learning and sentiment analysis. Humanities and Social Sciences Communications, 11(1), 1-18.

[29] Liu, H., Lou, T., Zhang, Y., Wu, Y., Xiao, Y., Jensen, C. S., & Zhang, D. (2024). EEG-based multimodal emotion recognition: a machine learning perspective. IEEE Transactions on Instrumentation and Measurement.

[30] Goshvarpour, A. (2024). Emerging Trends of Biomedical Signal Processing in Intelligent Emotion Recognition. Brain Sciences, 14(7), 628.

[31] Li, Z. L., Cao, H., & Zhang, J. S. (2024). Emotion Recognition in EEG Based on Multilevel Multidomain Feature Fusion. IEEE Access.

13

Neuroinformatics in the Era of Personalized Neuroscience

Farhana Yasmin[1], Mahade Hasan[2], Md. Mehedi Hassan[3], and Xue Yu[2]

[1]School of Computer and Software, Nanjing University of Information Science and Technology, China
[2]School of Software, Nanjing University of Information Science and Technology, China
[3]Computer Science and Engineering Discipline, Khulna University, Bangladesh
E-mail: farhana005.cse@gmail.com, farabi.eu.bd@gmail.com, mehedihassan@ieee.org, xueyu@nuist.edu.cn

Abstract

Integrating neuroinformatics with personalized neuroscience is crucial for advancing our understanding and treatment of brain functions and disorders. This chapter explores the definitions and scopes of both fields and underscores the importance of their convergence. It provides a historical perspective on the evolution of neuroinformatics and highlights key milestones that have paved the way for personalized approaches in neuroscience. The chapter delves into essential technologies and tools in neuroinformatics, including neuroimaging, data acquisition, preprocessing methods, databases, computational modeling, and the use of ML and AI. It also examines personalized neuroscience concepts, such as genetic and epigenetic influences, brain connectivity, cognitive and behavioral phenotyping, and treatment strategies. The discussion includes the role of big data, multi-modal analysis, and predictive modeling in developing precise diagnostic tools and interventions. Challenges like data heterogeneity, ethical considerations, and technological barriers are addressed, alongside opportunities for innovation and interdisciplinary collaboration. Future directions focus on emerging trends, advancements, and the impact of AI and ML on clinical practice. The conclusion emphasizes the importance of ongoing integration and collaboration in advancing neuroinformatics and personalized neuroscience.

Keywords: Neuroinformatics, personalized neuroscience, neuroimaging, predictive modeling, precision diagnostics.

13.1 Introduction

Neuroinformatics is a specialized field that merges neuroscience with information technology, focusing on the organization and analysis of complex neuroscience data. It involves developing computational models, creating databases, and applying data mining and machine learning techniques to manage vast volumes of data from neuroimaging, electrophysiological recordings, genomics, and behavioral studies [1]. Personalized neuroscience, or precision neuroscience, modifies interventions based on individual genetic, environmental, and lifestyle factors, moving beyond the traditional approach. Integrating neuroinformatics with personalized neuroscience is essential for understanding brain function at an individual level, enhancing diagnostic accuracy, and improving treatment efficacy [2]. This integration fosters innovative research, promotes data sharing, and accelerates advancements in personalized therapeutic interventions.

13.1.1 Scope of neuroinformatics

Neuroinformatics is a specialized field at the intersection of neuroscience and information technology, dedicated to the organization and analysis of diverse and complex neuroscience data. It encompasses a range of activities, including developing computational models and analytical tools, creating and maintaining databases, and applying data mining and machine learning techniques [3]. Neuroinformatics aims to facilitate the understanding of the nervous system by managing the large volumes of data generated by modern neuroscience research, which includes neuroimaging data, electrophysiological recordings, genomic data, and behavioral studies [4]. Neuroinformatics provides a cohesive framework that supports hypothesis generation, data sharing, and collaborative research by integrating data from multiple sources and formats.

13.1.2 Overview of personalized neuroscience

Personalized neuroscience, also known as precision neuroscience, is an emerging approach that interventions and treatments based on individual variability in genes, environment, and lifestyle. This approach recognizes that each person's brain is unique, and shaped by a complex interplay of genetic, epigenetic, and environmental factors [5]. Personalized neuroscience leverages advancements in genomics, neuroimaging, and computational biology to develop precise diagnostic tools and personalized treatment strategies. This field aims to move away from the traditional "one-size-fits-all" approach in neuroscience and psychiatry, towards more customized interventions that can improve outcomes in mental health, neurodegenerative diseases, and neurological disorders.

13.1.3 Importance of integrating neuroinformatics with personalized neuroscience

The integration of neuroinformatics with personalized neuroscience is crucial for advancing our understanding and treatment of brain function and dysfunction. Neuroinformatics provides the necessary tools and methodologies [6] to integrate and analyze diverse datasets, such as genetic information, brain imaging, and clinical data. This comprehensive data integration is essential for unraveling the complex and multifaceted nature of brain function on an individual level. By leveraging neuroinformatics, researchers can develop highly precise models of brain function that account for individual variability, thereby enhancing the accuracy of diagnostic tools and the efficacy of personalized treatment plans. Moreover, neuroinformatics enhances the scalability and efficiency of managing large-scale datasets, enabling researchers to conduct extensive studies [7] that identify subtle patterns and correlations critical for personalized neuroscience. The field also promotes data sharing and collaborative research through shared databases and standardized data formats, ensuring that researchers worldwide can contribute to and benefit from collective knowledge. This collaborative approach accelerates advancements in personalized neuroscience. Furthermore, the integration of neuroinformatics with personalized neuroscience fosters innovative research and the development of novel therapeutic interventions [8]. By understanding the unique aspects of each individual's brain, researchers can design targeted therapies that are more effective and have fewer side effects, paving the way for personalized treatment strategies that significantly improve patient outcomes.

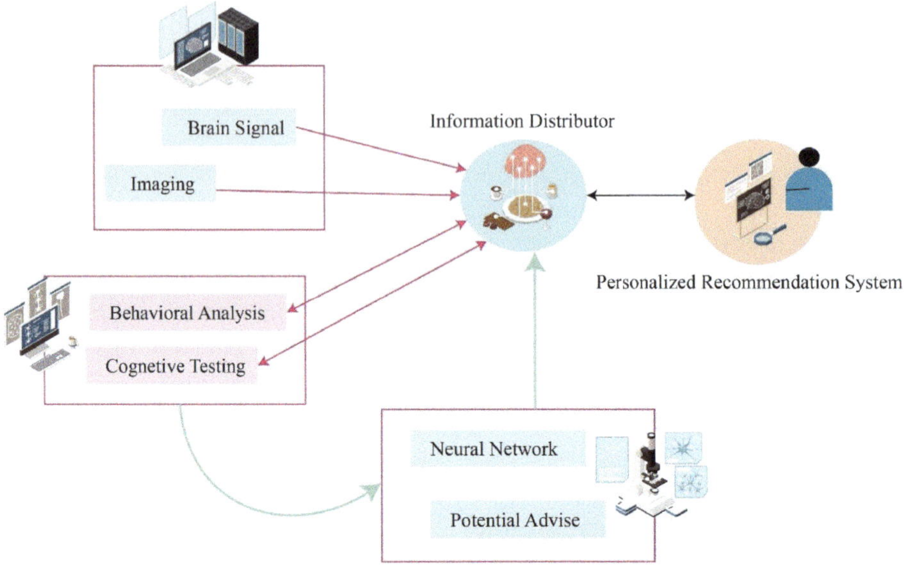

Figure 13.1 The fusion of neuroinformatics and personalized neuroscience.

As illustrated in Figure 13.1, the fusion of neuroinformatics and personalized neuroscience holds the promise of revolutionizing our understanding of the brain and transforming the landscape of neurological and psychiatric care. This integrated approach is essential for the development of next-generation diagnostic tools and treatments that are personalized to the individual, ultimately leading to better health outcomes and a deeper understanding of the human brain.

13.2 Evolution of Neuroinformatics

The field of neuroinformatics emerged from the convergence of computational technology and neuroscience. Initially, the digitization and analysis of neural data were made possible by the advent of computers, enabling the development of computational models like the Hodgkin-Huxley model [9]. This period marked the first attempts to simulate neural activity and manage the data produced by neurophysiological experiments. As neuroimaging techniques such as magnetic resonance imaging (MRI) [10] and positron emission tomography (PET) [11] developed, researchers were able to collect large volumes of brain data, necessitating the creation of databases for storage and management. This led to the establishment of the first neuroinformatics

Table 13.1 Summarized evolution of neuroinformatics.

Period	Achievements	Description
1985–1989	Foundational computational models	Development of initial computational models to simulate neural networks and basic brain functions.
1990–1994	Emergence of neuroinformatics	Formal recognition of neuroinformatics as a distinct field, focusing on data management and analysis.
1995–1999	Establishment of early databases	Creation of initial neuroscience databases and repositories, setting the groundwork for data sharing.
2000–2004	Advances in neuroimaging technologies	Introduction and widespread adoption of advanced neuroimaging techniques such as fMRI and DTI.
2005–2009	Integration of genomics and neuroimaging	Combining genetic data with neuroimaging to enhance understanding of brain function and disorders.
2010–2014	Rise of big data and machine learning	Utilization of big data analytics and machine learning techniques to manage and interpret complex datasets.
2015–2019	Development of comprehensive brain atlases	Launch of projects like the Human Connectome Project, providing detailed maps of brain connectivity.
2020–2023	Precision neuroscience and personalized approaches	Emphasis on personalized approaches in neuroscience, integrating multi-modal data for individualized analysis.
Present	AI and advanced computational models	Implementation of AI and sophisticated computational models for predictive diagnostics and therapeutic interventions.

databases and the development of tools to handle and analyze this new wealth of data. Table 13.1 summarizes the key achievements in the evolution of neuroinformatics in past periods to present, highlighting the significant developments and advancements that have contributed to the field.

Subsequent decades saw a growing emphasis on mapping the brain's structure and function, exemplified by initiatives aimed at comprehensive brain mapping [12]. The rise of the internet facilitated data sharing and collaboration, leading to the creation of extensive databases like the BrainMap and the Allen Brain Atlas. This period also saw the integration of genomics with neuroscience, giving rise to neurogenomics, and the establishment of organizations to promote global collaboration and standardization in neuroinformatics. In recent years, the era of big data and cloud computing has further expanded the capabilities of neuroinformatics. Major initiatives such as the BRAIN Initiative and the Human Connectome Project have generated massive datasets [13], requiring advanced computational methods for analysis. Neuroinformatics now includes a broad range of activities, from data integration and machine learning [14] to the development of sophisticated computational models of brain function.

13.2.1 Advances in neuroscience leading to personalization

Several key advances in neuroscience have paved the way for the emergence of personalized neuroscience, transforming how we understand and treat brain disorders. The sequencing of the human genome and discoveries in epigenetics have been particularly influential. These breakthroughs have highlighted the critical role that genetic and epigenetic factors play in brain function and disease, providing a foundation for personalized diagnostic and treatment approaches to individual genetic profiles [15]. Innovations in neuroimaging technologies, such as functional MRI (fMRI), diffusion tensor imaging (DTI), and electroencephalography (EEG), have also been essential [16]. These advanced imaging techniques offer detailed maps of brain structure and function, enabling the identification of individual variations in brain connectivity and activity. This level of detailed, personalized mapping is essential for developing customized approaches in neuroscience. Advances in computational neuroscience have further deepened our understanding of neural dynamics and brain function. Computational models and simulations can now be implemented to individual data, allowing for personalized predictions and interventions that are more accurate and effective. This capability is crucial for developing individualized treatment plans and understanding the unique aspects of each person's brain function.

The rise of big data and machine learning has revolutionized neuroscience research by enabling the analysis of large, complex datasets. Machine learning algorithms can detect patterns and correlations that are not apparent through traditional analysis methods [17], which is vital for developing personalized diagnostic tools and treatment plans. These technologies enhance the ability to process and interpret vast amounts of data quickly and accurately, driving forward the field of personalized neuroscience. The broader field of precision medicine has significantly influenced neuroscience by promoting medical treatment to individual characteristics. Precision medicine approaches in neurology and psychiatry now emphasize the development of treatments based on an individual's unique biological, psychological, and environmental factors [18]. This shift towards personalized treatment

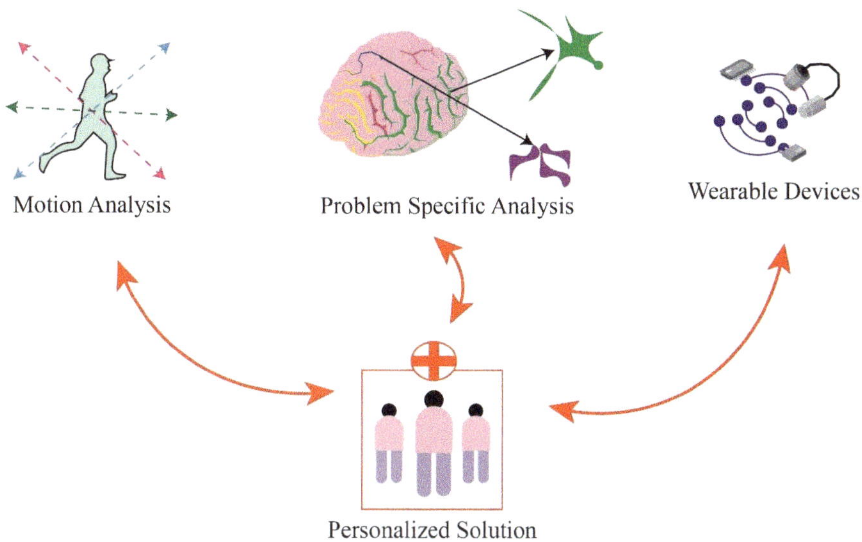

Figure 13.2 Advancements in neuroscience personalization.

strategies aims to improve the efficacy and outcomes of medical interventions, reflecting a more holistic and individualized approach to healthcare in neuroscience.

As shown in Figure 13.2, the integration of genomics, neuroimaging technologies, computational neuroscience, and machine learning have revolutionized neuroscience, enabling personalized diagnostic and treatment approaches based on individual genetic profiles, brain connectivity, and neural dynamics. Advances in big data analytics and precision medicine have further facilitated the development of interventions in neurology and psychiatry. These innovations underscore the critical role of personalized neuroscience in understanding and treating brain disorders on an individual basis.

13.2.2 Milestones in the integration of neuroinformatics and personalized neuroscience

Several significant milestones that have collectively propelled the field forward have marked the integration of neuroinformatics and personalized neuroscience. Table 13.2 outlines the significant innovations that have collectively propelled the field forward in the integration of neuroinformatics and personalized neuroscience. The completion of the Human Genome

Project provided critical insights into the genetic basis of brain function and dysfunction [19], laying the groundwork for personalized approaches in neuroscience. The launch of comprehensive brain atlases, such as the Allen Brain Atlas [20], set new standards for integrating genetic and anatomical data, offering detailed maps of gene expression that serve as essential models for human brain research. Major brain research [21] projects have generated vast amounts of data, necessitating the development of advanced neuroinformatics tools for their analysis and integration. Among these, the Human Connectome Project [22] stands out for its efforts to map the neural pathways underlying human brain function and behavior, producing datasets crucial for personalized models of brain connectivity. Precision medicine initiatives have underscored the importance of individualized treatment approaches, significantly influencing the strategies adopted in neuroscience and neuroinformatics [23].

Furthermore, advances in AI and machine learning have revolutionized the analysis of neural data [24]. These technologies enable more sophisticated

Table 13.2 Summarized findings in the integration of neuroinformatics and personalized neuroscience.

References	Findings	Description
Peter et al. [19]	Details of the Human Genome Project	Provided critical insights into the genetic basis of brain function and dysfunction, laying the groundwork for personalized approaches in neuroscience.
Gomes et al. [20]	Study of comprehensive brain Atlases	Allen Brain Atlas set new standards for integrating genetic and anatomical data, offering detailed maps of gene expression as essential models for human brain research.
Zhou et al. [21]	Compilation of major brain research projects	Generated vast amounts of data, necessitating the development of advanced neuroinformatics tools for analysis and integration.
Kruper et al. [22]	Study on Human Connectome Project	Aimed to map the neural pathways underlying human brain function and behavior, producing crucial datasets for personalized models of brain connectivity.
Galasso [23]	Highlights of precision medicine initiatives	Highlighted the importance of individualized treatment approaches, significantly influencing the strategies adopted in neuroscience and neuroinformatics.
Chen and Yadollahpour [24]	Advances in AI and machine learning	Revolutionized the analysis of neural data, enabling more sophisticated and comprehensive analyses, facilitating the development of personalized diagnostic and therapeutic tools.

and comprehensive analyses, facilitating the development of personalized diagnostic and therapeutic tools. Collectively, these milestones highlight the remarkable progress in integrating neuroinformatics with personalized neuroscience, connecting the way for individualized understanding and treatment of brain disorders.

13.3 Key Technologies and Tools in Neuroinformatics

Key technologies and tools in neuroinformatics are crucial for advancing our understanding of brain function and developing personalized treatments. These include advanced neuroimaging techniques, sophisticated data acquisition and preprocessing methods, comprehensive databases and repositories, computational modeling tools, and machine learning algorithms [25]. Together, they enable the collection, integration, and analysis of complex brain data, facilitating breakthroughs in personalized neuroscience as described in the following sections.

13.3.1 Neuroimaging and brain mapping techniques

Neuroimaging and brain mapping techniques are critical tools in neuroinformatics, providing detailed insights into the structure and function of the brain. Techniques such as magnetic resonance imaging (MRI) and its variant functional MRI (fMRI) allow for the detailed visualization of brain anatomy and activity, respectively, by utilizing magnetic fields and radio waves.

positron emission tomography (PET) imaging uses radioactive tracers to visualize metabolic processes, offering valuable data on brain metabolism and neurotransmitter activity. Electroencephalography (EEG) records electrical activity through scalp electrodes, providing high temporal resolution data on neural dynamics. Diffusion tensor imaging (DTI) maps the diffusion of water molecules in brain tissue, particularly useful for imaging white matter tracts, while magnetoencephalography (MEG) measures magnetic fields produced by neuronal activity, capturing real-time brain function and connectivity [26]. Figure 13.3 illustrates the neuroimaging and brain mapping techniques.

13.3.2 Data acquisition and preprocessing methods

The accuracy and utility of neuroinformatics analyses depend heavily on the quality of data acquisition and preprocessing methods. High-quality data acquisition protocols ensure that the collected neuroimaging, electrophysiological, or genomic data are accurate and reproducible.

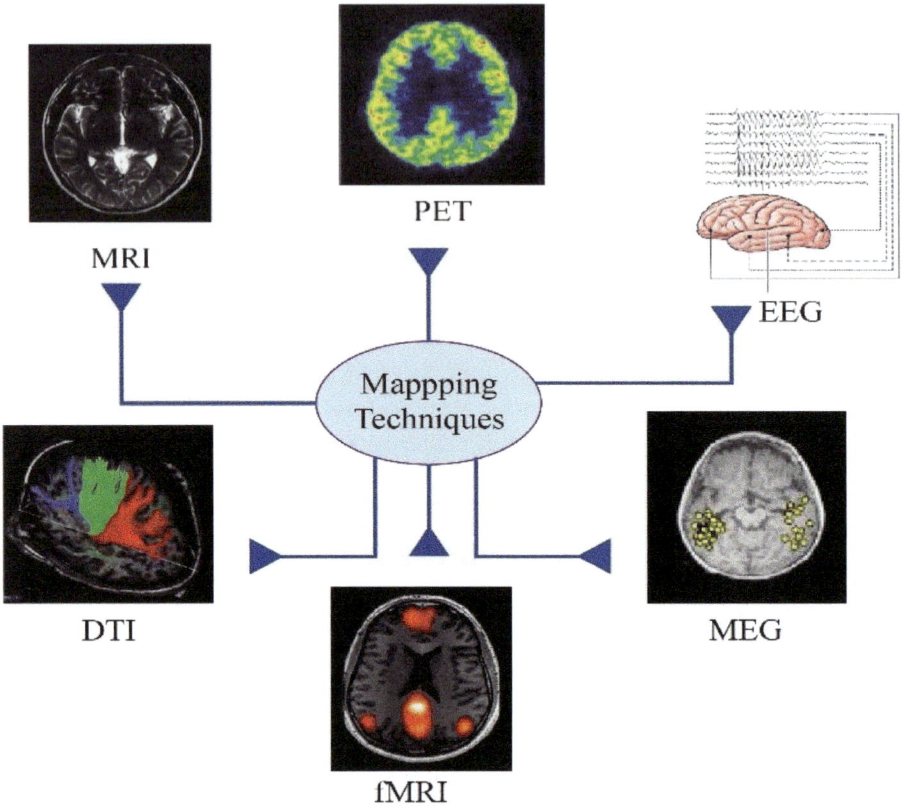

Figure 13.3 Illustration of neuroimaging and brain mapping techniques.

This involves standardized procedures for equipment calibration, participant preparation, and data collection, which are crucial for maintaining consistency across different studies and experiments. Preprocessing is another critical step, as raw data often contain noise and artifacts that need to be removed before analysis. Typical preprocessing steps include motion correction, normalization, artifact removal (such as removing noise from MRI or EEG data), and spatial smoothing. Techniques like independent component analysis (ICA) and principal component analysis (PCA) are commonly employed to decompose and clean the data, ensuring that the resulting dataset is suitable for subsequent analysis. Data transformation is also essential in neuroinformatics. Transforming data into a standard format or space, such as mapping individual brain scans to a common template, facilitates comparison and integration across studies. This step often involves feature extraction,

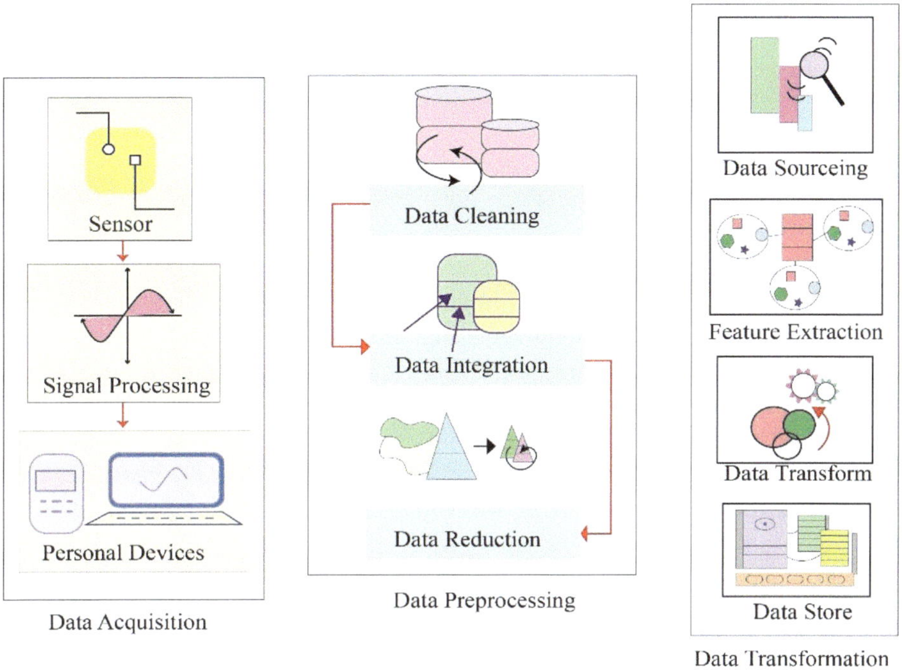

Figure 13.4 Illustration of data acquisition and preprocessing methods.

which highlights relevant aspects of the data for further analysis. By ensuring that data are accurately acquired, meticulously preprocessed, and properly transformed, researchers can maximize the reliability and validity of their neuroinformatics analyses.

As shown in Figure 13.4, the success of neuroinformatics analyses relies on careful data acquisition and preprocessing. Standardized protocols for equipment setup and participant preparation, along with advanced techniques for cleaning and normalizing data, ensure that the information gathered is accurate and reliable. Transforming this data into a common format and extracting key features make it easier to compare and integrate across studies, ultimately enhancing the insights and conclusions drawn from neuroinformatics research.

13.3.3 Neuroinformatics databases and repositories

Centralized databases and repositories are essential for the effective storage, sharing, and access of large-scale neuroscience data. The Human

Connectome Project (HCP) stands out by providing comprehensive datasets on human brain connectivity, encompassing structural and functional MRI data along with behavioral and genetic information, which are invaluable for research on brain function and behavior. The Allen Brain Atlas offers detailed maps of gene expression in both mouse and human brains, integrating anatomical and genetic data that serve as crucial references for various research studies. The neuroimaging informatics tools and resources clearinghouse (NITRC) supports the neuroinformatics community by offering access to a broad range of neuroimaging tools, resources, and datasets, facilitating advancements in neuroimaging analysis [27]. OpenNeuro, an open-access platform, enables researchers to share and access neuroimaging data freely, fostering transparency and collaboration across the neuroscience research community. Lastly, the gene expression omnibus (GEO) repository supports high-throughput gene expression research, providing extensive data that aid in the integration of genetic information with neuroimaging studies [28]. Together, these resources enhance the ability of researchers to conduct comprehensive and collaborative studies in neuroscience.

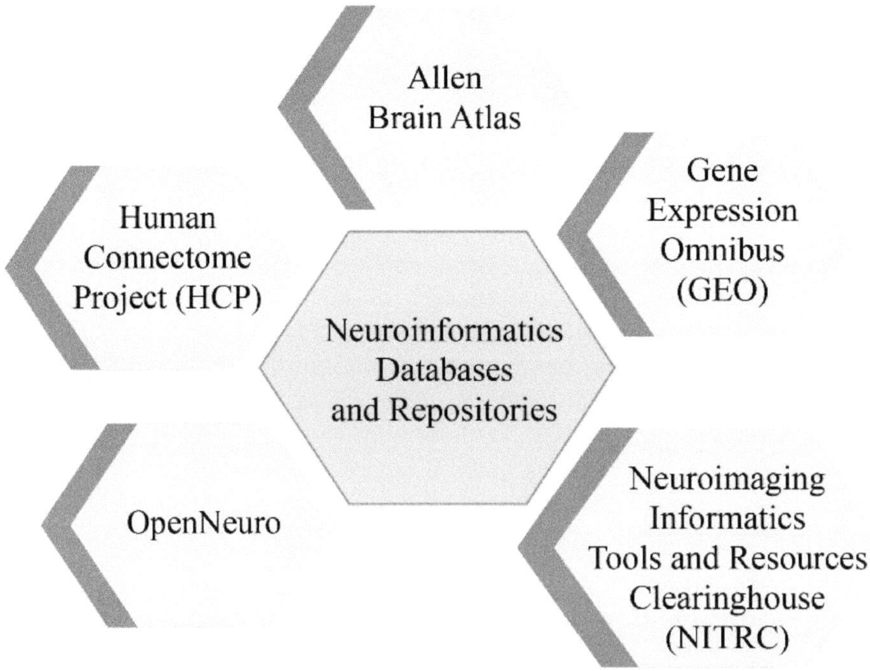

Figure 13.5 Illustration of neuroinformatics databases and repositories.

As shown in Figure 13.5 HCP, Allen Brain Atlas, NITRC, OpenNeuro, and GEO, are vital for storing, sharing, and accessing large-scale neuroscience data. These resources provide comprehensive datasets, tools, and platforms that support research in brain connectivity, gene expression, neuroimaging, and neurogenomics. Collectively, they enhance collaboration, transparency, and integration in neuroscience research, driving advancements in understanding and treating brain disorders.

13.3.4 Machine learning and AI in neuroinformatics

Machine learning (ML) and artificial intelligence (AI) techniques have become indispensable in neuroinformatics, enabling the analysis of complex, high-dimensional data that traditional methods struggle to handle. One of the key applications of ML in neuroinformatics is pattern recognition. Algorithms such as support vector machines (SVM) [29] and deep learning models are employed to identify patterns in neuroimaging data that correlate with specific brain states or diseases, allowing for early detection and diagnosis. Predictive modeling is another critical application of AI in neuroinformatics. AI techniques help develop models that predict the onset progression of neurological and psychiatric disorders. These models are also used to personalized treatment plans based on individual patient data, improving the effectiveness of therapeutic interventions. Machine learning also plays a vital role in data integration [30], facilitating the combination of multimodal data, such as neuroimaging, genetic, and clinical information. This comprehensive approach provides a more complete understanding of brain function and pathology, which is essential for advancing personalized neuroscience. Automated analysis powered by AI significantly reduces the time and effort required to process large datasets. AI-driven tools can analyze vast amounts of data quickly and accurately, increasing the reproducibility of results and allowing researchers to focus on interpreting the findings rather than on data processing. Feature extraction is another area where advanced ML algorithms excel. These algorithms can extract relevant features from complex datasets, highlighting important aspects of the data that might be overlooked by traditional analysis methods [31]. This capability is crucial for identifying subtle but significant patterns that contribute to our understanding of brain function and disease.

In summary, as illustrated in Figure 13.6, the key technologies and tools in neuroinformatics-neuroimaging techniques, data acquisition and preprocessing methods, databases, and repositories, computational modeling tools,

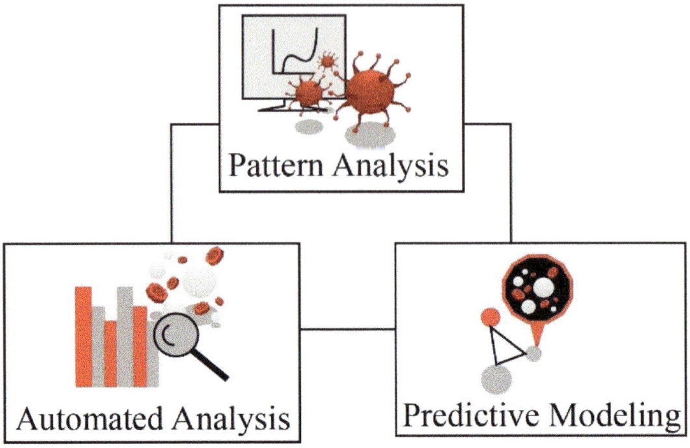

Figure 13.6 Illustration of ML and AI in neuroinformatics.

and AI techniques – are crucial for advancing the field of personalized neuroscience. These tools enable the collection, integration, and analysis of complex brain data, facilitating a deeper understanding of individual brain function and the development of personalized diagnostic and therapeutic strategies.

13.4 Personalized Neuroscience Concepts and Applications

13.4.1 Genetic and epigenetic influences on brain function

Genetic and epigenetic factors play a critical role in shaping brain function and behavior, Genetic and epigenetic factors play a crucial role in shaping brain function and behavior. Genetics refers to the hereditary information encoded in an individual's deoxyribonucleic acid (DNA), which influences the development and function of neural circuits [32]. Genetic variations can affect neurotransmitter systems, receptor sensitivity, and synaptic plasticity, contributing to individual differences in cognition, emotion, and susceptibility to neurological disorders. Epigenetics, on the other hand, involves changes in gene expression that do not alter the underlying DNA sequence but are influenced by environmental factors, lifestyle, and experiences. Epigenetic modifications, such as DNA methylation and histone acetylation, can impact brain function by regulating the activity of genes involved in neuronal development and plasticity [33].

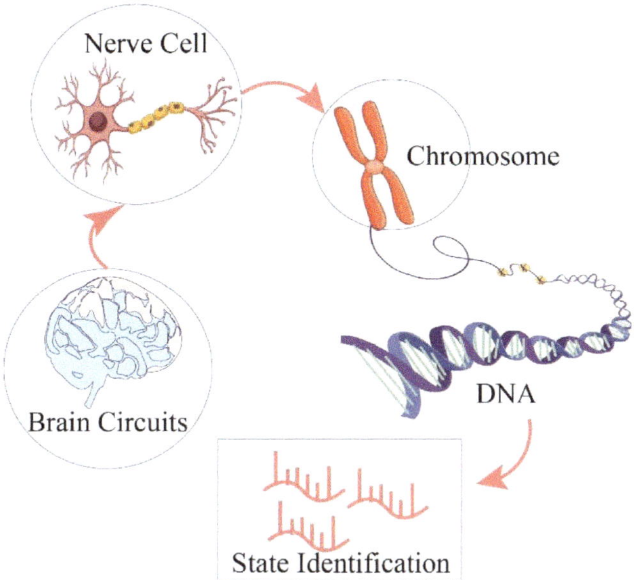

Figure 13.7 Illustration of brain influences analysis.

As shown in Figure 13.7, understanding these genetic and epigenetic mechanisms is essential for identifying biomarkers of brain disorders and developing personalized therapeutic approaches that target these specific pathways.

13.4.2 Brain connectivity and individual variability

Brain connectivity refers to the intricate network of neural connections that facilitate communication between different brain regions. This connectivity underlies all cognitive processes and behaviors, and its patterns can vary significantly between individuals. Techniques such as functional MRI (fMRI) and diffusion tensor imaging (DTI) allow researchers to map these connections and study how they differ across people [34]. Individual variability in brain connectivity can influence how information is processed and integrated with the brain, affecting everything from memory and attention to emotional regulation and social interactions. In the context of personalized neuroscience, understanding these unique connectivity patterns is crucial for interventions. For instance, customized brain stimulation techniques can be designed to target specific neural networks that are disrupted in a particular individual, thereby enhancing therapeutic outcomes.

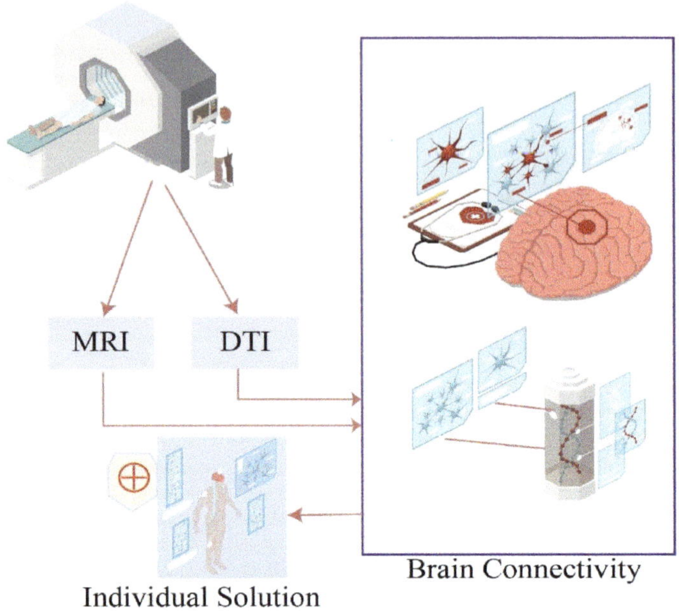

Figure 13.8 Illustration of brain connectivity and individual variability.

As shown in Figure 13.8, describing individual variability in brain connectivity is essential for developing interventions in personalized neuroscience. Techniques like fMRI and DTI allow researchers to map these unique neural connections, revealing how differences in connectivity influence cognitive processes and behaviors. This knowledge enables the design of personalized brain stimulation techniques targeting specific disrupted networks, thereby enhancing therapeutic outcomes.

13.4.3 Personalized treatment strategies and interventions

Personalized treatment strategies in neuroscience involve designing interventions that are specific to the unique characteristics of an individual's brain and behavior. These strategies can encompass a wide range of approaches, including pharmacological treatments, cognitive-behavioral therapies, neurostimulation techniques, and lifestyle modifications [35]. Pharmacological treatments can be personalized based on an individual's genetic makeup, which affects drug metabolism and response. For instance, pharmacogenomic testing can identify genetic variations that influence how a person responds to certain medications, allowing for the selection of the most effective drugs with the least side effects. Cognitive-behavioral therapies can be customized to address the specific cognitive distortions and behavioral patterns identified through phenotyping. Personalized neurostimulation techniques, such as transcranial magnetic stimulation (TMS) or transcranial direct current stimulation (tDCS), can target specific brain regions or networks that are implicated in an individual's condition [36]. Additionally, personalized interventions can include lifestyle and environmental modifications that take into account the individual's unique needs and preferences. This holistic approach ensures that all aspects of the person's life are considered in the treatment plan, leading to more comprehensive and effective care [37]. By integrating genetic and epigenetic insights, understanding individual brain connectivity, conducting detailed cognitive and behavioral phenotyping, and developing personalized treatment strategies, personalized neuroscience aims to provide more precise and effective interventions for neurological and psychiatric disorders. This approach holds the promise of improving patient outcomes by modifying treatments to the unique needs of each individual.

13.5 Integrating Neuroinformatics and Personalized Neuroscience

Big data plays a fundamental role in advancing personalized neuroscience by enabling the collection, storage, and analysis of vast amounts of diverse neurological data. This includes genetic information, neuroimaging data, electrophysiological recordings, and behavioral assessments. The sheer volume and variety of data facilitate the identification of subtle patterns and correlations that are not apparent through traditional analysis methods. For example, large-scale datasets allow researchers to explore the genetic underpinnings of brain disorders, identify biomarkers for early diagnosis, and understand the complex interactions between genetics and environmental factors [38]. Moreover, big data analytics can uncover insights into individual differences in brain structure and function, which are crucial for developing personalized treatment strategies. In addition, data integration and multi-modal analysis are fundamental to personalized neuroscience, as they enable a comprehensive understanding of brain function and pathology. By combining data from multiple sources – such as genomics, neuroimaging, and clinical records, researchers can create a holistic view of an individual's brain health. Multi-modal analysis leverages advanced computational techniques to integrate and analyze these diverse datasets, revealing complex interactions and providing deeper insights into the neural mechanisms underlying various conditions [39]. For instance, integrating genetic data with neuroimaging findings can help identify how specific genetic variations influence brain connectivity and function. This integrated approach is essential for identifying individualized biomarkers and adapting interventions to the unique neurobiological profile of each person.

Following that predictive modeling and precision diagnostics are key components of personalized neuroscience, aiming to predict disease onset, progression, and treatment response at an individual level. Using advanced ML and AI techniques, predictive models can analyze multi-dimensional data to forecast outcomes and identify the most effective treatments for each patient [40]. These models are trained on large datasets to recognize patterns associated with specific neurological or psychiatric conditions, enabling early and accurate diagnosis. Precision diagnostics leverage these predictive capabilities to fit diagnostic assessments and interventions based on an individual's unique genetic, neurobiological, and clinical profile. For example, machine learning algorithms can predict the likelihood of developing Alzheimer's disease based on genetic risk factors, neuroimaging abnormalities, and cognitive performance, facilitating early intervention and personalized care plans [41].

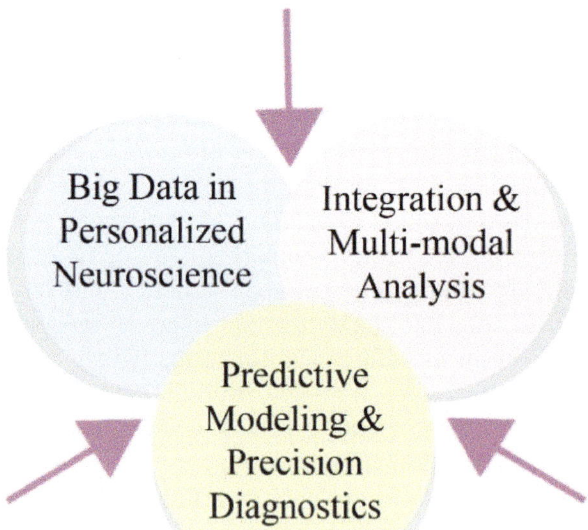

Figure 13.9 Illustration of integrating neuroinformatics and personalized neuroscience.

Therefore as shown in Figure 13.9, the integration of neuroinformatics with personalized neuroscience harnesses the power of big data, multi-modal analysis, and predictive modeling to revolutionize our understanding of brain health. These approaches enable the development of precision diagnostics and personalized treatments, ultimately improving outcomes and quality of life for individuals with neurological and psychiatric disorders.

13.6 Challenges and Opportunities

One of the major challenges in neuroinformatics and personalized neuroscience is the heterogeneity of data sources. Different studies often use varying protocols, equipment, and data formats, making it difficult to compare and integrate findings across research efforts [42]. Standardization of data acquisition, storage, and processing methods is crucial to ensure consistency and reliability. Efforts are underway to develop common standards and frameworks, but achieving widespread adoption remains a significant hurdle. Moreover, the integration of large-scale neurobiological and clinical data raises important ethical and privacy concerns. Protecting the confidentiality of sensitive information is paramount, particularly when dealing with genetic and health data. Informed consent, data anonymization, and secure data storage practices are essential to maintain trust and protect individuals' rights [43]. Additionally, there is a need to address issues related to the potential misuse of data and ensure that the benefits of personalized neuroscience are equitably distributed [44]. In addition, the sheer volume and complexity of neuroinformatics data present significant technological and computational challenges. Advanced computing infrastructure and sophisticated algorithms are required to process, analyze, and interpret large datasets. Additionally, there is a need for scalable and efficient data storage solutions [45]. Overcoming these challenges necessitates ongoing investment in technology and the development of new computational methods to the unique demands of neuroinformatics.

Despite the challenges, the field of neuroinformatics presents numerous opportunities for innovation and interdisciplinary collaboration. By bringing together experts from neuroscience, computer science, engineering, and other disciplines, new tools and approaches can be developed to address complex questions about brain function and disease. Collaborative efforts can lead to breakthroughs in understanding individual variability in brain health, developing personalized treatments, and improving diagnostic accuracy. The integration of diverse perspectives and expertise is key to driving progress in this rapidly evolving field.

13.7 Future Directions

Emerging trends in neuroinformatics include the increasing use of ML and AI to analyze complex data, the development of more sophisticated brain imaging techniques, and the integration of multi-omics data (genomics, proteomics, etc.) to provide a comprehensive view of brain function. Advances in computational neuroscience are also expected to enhance our understanding of neural dynamics and facilitate the creation of more accurate models of brain function. Personalized neuroscience is poised to benefit from several potential advancements, including the development of more precise diagnostic tools and targeted therapies. Enhanced understanding of genetic and epigenetic factors will enable the identification of new biomarkers for brain disorders, leading to earlier and more accurate diagnoses. Additionally, personalized interventions to an individual's unique neurobiological profile will improve treatment outcomes and reduce adverse effects. AI and ML are set to play a central role in the future of neuroinformatics and personalized neuroscience. These technologies can handle large and complex datasets, uncovering patterns and insights that would be difficult to detect through traditional methods. AI and ML algorithms can improve predictive modeling, optimize treatment plans, and support real-time analysis of brain activity, contributing to more effective and personalized healthcare solutions. The integration of neuroinformatics into clinical practice has the potential to transform healthcare. By providing clinicians with more detailed and accurate information about brain health, neuroinformatics can enhance diagnostic precision, guide treatment decisions, and monitor disease progression. Personalized neuroscience approaches can lead to more effective and individualized patient care, improving outcomes and quality of life. The use of advanced data analytics in clinical settings will also support the development of new therapies and interventions.

13.8 Conclusion

Neuroinformatics and personalized neuroscience are intertwined fields that leverage advanced data analytics, machine learning, and interdisciplinary collaboration to enhance our understanding of brain function and improve patient care. Key challenges include data heterogeneity, ethical considerations, and technological demands, while opportunities for innovation and collaboration abound. Continued integration of neuroinformatics and personalized neuroscience, supported by collaboration across disciplines, is essential for overcoming current challenges and achieving future breakthroughs. Standardization of data practices, ethical handling of sensitive information, and investment in advanced technologies will drive progress in these fields. The future of neuroinformatics and personalized neuroscience holds immense promise for transforming our understanding of the brain and revolutionizing clinical practice. As we continue to harness the power of big data, AI, and interdisciplinary research, we move closer to a future where personalized, precise, and effective brain health solutions are a reality. The ongoing integration of these fields will not only enhance scientific knowledge but also improve the lives of individuals affected by neurological and psychiatric disorders.

Acknowledgments

We gratefully acknowledge the editors for their invaluable contributions. Their expert guidance and keen insights have enriched the fabric of this chapter, elevating its impact and narrative coherence.

References

[1] J. Royer et al., "Gradients of Brain Organization: Smooth Sailing from Methods Development to User Community," Neuroinform, Apr. 2024, doi: 10.1007/s12021-024-09660-y.

[2] C. L. Adamson et al., "Updates to the Melbourne Children's Regional Infant Brain Software Package (M-CRIB-S)," Neuroinform, vol. 22, no. 2, pp. 207–223, Apr. 2024, doi: 10.1007/s12021-024-09656-8.

[3] J. D. Van Horn, "Editorial: On the Economics of Neuroscientific Data Sharing," Neuroinform, vol. 22, no. 1, pp. 1–4, Jan. 2024, doi: 10.1007/s12021-023-09649-z.

[4] D. Aquilué-Llorens, J. S. Goldman, and A. Destexhe, "High-Density Exploration of Activity States in a Multi-Area Brain Model," Neuroinform, vol. 22, no. 1, pp. 75–87, Jan. 2024, doi: 10.1007/s12021-023-09647-1.

[5] R. J. M. Hermosillo et al., "A precision functional atlas of personalized network topography and probabilities," Nat Neurosci, vol. 27, no. 5, Art. no. 5, May 2024, doi: 10.1038/s41593-024-01596-5.

[6] Y. K. Choi, L. Feng, W.-K. Jeong, and J. Kim, "Connecto-informatics at the mesoscale: current advances in image processing and analysis for mapping the brain connectivity," Brain Inf., vol. 11, no. 1, p. 15, Jun. 2024, doi: 10.1186/s40708-024-00228-9.

[7] A. Serra, P. Galdi, and R. Tagliaferri, "16 - Multiview learning in biomedical applications," in Artificial Intelligence in the Age of Neural Networks and Brain Computing (Second Edition), R. Kozma, C. Alippi, Y. Choe, and F. C. Morabito, Eds., Academic Press, 2024, pp. 307–324. doi: 10.1016/B978-0-323-96104-2.00010-5.

[8] A. Rokem and N. C. Benson, "Hands-On Neuroinformatics Education at the Crossroads of Online and In-Person: Lessons Learned from NeuroHackademy," Neuroinform, May 2024, doi: 10.1007/s12021-024-09666-6.

[9] Q. Xu, Y. Wang, H. Wu, M. Chen, and B. Chen, "Periodic and chaotic spiking behaviors in a simplified memristive Hodgkin-Huxley circuit," Chaos, Solitons & Fractals, vol. 179, p. 114458, Feb. 2024, doi: 10.1016/j.chaos.2024.114458.

[10] A. Nomani, S. Yousefi, D. Sargsyan, and A. Hatefi, "A quantitative MRI-based approach to estimate the permeation and retention of nanomedicines in tumors," Journal of Controlled Release, vol. 368, pp. 728–739, Apr. 2024, doi: 10.1016/j.jconrel.2024.03.019.

[11] L. Zimmer, "Recent applications of positron emission tomographic (PET) imaging in psychiatric drug discovery," Expert Opinion on Drug Discovery, vol. 19, no. 2, pp. 161–172, Feb. 2024, doi: 10.1080/17460441.2023.2278635.

[12] Y. Lu et al., "Macaque Brainnetome Atlas: A multifaceted brain map with parcellation, connection, and histology," Science Bulletin, Mar. 2024, doi: 10.1016/j.scib.2024.03.031.

[13] G. J. B. Elias et al., "A large normative connectome for exploring the tractographic correlates of focal brain interventions," Sci Data, vol. 11, no. 1, p. 353, Apr. 2024, doi: 10.1038/s41597-024-03197-0.

[14] F. Yasmin et al., "PoxNet22: A Fine-Tuned Model for the Classification of Monkeypox Disease Using Transfer Learning," IEEE Access, vol. 11, pp. 24053–24076, 2023, doi: 10.1109/ACCESS.2023.3253868.

[15] E. Vivi and B. Di Benedetto, "Brain stars take the lead during critical periods of early postnatal brain development: relevance of astrocytes in

health and mental disorders," Mol Psychiatry, pp. 1–13, Mar. 2024, doi: 10.1038/s41380-024-02534-4.

[16] N. Naghibi et al., "Embodying Time in the Brain: A Multi-Dimensional Neuroimaging Meta-Analysis of 95 Duration Processing Studies," Neuropsychol Rev, vol. 34, no. 1, pp. 277–298, Mar. 2024, doi: 10.1007/s11065-023-09588-1.

[17] F. Yasmin, M. M. Hassan, S. Zaman, S. T. Aung, A. Karim, and S. Azam, "A Forecasting Prognosis of the Monkeypox Outbreak Based on a Comprehensive Statistical and Regression Analysis," Computation, vol. 10, no. 10, Art. no. 10, Oct. 2022, doi: 10.3390/computation10100177.

[18] Z. Wang et al., "Spatiotemporal manipulation metal–organic frameworks as oral drug delivery systems for precision medicine," Coordination Chemistry Reviews, vol. 502, p. 215615, Mar. 2024, doi: 10.1016/j.ccr.2023.215615.

[19] R. Peter et al., "AnNoBrainer, an Automated Annotation of Mouse Brain Images using Deep Learning." bioRxiv, p. 2024.01.12.575415, Jan. 16, 2024. doi: 10.1101/2024.01.12.575415.

[20] I. J. L. Gomes et al., "Exploring the Allen Brain Atlas with medical students in an official program of scientific training in neuropsychiatric genetics," Brazilian Journal of Health Review, vol. 7, no. 1, pp. 5805–5812, Feb. 2024, doi: 10.34119/bjhrv7n1-470.

[21] Y. Zhou, H. Song, and G. Ming, "Genetics of human brain development," Nat Rev Genet, vol. 25, no. 1, Art. no. 1, Jan. 2024, doi: 10.1038/s41576-023-00626-5.

[22] J. Kruper et al., "Tractometry of the Human Connectome Project: Resources and Insights," Front. Neurosci., vol. 18, May 2024, doi: 10.3389/fnins.2024.1389680.

[23] I. Galasso, "Precision Medicine for Whom? Public Health Outputs from 'Genomics England' and 'All of Us' to Make Up for Upstream and Downstream Exclusion," The American Journal of Bioethics, vol. 24, no. 3, pp. 71–85, Mar. 2024, doi: 10.1080/15265161.2023.2180108.

[24] Z. Chen and A. Yadollahpour, "A new era in cognitive neuroscience: the tidal wave of artificial intelligence (AI)," BMC Neurosci, vol. 25, no. 1, p. 23, May 2024, doi: 10.1186/s12868-024-00869-w.

[25] F. Yasmin et al., "GastroNet: Gastrointestinal Polyp and Abnormal Feature Detection and Classification With Deep Learning Approach," IEEE Access, vol. 11, pp. 97605–97624, 2023, doi:10.1109/ACCESS.2023.3312729.

[26] A. G. Andersen, A. C. Riparbelli, H. R. Siebner, L. Konge, and F. Bjerrum, "Using neuroimaging to assess brain activity and areas associated with surgical skills: a systematic review," Surg Endosc, vol. 38, no. 6, pp. 3004–3026, Jun. 2024, doi: 10.1007/s00464-024-10830-x.

[27] P. Dzianok and E. Kublik, "PEARL-Neuro Database: EEG, fMRI, health and lifestyle data of middle-aged people at risk of dementia," Sci Data, vol. 11, no. 1, p. 276, Mar. 2024, doi: 10.1038/s41597-024-03106-5.

[28] M. E. Martone, "The past, present and future of neuroscience data sharing: a perspective on the state of practices and infrastructure for FAIR," Front. Neuroinform., vol. 17, Jan. 2024, doi:10.3389/fninf.2023.1276407.

[29] M. Hasan, F. Yasmin, and L. Deng, "Rule Mining of Early Diabetes Symptom and Applied Supervised Machine Learning and Cross Validation Approaches based on the Most Important Features to Predict Early-stage Diabetes," IJIRMPS - International Journal of Innovative Research in Engineering & Multidisciplinary Physical Sciences, vol. 11, no. 3, Jun. 2023, doi: 10.37082/IJIRMPS.v11.i3.230225.

[30] F. Yasmin et al., "AQIPred: A Hybrid Model for High Precision Time Specific Forecasting of Air Quality Index with Cluster Analysis," Hum-Cent Intell Syst, vol. 3, no. 3, pp. 275–295, Sep. 2023, doi:10.1007/s44230-023-00039-x.

[31] E. J. Bacon et al., "Neuroimage analysis using artificial intelligence approaches: a systematic review," Med Biol Eng Comput, Apr. 2024, doi: 10.1007/s11517-024-03097-w.

[32] D. J. Lin et al., "Transforming modeling in neurorehabilitation: clinical insights for personalized rehabilitation," J NeuroEngineering Rehabil, vol. 21, no. 1, p. 18, Feb. 2024, doi: 10.1186/s12984-024-01309-w.

[33] G. C. Kretzschmar, A. B. W. Boldt, and A. D. S. Targa, "Editorial: The genetics and epigenetics of mental health," Front. Genet., vol. 15, Apr. 2024, doi: 10.3389/fgene.2024.1402495.

[34] S. Pal, J. W. C. Lim, and L. J. Richards, "Diverse axonal morphologies of individual callosal projection neurons reveal new insights into brain connectivity," Current Opinion in Neurobiology, vol. 84, p. 102837, Feb. 2024, doi: 10.1016/j.conb.2023.102837.

[35] S. Khalighi, K. Reddy, A. Midya, K. B. Pandav, A. Madabhushi, and M. Abedalthagafi, "Artificial intelligence in neuro-oncology: advances and challenges in brain tumor diagnosis, prognosis, and precision treatment," npj Precis. Onc., vol. 8, no. 1, pp. 1–12, Mar. 2024, doi: 10.1038/s41698-024-00575-0.

[36] R. F. H. Cash and A. Zalesky, "Personalized and Circuit-Based Transcranial Magnetic Stimulation: Evidence, Controversies, and Opportunities," Biological Psychiatry, vol. 95, no. 6, pp. 510–522, Mar. 2024, doi:10.1016/j.biopsych.2023.11.013.

[37] G. Drossel and A. Zilverstand, "Towards personalized medicine: subtyping using functional profiles," Neuropsychopharmacology, vol. 49, no. 1, pp. 347–348, Jan. 2024, doi: 10.1038/s41386-023-01704-2.

[38] A.-K. Deisenhofer et al., "Implementing precision methods in personalizing psychological therapies: Barriers and possible ways forward," Behaviour Research and Therapy, vol. 172, p. 104443, Jan. 2024, doi: 10.1016/j.brat.2023.104443.

[39] Z. Cao, X. Xiao, C. Xie, L. Wei, Y. Yang, and C. Zhu, "Personalized connectivity-based network targeting model of transcranial magnetic stimulation for treatment of psychiatric disorders: computational feasibility and reproducibility," Front. Psychiatry, vol. 15, Feb. 2024, doi: 10.3389/fpsyt.2024.1341908.

[40] B. Wang et al., "Aptamer Technologies in Neuroscience, Neuro-Diagnostics and Neuro-Medicine Development," Molecules, vol. 29, no. 5, Art. no. 5, Jan. 2024, doi: 10.3390/molecules29051124.

[41] M. Z. Zhong, T. Peng, M. L. Duarte, M. Wang, and D. Cai, "Updates on mouse models of Alzheimer's disease," Mol Neurodegeneration, vol. 19, no. 1, p. 23, Mar. 2024, doi: 10.1186/s13024-024-00712-0.

[42] R. Bhidayasiri, "The grand challenge at the frontiers of neurotechnology and its emerging clinical applications," Front. Neurol., vol. 15, Jan. 2024, doi: 10.3389/fneur.2024.1314477.

[43] S. Voigtlaender et al., "Artificial intelligence in neurology: opportunities, challenges, and policy implications," J Neurol, vol. 271, no. 5, pp. 2258–2273, May 2024, doi: 10.1007/s00415-024-12220-8.

[44] G. Luk and J. A. Christodoulou, "Cognitive Neuroscience and Education," in Handbook of Educational Psychology, 4th ed., Routledge, 2023.

[45] J. F. Ribeiro, K. L. Shepard, and P. Ruther, "Editorial: Problems, strategies, and developments for high-density long-term chronic intracortical neural interfaces and their application," Front. Neurosci., vol. 18, Feb. 2024, doi: 10.3389/fnins.2024.1373451.

14

Deciphering Minds and Motion: A Unified Exploration of Brain Signal Decoding and Activity Recognition Through AI-driven Analysis

Arafat Rahman[1], Ayontika Das[2], Anindya Nag[3], and Tamanna Zubairi Sana[4]

[1,3]Computer Science and Engineering Discipline, Khulna University, Bangladesh
[2]Department of Computer Science and Engineering, Adamas University, India
[4]Department of Computer Science and Engineering, East Delta University, Bangladesh
E-mail: shawonsr54@gmail.com, anindyanag@ieee.org, ayontikadas43@gmail.com, tamannazubairisana@gmail.com

Abstract

Human activity recognition (HAR) and machine learning approaches for brain signal decoding represent key intersections of technology and human behavior analysis. HAR involves automated detection of human actions through sensor data, with applications in healthcare, sports, and smart homes. Machine learning for brain signal decoding interprets brain signals, providing insights into cognition and neurological processes. Both fields face challenges due to the complexity of human movement and brain activity, requiring sophisticated algorithms and sensors for accurate analysis. Despite significant progress, selecting optimal algorithms and sensors for specific applications remains challenging. This study aims to bridge this gap by investigating effective machine learning strategies for accurately identifying human activities and decoding brain signals. It will examine the use of vision-based and non-vision-based acquisition devices in various contexts. The applications of HAR extend to environmental surveillance, security, and beyond, where accurate human activity detection is crucial. Similarly, brain signal decoding has profound implications for understanding and treating neurological conditions. This comprehensive analysis will highlight the current state and future prospects of these dynamic fields, contributing to advancements in neuroscience and artificial intelligence.

Keywords: Human activity recognition, machine learning, deep learning, convolutional neural network, long short-term memory network.

14.1 Introduction

Human activity recognition (HAR) is a specialized domain within computer science and artificial intelligence focused on developing algorithms and methods to automatically detect and categorize human actions using data from various sensors. These activities range from routine actions like walking, running, sitting, and standing to more complex behaviors such as cooking, driving, or exercising [1]. The primary goal of HAR is to analyze sensor data – commonly gathered from wearable devices like smartwatches or smartphones, as well as environmental sensors like cameras or accelerometers – to accurately identify the activity being performed [2]. HAR has diverse applications across several domains, including healthcare, fitness tracking, sports analytics, security monitoring, and human–computer interfaces [3]. In healthcare, HAR systems can monitor and analyze patients' movements and behaviors, providing insights into their overall well-being or detecting potential medical issues. In sports analytics, HAR offers valuable insights into athletic performance, aiding in the optimization and enhancement of training programs. Research leveraging sensors in wearable and mobile devices has significantly advanced the understanding of human behavior and the ability to predict human intentions. A key objective for many researchers is to develop systems that can accurately detect user behavior from raw data while minimizing resource consumption [4].

In the fascinating intersection of neuroscience and artificial intelligence, machine learning approaches have emerged as powerful tools for decoding brain signals, offering unprecedented insights into the workings of the human mind. By harnessing advanced algorithms and computational techniques, researchers delve deep into the complexities of neural activity, unraveling the intricate patterns hidden within brain signals. Through innovative AI-driven analysis, these decoded signals illuminate the underlying neural processes, shedding light on cognitive functions, emotions, and behaviors. This synergy between machine learning (ML) and neuroscience holds promise for transformative breakthroughs in understanding brain dynamics and unlocking the mysteries of consciousness. As we embark on this journey of exploration, the convergence of ML approaches and brain signal decoding opens new vistas of discovery, paving the way for revolutionary advancements in both scientific research and clinical applications.

The origins of human action recognition in HAR can be traced to fields such as human–computer interaction (HCI), computer vision, video surveillance, and automated observation. Research groups have explored various HAR system applications and challenges, enhancing hardware capabilities, introducing new performance metrics, and training classification models on diverse datasets [5]. While periodic surveys and review articles have highlighted research advancements in HAR, there has been a lack of a comprehensive recent study covering architecture, application areas, techniques/algorithms, evaluation methods, problems, and challenges [6]. These findings are particularly relevant to researchers working on HAR systems. Figure 14.1 illustrates the different types of activities. Human activities are classified into four groups based on the body components involved and the level of complexity:

Figure 14.1 Different types of activities.

- **Gesture**: Gestures are non-verbal forms of communication where individuals use their hands, face, or other body parts to convey a message. Examples include the "OK" gesture or an apologetic hand gesture.
- **Activity**: Activities involve the physical movement of the body, such as swimming or jumping.
- **Interactivities**: Interactivities require at least two activities involving multiple actors, such as hugging or chatting.
- **Group activities**: Group activities involve the collective participation of multiple individuals in gestures, actions, and interactions, such as playing cricket or football.

The development of a novel activity detection system is driven by various challenges and concerns, particularly those related to its real-time implications and the complexity of the operating environment. The literature addresses numerous issues associated with activity recognition. When designing these systems, it is crucial to emphasize simplicity in learning and effectiveness, while ensuring the security of sensitive user data and respecting personal privacy [7]. The accuracy of activity recognition significantly depends on the individuals involved in the training and testing phases. Flexibility is a key characteristic, reflecting the system's ability to efficiently adapt and respond to different contexts and conditions.

To achieve accurate activity recognition, HAR algorithms frequently employ ML pattern recognition techniques. These algorithms analyze sensor data to extract meaningful features, which are then used to train models capable of classifying various actions [8]. The common ML methods utilized in HAR include supervised learning, where models are trained with labeled data, and unsupervised learning, where models identify patterns in unlabeled data [9]. HAR is vital for understanding human behavior and supporting various applications aimed at improving health, safety, and overall quality of life. The following are the primary contributions of this chapter.

- This chapter introduces an effective and efficient HAR model based on different algorithms, capable of classifying human activities with high accuracy and precision.
- We validate the quality of the proposed models through extensive experiments and comparisons.
- In the captivating fusion of neuroscience and artificial intelligence, ML techniques have arisen as potent instruments for deciphering brain signals and we are providing unparalleled glimpses into the intricacies of human cognition.

- Demonstrated advancements in brain signal decoding and its relationship to human activity recognition.
- The chapter provides a concise summary of HAR and its fundamental components, illustrating the operation of these elements and presenting simulations.
- Additionally, this chapter emphasizes the importance of HAR in the processes of discovery and development.

The organization of this chapter includes a concise overview of HAR and the application of ML techniques for HAR in Section 14.2. This is followed by an exploration of HAR in Section 14.3 and a comprehensive analysis of HAR in Section 14.4. Section 14.5 provides a comparison of HAR methodologies. Chapter 14.6 explores the correlation between brain signals and HAR. Chapter 14.7 outlines the future scope of HAR, while Section 14.8 presents the conclusions.

14.1.1 Basic concept of HAR

HAR is a dynamic and complex field within computer science. Due to the intricate and detailed nature of human movements, accurately detecting human activity for various applications is challenging. Typically, activities are identified based on sequences of actions performed by individuals using either vision-based or non-vision-based sensors. HAR's applications span numerous fields, including healthcare, sports, smart home systems, and other sectors. Moreover, automating systems for environmental monitoring and identifying suspicious activities necessitates accurate human activity detection. In pervasive computing, providing precise information on individuals is crucial. However, recognizing human activities and movements is difficult due to the complexity of the activities, the rapidity of movements, constantly changing recording conditions, and the wide range of application domains. Additionally, all actions and activities occur in distinct settings and contexts. Figure 14.2 illustrates the basic concept of HAR, which involves several steps outlined below:

HAR is a discipline within computer vision and ML that aims to detect and categorize human behaviors or activities using data from various sensors. These sensors may include accelerometers, gyroscopes, magnetometers, and sometimes even cameras or microphones.

- **Data collection**: Sensor data is gathered from diverse sources, such as wearable devices (e.g., smartwatches and fitness trackers), smartphones, and Internet of Things (IoT) sensors embedded in the environment.

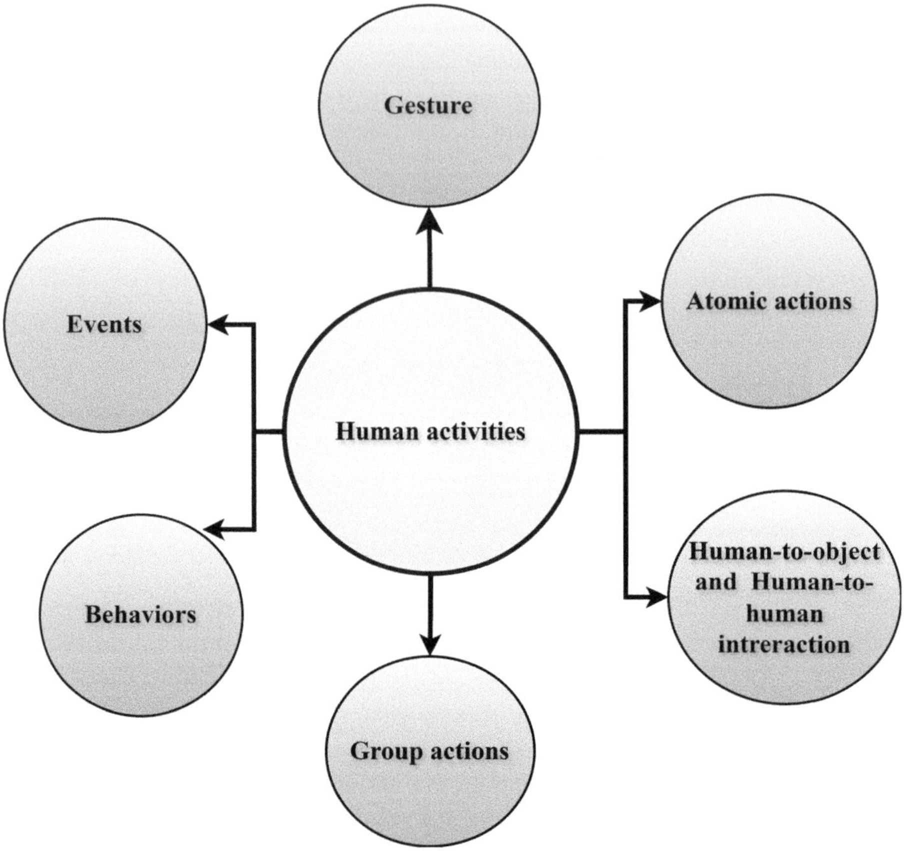

Figure 14.2 Actions involving HAR.

- **Preprocessing**: Raw sensor data often requires preprocessing to remove noise and irrelevant information, preparing it for analysis. This step may involve techniques such as signal filtering, normalization, and feature extraction.
- **Feature extraction**: Significant features are extracted from the preprocessed data. These features can include statistical metrics like mean, variance or frequency domain characteristics obtained using techniques like Fourier transform. Feature selection methods can reduce dimensionality and computational complexity.
- **Model training**: The extracted features and labeled data are used to train ML models. Popular models in ML include decision trees (DT), support vector machines (SVM), k-nearest neighbors (KNN), and deep

learning (DL) architectures such as recurrent neural networks (RNNs) and convolutional neural networks (CNNs).

- **Classification**: Once the model is trained, it can classify new sensor data instances into predefined activity categories. The model assigns a probability or confidence score to each class, indicating the likelihood of a specific activity being performed.
- **Evaluation**: The HAR system is evaluated using metrics such as accuracy, precision, recall, and F1-score. This step facilitates the assessment of the system's ability to generalize to unfamiliar data and identifies potential areas for improvement.
- **Deployment**: After training and evaluation, the model can be deployed in practical applications such as fitness tracking, healthcare monitoring, gesture recognition, security systems, and human–computer interaction interfaces.

HAR is crucial in various fields, including healthcare, sports, smart homes, and surveillance, enabling the automated analysis and understanding of human behavior.

14.2 Usages of Machine Learning Technique for HAR

The significance of ML techniques lies in their ability to extract valuable information from large datasets, facilitate predictive analysis, customize experiences, automate procedures, enhance decision-making, uncover hidden patterns, adapt to changes, and leverage technological advancements to promote innovation and efficiency across various domains.

14.2.1 Neural network with dynamic edges-based HAR

Recognizing human actions using skeletal frameworks with a dynamic edge convolutional neural network involves the use of monitoring devices to bridge the digital and physical realms. The core idea of HAR is its ability to enable machines to observe, collect data, and respond to human activities [10]. HAR has extensive and significant applications across various fields, including healthcare, smart home technologies, sports, and security, among others.

In healthcare, HAR is crucial as it provides valuable insights into the overall health condition of patients, aiding in better risk management and patient care. Similarly, in the security sector, HAR enhances the ability to monitor and recognize potential threats. Figure 14.3 illustrates the user and the block diagram for HAR, highlighting its fundamental components and workflow.

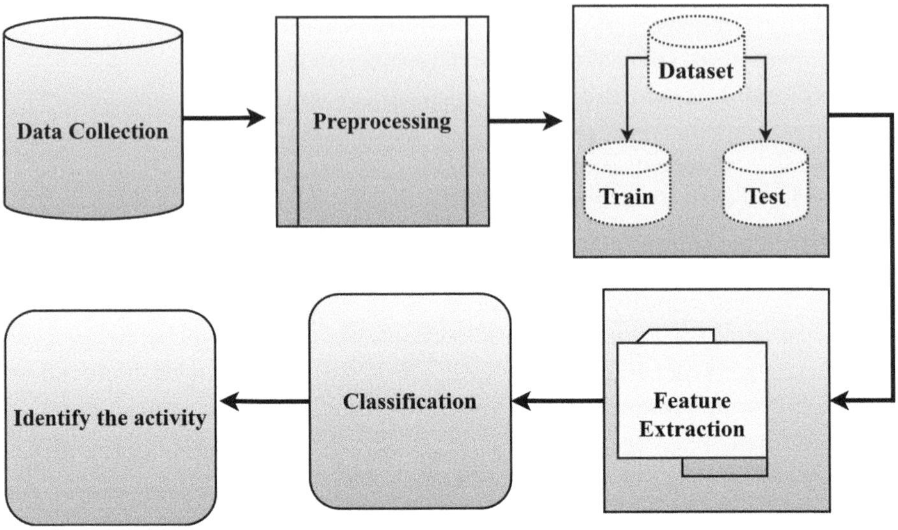

Figure 14.3 Block diagram of HAR system.

This HAR algorithm is based on a 3D skeleton joint model, contrasting it with two alternative approaches also utilizing a 3D skeleton joint model. Our findings reveal that the proposed algorithm excels particularly in facial recognition tasks. While this algorithm may be more time-consuming compared to other ML approaches, it delivers notably superior accuracy and value in its results.

14.2.2 HAR with self-attention

Spatio-temporal action representation recognition model, designed for the recognition and categorization of human actions. The approach utilizes hierarchical pyramid depth motion images (HP-DMI) and an ST-GCN extractor to compress skeletal information into spatio-temporal joint descriptors. In [11], the incorporation of self-attention for human action analysis was proposed. The neural network architecture employs multi-layer and multi-head self-attention, along with two robust baselines, to recognize human activities. The findings revealed that the neural network architecture achieved a test accuracy of 91.75%.

14.2.3 Multiple classifier ensemble to enhance HAR

The vision and inertial sensing fusion for HAR was developed to amalgamate activity data from wearable inertial sensors, depth cameras, RGB cameras,

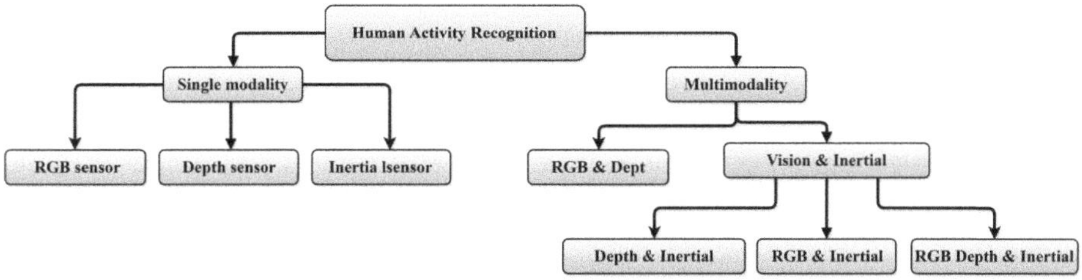

Figure 14.4 Different sensing methods for HAR.

and other sources for the recognition of human activities. Human actions encompass a range of behaviors involving their bodies, interactions with others, objects, and group activities. This review article specifically delves into human signal detection, emphasizing its diverse applications. These include smart video surveillance, home monitoring, human–machine interaction (HMI), video storage and retrieval, assistive living, and companion robots, among others. The discussion in this article covers various research topics in computer vision, such as human activity analysis and interpretation, tracking, location estimation, and video identification [12].

Recent years have witnessed significant advancements in HAR research, leading to the development of practical products. Numerous strategies have been employed to integrate detection modalities, with basic data combining methods including level combination, score/decision level combination, and others. Integrating homogeneous RGB or depth camera data necessitates consideration of multiple information levels. However, when dealing with heterogeneous sensors, feature-level or decision-level fusion is typically preferred. Figure 14.4 visually illustrates these combined methods, offering insights into their application within the context of HAR.

Utilizing optical and inertial sensing techniques, our research aims to identify human actions by integrating visual perception and inertial sensing modalities. This compilation of papers encompasses collaborative endeavors focused on merging these two sensing modalities. Additionally, we address the challenges encountered and propose solutions for effectively integrating optical and inertial sensing in practical applications.

14.2.4 HAR in multi-view binary coded images with GLAC features

Our research focuses on activity recognition utilizing self-relationship attributes derived from temporal and static depth data extracted from video

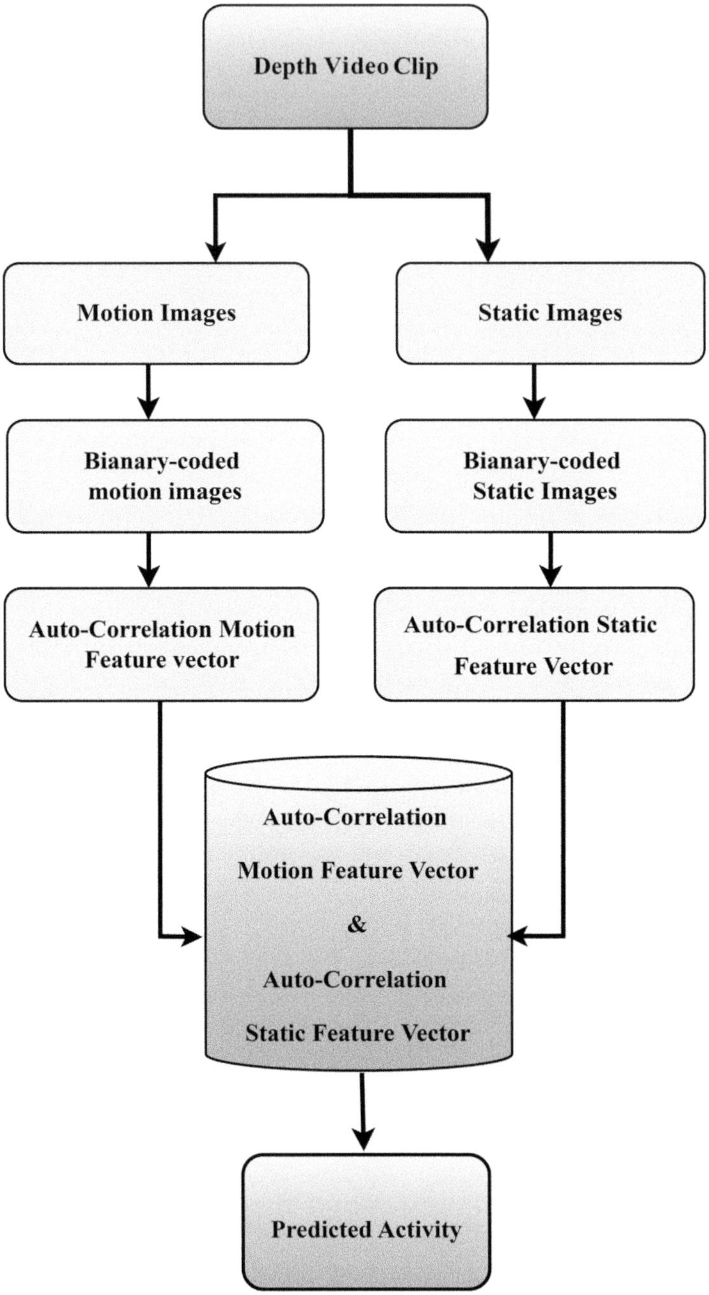

Figure 14.5 Architecture of the HAR system.

clips. The proposed 3D motion trail model (3DMTM) swiftly generates 2D motion and static data images immediately following activity video segmentation. The 3DMTM produces three sets of images – each representing the top, side, and front views of the activity – and one set of static images to comprehensively capture the complexity of the video clip. Subsequently, the images undergo double-encoding using the local binary pattern (LBP) operator. The gradient nearby auto-correlation (GLAC) calculation is then applied to the parallel-coded images to assess depth movements through auto-correlation analysis. The extreme learning machine (ELM) orchestrates operations by utilizing the extracted features in conjunction with the piece stunt. The MSRAction3D dataset has proven to be highly conducive to the effectiveness of the technique. Our exploratory analysis reveals that our framework exhibits performance comparable to top-tier frameworks in activity recognition. Figure 14.5 illustrates the flowchart of the HAR system, depicting the sequential processes involved.

This presents a method leveraging depth sensor data for movement detection. It involves creating three dynamic activity edges and three static ones, each with two layers [13]. A paired structure is established by utilizing both movement and static cues through LBP. The gradient local auto-correlation (GLAC) technique employs binary-coded patterns to characterize motion via auto-correlation analysis. GLAC highlights vectors derived from static images, which are then combined with double-coded movement to provide a more comprehensive representation of motion. The approach surpasses the performance of the 3D-CNN with the DMM pyramid algorithm in DL for motion image analysis by 91.31%. Consequently, integrating dynamic and static data images enhances the effectiveness of the recognition system when identifying combined activities.

14.2.5 Uncontrolled settings using machine learning

The focus of ML research is on uncontrolled human action recognition, particularly in video-based environments, where the goal is to detect and categorize human activities within video footage. Successfully identifying human actions in videos represents a significant advancement towards real-life activity detection. While various computations and approaches have been proposed to enhance the accuracy of the HAR process, ongoing refinement is still necessary. Challenges arise due to random variations in human appearance, clothing, lighting conditions, and background elements, making the detection and characterization of human behaviors a complex task [14]. A robust strategy for analyzing human behavior is employing a multi-stage method. The process begins with the extraction of the segmentation of interest (SoI), followed by joint motion capture (JMOCAP) and the mining of geodesic distance (GD) highlight descriptors. Subsequently, redundant sections are removed from the recordings. The classification phase culminates with the utilization of a neuro fuzzy classifier (NFC). Figure 14.6 illustrates the structural framework of the HAR model, showcasing the sequential stages involved in the process.

HAR entails the analysis of video or still images, which consist of a continuous sequence of frames, to achieve real-time recognition of human actions. The development of the HAR model involves leveraging methods such as GD, 3D convolutional features (3D-CF), JMOCAP, and non-parametric temporal grouping (nPTG) descriptor mining. Additionally, redundant frames are eliminated from recordings, and significant points of interest are extracted. The effectiveness of this approach was validated on HMDB-51 and Hollywood2, two widely used open-source datasets.

The primary objective of this research is to identify the actions being performed by two individuals in a given photo. However, overcrowding within a frame diminishes efficiency. Enhancements to the proposed approach can be achieved by incorporating a greater number of human joints and adding additional layers to the NFC to optimize the arrangement process. Increasing the number of joints facilitates easier human discernment in still images. Furthermore, testing the technique across multiple contexts and settings serves as another avenue for evaluation.

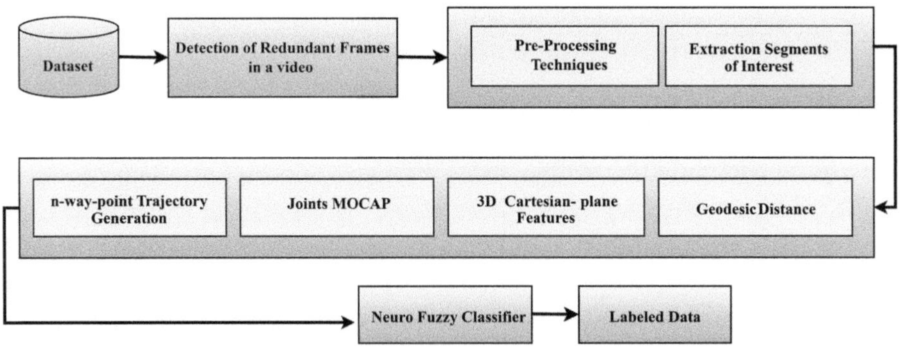

Figure 14.6 Structure of the HAR model for uncontrolled settings using ML.

14.3 Application of HAR

HAR stands at the intersection of ML and pattern recognition, focusing on detecting and analyzing human behavior using data from various sensors. These sensors can be integrated into wearable devices, smartphones, and cameras, or deployed across different locations. HAR boasts a plethora of applications across diverse fields, leveraging data to enhance safety, health, efficiency, and user experiences. Here's a detailed exploration of HAR and its myriad uses:

- **Healthcare and wellness**: HAR aids in monitoring and assessing individuals' health and well-being. It tracks physical activity levels to gauge fitness, identifies sedentary behavior, and encourages exercise. Additionally, it evaluates gait patterns to detect abnormalities indicative of neurological or musculoskeletal conditions. Surveillance of elderly individuals or patients helps identify falls, changes in behavior, or health-related abnormalities.
- **Assistive technologies**: HAR facilitates the implementation of assistive technologies to enhance accessibility and promote self-sufficiency for individuals with disabilities or special needs. Gesture recognition systems enable hands-free control of devices or interfaces, benefiting those with limited mobility. It also provides surveillance and reminders for individuals with cognitive impairments or memory issues, aiding in daily activities and medication management.
- **Smart homes and home automation**: HAR automates smart home devices based on user activity, optimizing energy consumption and enhancing convenience. It enhances home security by identifying abnormal activities or intrusions through pattern recognition.
- **Workplace safety and productivity**: In workplace settings, HAR ensures ergonomic practices to prevent injuries by monitoring employees' body positions and movements. It also detects signs of fatigue or strain in high-risk occupations like construction or manufacturing, thereby preventing accidents.
- **Sport and performance analysis**: HAR optimizes athletes' techniques and prevents injuries by analyzing movement patterns. It provides feedback to avoid hazardous movements and enhance performance.
- **Transportation**: In transportation, HAR identifies driver fatigue, distraction, or unusual driving behaviors to improve road safety. It also analyzes passenger activities in autonomous vehicles to enhance comfort and offer tailored services.

- **Entertainment and gaming**: HAR enhances gaming experiences by using body movements as input and provides immersive experiences in virtual reality (VR) environments by accurately tracking users' motions.
- **Education and special education**: HAR aids in monitoring student activities in e-learning environments, personalizing educational content, and improving engagement. It also facilitates the creation of tailored educational programs for students with special needs.
- **Retail and marketing**: HAR analyzes customer behaviors to optimize store layouts and marketing strategies. It delivers personalized product suggestions based on clients' actions and preferences.
- **Military and defense**: HAR monitors soldiers' physical condition and actions to ensure optimal performance. It also improves military training programs by analyzing troops' movements and techniques.
- **Agriculture**: HAR ensures the safety and efficiency of agricultural workers by monitoring their actions. It also tracks livestock movements and behaviors to ensure their health and welfare.
- **Urban planning and public safety**: HAR aids in crowd management and enhances public safety by detecting abnormal behaviors or events in public spaces, thus improving security measures and response times to emergencies.

HAR has the potential to revolutionize numerous domains, enhancing safety, quality of life, efficiency, and personalized experiences. Its influence will continue to expand with advancements in sensor technology, ML algorithms, and data analytics.

14.4 Comparative Study

In this era marked by extraordinary technological advancements, the ability to comprehend and assess human actions has become a significant challenge and avenue for development. Referred to as "human-assisted reasoning," this entails utilizing sensing technology, ML, and AI to decipher the complexities of human behavior [14]. In the realm of HAR, KUN XIA, and colleagues proposed an innovative long short-term memory (LSTM)-CNN architecture. This unique deep neural network design integrates convolutional layers with LSTM, effectively extracting activity features and classifying them with minimal model parameters. LSTM networks, tailored for handling temporal sequences, comprise two layers in the suggested architecture, followed by convolutional layers [15]. Raw data collected from mobile sensors is fed into this design, with a global average pooling layer replacing the fully connected layer post-convolution to reduce model parameters. Further enhancement is achieved through the incorporation of a batch normalization layer, which enhances convergence rates and overall performance. Evaluation of the model's efficacy was conducted on publicly available datasets including OPPORTUNITY, UCI, and WISDM. Another notable proposition by Dong Seog Han et al. [16] introduces a CNN-LSTM approach for HAR, aimed at better understanding human behavior and forecasting human intentions. This approach, leveraging sensors in wearable and handheld devices, reduces reliance on extensive feature engineering by utilizing raw data for improved action prediction accuracy. The depth of the CNN-LSTM network, both spatially and temporally, enhances its effectiveness in activity identification [17]. Rahman et al. [18] proposed a hybrid model for HAR using an extreme learning machine (ELM) based on local binary patterns (LBP). Additionally, HAR utilizing an inertial measurement unit has become the gold standard for continuous tracking of human, machine, and pet movements. However, despite advancements in correctly classifying user actions based on IMU sensor data, challenges persist, including resource constraints and specialized knowledge requirements [19]. Recent progress in HAR has been propelled by the application of DL techniques, facilitating easier action recognition and classification [20]. This project aims to develop a HAR model capable of accurately identifying and categorizing various human behaviors from video sequences. These advancements underscore HAR's significance across diverse domains, from healthcare to security and human–computer interface applications [21–22]. Table 14.1 provides a summary of these works, highlighting their contributions and methodologies in advancing the field of HAR.

Table 14.1 provides a comprehensive comparison of various models and algorithms across multiple dimensions, including complexity, handling of non-linear data, overfitting, memory and training time, interpretability, and robustness to outliers and irrelevant features. Depending on the specific characteristics of the data and the problem at hand, certain options may be more suitable than others. Notably, [18] achieves the highest accuracy among the models, while [23] utilizes the NFC model and achieves an accuracy of 82.55%. In [13], the use of LBP is highlighted as a highly efficient texturing operator. LBP assigns labels to pixels in an image by applying a threshold to their surrounding area and translating the outcome into a binary integer. This technique combines statistical and structural models of texture analysis, offering high discriminatory ability and computational efficiency.

Table 14.1 Summary of related work.

References	**Method**	**Dataset**	**Accuracy (%)**
Mohsen et al. [14]	KNN	MotionSense	96.50
Wang et al. [15]	DMMt-HOG	MSRAction3D	90.20
Singh et al. [16]	Deep neural network	UCI-HAR	95.78
Xia et al. [17]	CNN-LSTM	UCI-HAR	92
Rahman et al. [18]	LBP based ELM	MotionSense	98.96
Tasnim et al. [19]	3D skeleton joint model	MSR-Action3D	95.64
Tan et al. [20]	3D motion trail model (3DMTM).	MSR-Action3D	94.87
Bulbul et al. [21]	3D skeleton joint model	MSR-Action3D	91.21
Bulbul et al. [22]	3DMTM	MSR-Action3D	97.41
Nasir et al. [23]	NFC	Hollywood2 & HMDB-51	82.55

Its resilience to fluctuations in illumination, crucial in practical scenarios, makes it widely applicable. LBP transforms RGB images into grayscale and is compatible with grayscale images in Python applications. It is commonly used for picture representation, particularly in the field of HAR [22]. The LBP algorithm analyzes 3×3 windows of images, comprising nine-pixel values. From each window, the LBP code is extracted using a processing approach. The ELM classifier then generates a confusion matrix using the pixel values of the images. Employing the ELM approach reduces processing complexity and noise, thereby enhancing result accuracy. The ELM structure utilizes hidden layer weights and calculates optimal weights for a single layer feedforward network (SLFN) to address the limitations of gradient-based learning. Arbitrary initial weights are assigned between the input and hidden layers, facilitating efficient learning [18].

14.5 Result Analysis

Our study aimed to develop a HAR system capable of accurately and reliably categorizing human actions in real-world scenarios. To achieve this, we conducted a thorough analysis comparing and evaluating various ML models, including decision trees, SVMs, and DL architectures like CNNs and RNNs. Our research dataset encompassed diverse everyday behaviors such as walking, running, sitting, and standing. We observed that the performance of the HAR system was significantly influenced by preprocessing techniques such as windowing, data normalization, and feature extraction. Through meticulous selection and engineering of informative features, we were able to enhance the discriminative power and overall recognition accuracy of the models. Our study advances the field of HAR by shedding light on the efficacy of different preprocessing and machine learning methods. The developed HAR system exhibits promising potential across various domains, including healthcare, sports analytics, assisted living, and human–computer interface design. Future research directions may include exploring multi-sensor fusion techniques, refining model architectures, and conducting real-world evaluations to assess system performance. While there hasn't been a comprehensive survey covering algorithms and sensors for diverse HAR applications, existing surveys in the field have provided valuable insights. Our survey aims to identify the most effective machine learning methods for reliably detecting human actions in the HAR domain. We provide an in-depth analysis of algorithms tailored to specific application areas. Additionally, we investigate whether vision-based or non-vision-based acquisition devices are more prevalent in current HAR applications. Figure 14.7 illustrates the accuracy comparison across different models.

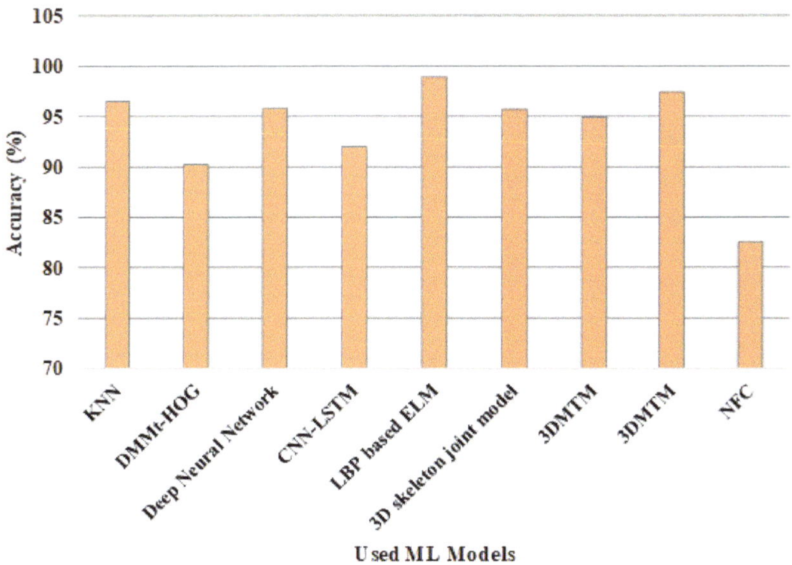

Figure 14.7 Comparing the accuracy of several ML models.

14.6 Connection between Decoding Brain Signals and HAR

Both brain signals and HAR rely on state-of-the-art ML methods to decipher intricate biological signals. At their core, both domains aim to extract meaningful patterns from data to understand and predict human behavior [24]. Brain signal decoding focuses on interpreting neural activity through techniques like EEG, fMRI, and MEG, translating brain signals into actionable information such as control commands for brain–computer interfaces or insights into cognitive states. Similarly, HAR employs sensors to monitor and classify physical actions, utilizing algorithms like CNNs and RNNs to recognize activities such as walking, running, or sitting. The convergence of these fields lies in their shared challenges and technological approaches. Both require sophisticated data processing and feature extraction methods to handle noisy and variable signals. They also face the task of developing models that are not only accurate but also interpretable. Advances in one field can inform the other; for instance, improvements in DL algorithms for brain signal decoding can enhance HAR by providing better models for pattern recognition. Moreover, the integration of neural and physical activity data holds potential for comprehensive systems that monitor both mental and physical states, benefiting applications in healthcare, neurorehabilitation, and smart environments. Through the synergistic development of techniques and technologies, brain signal decoding and HAR can jointly advance our understanding and interaction with human behaviors and health [25].

14.6.1 Brain signal decoding using ML techniques

- **Functional MRI (fMRI):** fMRI is a neuroimaging technique that measures and maps brain activity by detecting changes associated with blood flow. When brain regions are more active, they receive more blood, which fMRI can capture. In the context of brain signal decoding, DL models are employed to interpret the complex data generated by fMRI scans. These models can analyze patterns of brain activity to decode cognitive states, predict mental tasks, and understand brain function. The high spatial resolution of fMRI allows researchers to pinpoint specific brain regions involved in various cognitive processes.
- **Brain–computer interfaces (BCIs):** BCIs are systems that enable direct communication between the brain and external devices. By interpreting brain signals, BCIs allow users to control computers, prosthetics,

or other devices using their thoughts. Initially developed for basic command-based operations, BCIs have advanced to recognize more complex thought patterns and intentions. This evolution has expanded their applications from simple tasks like cursor movement to more sophisticated functions such as speech synthesis for individuals with communication impairments and complex device control [26].

- **Magnetoencephalography (MEG):** MEG is a non-invasive imaging technique that records magnetic fields produced by neural activity in the brain. MEG offers high temporal resolution, capturing rapid changes in brain activity in real time. This makes it particularly useful for real-time decoding of brain signals. MEG's ability to track the timing of neural events with precision enables researchers to study dynamic brain processes and understand how different brain regions interact over time.
- **Deep learning (DL):** DL involves using artificial neural networks to model and understand complex data. In brain signal decoding, DL techniques, such as long short-term memory fully convolutional networks (LSTM-FCN), are employed to classify electroencephalogram (EEG) signals. EEG measures electrical activity in the brain, and LSTM-FCNs are well-suited for handling the sequential nature of EEG data. These networks can learn to recognize patterns in EEG signals associated with different cognitive states or tasks, improving the accuracy of brain signal classification.
- **Convolutional neural networks (CNNs):** CNNs are a class of DL algorithms particularly effective for analyzing structured data like images and videos. In brain signal decoding, CNNs can process complex neural data to decode behaviors and stimuli from various brain regions. CNNs have been used across different species, including humans, to identify patterns in brain activity related to specific behaviors or sensory experiences. Their ability to learn hierarchical features from data makes CNNs powerful tools for understanding the neural basis of behavior and cognition [27].
- **Autoencoders:** Autoencoders are a type of neural network used for unsupervised learning, where the network learns to encode input data into a lower-dimensional representation and then reconstruct it. In EEG signal classification, autoencoder-based approaches can enhance performance by learning efficient representations of the EEG data. These representations can capture essential features that differentiate between various cognitive states, leading to improved classification accuracy.

Table 14.2 Brain signal decoding using ML techniques.

Method	Description
fMRI	Decoding brain activity from fMRI data with DL models.
BCIs	Bridging the gap between the brain and external devices by interpreting brain signals for control or communication. BCIs have evolved from simple command-based systems to recognizing complex thought patterns.
MEG	High-temporal-resolution neuroimaging for real-time decoding.
DL	Utilizing neural networks, such as LSTM-FCN, for EEG signal classification.
CNNs	AI-driven algorithms are capable of decoding various behaviors and stimuli from different brain regions across species, including humans.
Autoencoders	Improved EEG signal classification using autoencoder-based approaches.

Autoencoders can also help in denoising EEG signals, making the decoding process more robust [28].

Brain signal decoding using ML techniques has made significant progress in recent years. Here's a concise Table 14.2 summarizing ML approaches for brain signal decoding in the context of HAR.

These methodologies highlight the intersection of advanced neuroimaging techniques and ML models in brain signal decoding. By leveraging the strengths of fMRI, MEG, deep learning, and autoencoders, researchers can gain deeper insights into brain function, leading to potential breakthroughs in neuroscience and related fields.

14.6.2 Comparative analysis of brain signal decoding and human activity recognition

In the realm of ML approaches for brain signal decoding and interpreting brain signals: Insights from AI-driven Analysis, several avenues for future research present themselves. Firstly, there is a need to explore novel ML algorithms and techniques tailored specifically for decoding complex brain signals. DL models such as CNNs, RNNs, and attention mechanisms offer promising avenues for advancing brain signal analysis. Investigating the fusion of multiple modalities, such as EEG, fMRI, and MEG, could enhance the accuracy and robustness of brain signal decoding systems [29]. Additionally, future research should focus on developing interpretable machine learning models for brain signal analysis [30]. Interpretability is crucial for understanding how ML algorithms arrive at their predictions and can provide

Table 14.3 Comparative analysis of brain signal decoding and HAR.

Aspect	ML approaches for brain signal decoding	Interpreting brain signals insights from AI-driven analysis	HAR
Objective	Decode and interpret brain signals using ML techniques.	Gain insights into brain activity patterns and behaviors using AI-driven analysis.	Recognize and classify human activities based on sensor data.
Techniques	EEG, fMRI, and MEG signal processing with ML algorithms such as SVM, CNN, and RNN.	Dl models, neural networks, and pattern recognition methods.	CNNs, RNNs, decision trees, SVMs, feature extraction, and data preprocessing.
Data sources	EEG, fMRI, MEG.	EEG, fMRI, brain imaging data, neural activity recordings.	Wearable sensors, IoT devices, visual sensors, non-visual sensors.
Applications	BCI, neurofeedback, clinical diagnosis, cognitive neuroscience research.	Cognitive neuroscience, mental health diagnostics, brain–machine interfaces.	Healthcare monitoring, sports analytics, smart home systems, security.
Challenges	Noise in brain signals, signal variability, feature selection, interpretability of ML models.	Complex neural dynamics, signal processing complexity, data interpretation, ethical considerations.	Sensor noise, diverse activity patterns, real-time processing, and model interpretability.
Future directions	Multi-modal fusion, real-time decoding, interpretability enhancements, clinical translation.	Advanced AI algorithms, integration with neuroimaging techniques, personalized medicine.	Multi-sensor fusion, real-world application testing, adaptive learning systems.

valuable insights into the underlying cognitive processes. Techniques such as attention mechanisms, explainable AI, and model visualization methods can aid in elucidating the neural mechanisms underlying brain function. Here's a concise analysis of brain signal decoding and human activity recognition (Table 14.3).

By comparing these aspects, it's evident that both brain signal decoding and HAR leverage sophisticated machine learning techniques to interpret complex data. While brain signal decoding focuses on understanding neural activities for applications like BCIs and clinical diagnostics, HAR aims to identify human actions for applications in healthcare, sports, and security. Both fields face unique challenges but hold significant promise for future advancements through continued research and development in ML and sensor technologies [31]. Moreover, there is a need to explore the integration of brain signal decoding technologies into practical applications across various domains. This includes healthcare, where brain signal decoding can aid in the diagnosis and treatment of neurological disorders, as well as assistive technologies for individuals with disabilities. Furthermore, brain signal decoding has the potential to revolutionize human–computer interaction, enabling seamless communication between humans and machines through BCIs). Finally, comprehensive surveys that investigate the efficacy of different machine learning methods and sensor technologies for brain signal decoding are warranted. Such surveys can provide valuable insights into state-of-the-art techniques and guide future research directions in this rapidly evolving field [32]. Overall, future research in ML approaches for brain signal decoding and interpreting brain signals holds immense potential for advancing our understanding of the human brain and enhancing human-machine interaction.

14.7 Future Scope

HAR is a burgeoning field within computational science and engineering, dedicated to devising systems and methodologies capable of autonomously identifying and categorizing human behavior based on sensor data. The prospects of HAR are wide-ranging and promising, with significant potential impact across various domains:

- **Healthcare and well-being**: HAR holds considerable promise in healthcare analytics and monitoring, where it can play a pivotal role in observing and analyzing human activities to detect patterns indicative of health issues or changes in overall well-being. For instance, it could aid in elderly care by monitoring daily activities and promptly identifying anomalies signaling potential health problems or emergencies.
- **Fitness and sports**: Integration of HAR technologies into wearable devices can revolutionize fitness and sports monitoring, providing users with valuable insights into their physical activity, performance metrics, and personalized recommendations for improvement.
- **Smart environments**: HAR stands to enhance the development of intelligent environments that adapt to human behaviors and preferences.

This encompasses smart homes, workplaces, and public spaces that adjust lighting, climate control, and other environmental factors based on occupants' activities and needs.

- **Safety and security**: HAR has the potential to enhance the effectiveness of safety and security systems by detecting and flagging abnormal or suspicious behaviors in public places, transportation hubs, or critical infrastructure sites. It can also be utilized for access control, authentication, and fraud detection through behavioral biometrics.
- **HCI**: The burgeoning field of human–AI relationships HAR can revolutionize human–computer interaction, enabling the creation of interfaces that are more natural and intuitive. This includes gesture recognition, facial expression analysis, and other forms of non-verbal communication for manipulating devices or providing responses in virtual and augmented reality environments.
- **Workforce productivity**: HAR can aid organizations in optimizing workflow efficiency and enhancing workforce productivity by assessing employees' task performance and identifying areas for automation, efficiency improvements, or ergonomic adjustments.
- **Education and training**: In the realm of education, HAR can enhance personalized learning experiences by monitoring student engagement and learning styles. This data can provide valuable feedback to educators and dynamically adjust educational content to meet the specific needs and preferences of individual learners.
- **Consumer electronics**: HAR features are increasingly being integrated into consumer electronics products such as smartphones, smartwatches, and home assistants, enabling novel capabilities and enriching user experiences.
- **Urban planning and transportation**: HAR data collection offers valuable insights for urban planners and transportation authorities, enabling a better understanding of movement patterns and social interactions within cities. This information can inform the design of infrastructure, public transportation systems, and urban areas.
- **Social sciences and market research**: HAR holds potential in social science research and market analysis, facilitating the study of human behavior, preferences, and trends. This data can inform policymaking, marketing strategies, and product development initiatives.

HAR has emerged as a multifaceted discipline with far-reaching applications, poised to revolutionize various aspects of human life and societal interactions through its advanced capabilities in behavior recognition and analysis.

14.8 Conclusion

In our research, we delve into the exploration of ML methods for decoding brain signals, aiming to develop a robust system capable of accurately categorizing neural activities in real-world scenarios. We conduct a thorough investigation comparing and analyzing various ML approaches, including decision trees, SVMs, and DL models such as CNNs and RNNs. Our dataset encompasses a spectrum of everyday behaviors, ranging from walking and running to sitting and standing, to comprehensively assess the performance of the HAR system. Additionally, we recognize the significant impact of preprocessing techniques, such as windowing, data normalization, and feature extraction, on the system's efficacy. By meticulously selecting and engineering informative features, we enhance the models' discriminative power and overall recognition accuracy. This study contributes to the advancement of brain signal decoding systems by shedding light on the effectiveness of various preprocessing and ML methods. Our developed system exhibits promising potential across multiple domains, including healthcare, sports analytics, assisted living, and HCI. Moving forward, avenues for further research include exploring multi-sensor fusion techniques, refining model topologies, and conducting real-world testing to validate the system's performance. While existing surveys in the field of brain signal decoding offer valuable insights, this study aims to provide a comprehensive analysis of the most effective ML methods and techniques for reliably interpreting brain signals. Through an in-depth examination of algorithms tailored to different applications, we strive to advance the understanding and utilization of AI-driven analysis in decoding brain signals.

References

[1] Nasir, I. M., Raza, M., Shah, J. H., Attique Khan, M., & Rehman, A. (2021). Human action recognition using machine learning in uncontrolled environment. 2021 1st International Conference on Artificial Intelligence and Data Analytics (CAIDA). https://doi.org/10.1109/caida51941.2021.9425202.

[2] Bulbul, M. F., Islam, S., Zhou, Y., & Ali, H. (2019a). Improving human action recognition using hierarchical features and multiple classifier ensembles. The Computer Journal, 64(11), 1633–1655. https://doi.org/10.1093/comjnl/bxz123

[3] Majumder, S., & Kehtarnavaz, N. (2021). Vision and inertial sensing fusion for human action recognition: A Review. IEEE Sensors Journal, 21(3), 2454–2467. https://doi.org/10.1109/jsen.2020.3022326

[4] Varges da Silva, M., & Nilceu Marana, A. (2020). Human action recognition in videos based on spatiotemporal features and bag-of-poses. Applied Soft Computing, 95, 106513. https://doi.org/10.1016/j.asoc.2020.106513.

[5] Chaudhuri, S., & Bhattacharya, S. (2023). VILP: Knowledge Exploration using vision, language, and pose embeddings for Video action recognition. Proceedings of the Fourteenth Indian Conference on Computer Vision, Graphics and Image Processing. https://doi.org/10.1145/3627631.3627637

[6] Ishrak, S., Munir, M. B., & Kabir, M. H. (2023). Dynamic hand gesture recognition using sequence of human joint relative angles. 2023 International Conference on Electrical, Computer and Communication Engineering (ECCE). https://doi.org/10.1109/ecce57851.2023.10101606

[7] Shi, X., Li, Y., Zhou, F., & Liu, L. (2018). Human Activity Recognition based on Deep Learning Method. 2018 International Conference on Radar (RADAR). https://doi.org/10.1109/radar.2018.8557335

[8] M. V. da Silva and A. N. J. A. S. C. Marana, "Human action recognition in videos based on spatiotemporal features and bag-of-poses," p. 106513, 2020.

[9] I.-M. Nasir, M.-A. Khan, M. Alhaisoni, T. Saba, A. Rehman, and T. Iqbal, "A Hybrid Deep Learning Architecture for the Classification of Superhero Fashion Products: An Application for Medical-Tech Classification," p. {pages}.

[10] M. F. Bulbul, S. Islam, and H. Ali, "Human action recognition using mhi and shi based glac features and collaborative representation classifier," Multimedia Tools and Applications, vol. 78, no. 15, pp. 21 085–21 111, 2019.

[11] Xia, K., Huang, J., & Wang, H. (2020). LSTM-CNN Architecture for Human Activity Recognition. IEEE Access, 8, 56855–56866. https://doi.org/10.1109/access.2020.2982225.

[12] Ling, C., Xiaoze, L., Liangying, P., Menghan, W.: Deep learning based multimodal complex Human Activity Recognition using wearable devices. Appl. Intell. 1–14 (2020)

[13] Laksitamas, P., & Adekoya, D. O. (2023). The impact of gender on consumer buying decision process towards online shopping: A study of Thai consumers. International Journal of Business Performance Management, 1(1), 1. https://doi.org/10.1504/ijbpm.2023.10052174

[14] Mohsen, S., Elkaseer, A., Scholz, S.G. (2022). Human Activity Recognition Using K-Nearest Neighbor Machine Learning Algorithm. In: Scholz, S.G., Howlett, R.J., Setchi, R. (eds) Sustainable Design and Manufacturing. KES-SDM 2021. Smart Innovation, Systems and Technologies, vol 262. Springer, Singapore. https://doi.org/10.1007/978-981-16-6128-0_29

[15] Wang, D., Ou, F., & Zhou, Y. (2018). Action recognition based on multi-feature depth motion maps. IECON 2018 - 44th Annual Conference of the IEEE Industrial Electronics Society, 2683–2688. https://doi.org/10.1109/iecon.2018.8591591

[16] Singh, A. K., Kumar, S., Kumar, B., & Chauhan, A. S. (2023). Face recognition using machine learning. 2023 International Conference on Sustainable Emerging Innovations in Engineering and Technology (ICSEIET). https://doi.org/10.1109/icseiet58677.2023.10303390

[17] Xia, K., Huang, J., & Wang, H. (2020). LSTM-CNN Architecture for Human Activity Recognition. IEEE Access, 8, 56855–56866. https://doi.org/10.1109/access.2020.2982225.

[18] Rahman, A., Nag, A., & Bairagi, A. K. (2023). A hybrid model for recognizing human activity using extreme learning. 2023 26th International Conference on Computer and Information Technology (ICCIT). https://doi.org/10.1109/iccit60459.2023.10441574.

[19] Tasnim, N., & Baek, J.-H. (2023). Dynamic edge convolutional neural network for skeleton-based human action recognition. Sensors, 23(2), 778. https://doi.org/10.3390/s23020778

[20] Tan, Y.-F., Poh, S.-C., Ooi, C.-P., & Tan, W.-H. (2023). Human Activity Recognition with self-attention. International Journal of Electrical and Computer Engineering (IJECE), 13(2), 2023. https://doi.org/10.11591/ijece.v13i2.pp2023-2029

[21] Bulbul, M. F., Islam, S., Zhou, Y., & Ali, H. (2019). Improving human action recognition using hierarchical features and multiple classifier ensembles. The Computer Journal, 64(11), 1633–1655. https://doi.org/10.1093/comjnl/bxz123

[22] M. Bulbul, M. F., Galib, S. Md., & Ali, H. (2019). Human action recognition using GLAC features on Multi-view binary coded images. 2019 UK/ China Emerging Technologies (UCET). https://doi.org/10.1109/ucet.2019.8881844

[23] Nasir, I. M., Raza, M., Shah, J. H., Attique Khan, M., & Rehman, A. (2021a). Human action recognition using machine learning in uncontrolled environment. 2021 1st International Conference on Artificial Intelligence and Data Analytics (CAIDA). https://doi.org/10.1109/caida51941.2021.9425202

[24] Nag, A., Mondal, H., Kabir, S. R. R., Islam, M. R., Ahmed, M. E., & Jamil, S. H. (2024, May). Emotion Decoding: An Extensive Examination of Electroencephalogram Signals Using Explainable Machine Learning. In 2024 6th International Conference on Electrical Engineering and Information & Communication Technology (ICEEICT) (pp. 604-609). IEEE.

[25] Dai, H. N., Kumar, J., & Pandey, O. J. (2023). AERed: An Autoencoder-decoder Dimensionality Reduction Method for Wearable-Based Human Activity Recognition. IEEE Sensors Journal.

[26] Gu, X., Cao, Z., Jolfaei, A., Xu, P., Wu, D., Jung, T. P., & Lin, C. T. (2021). EEG-based brain–computer interfaces (BCIs): A survey of recent studies on signal sensing technologies and computational intelligence approaches and their applications. IEEE/ACM transactions on computational biology and bioinformatics, 18(5), 1645-1666.

[27] Nag, A., Das, A., Sil, R., Kar, A., Mandal, D., & Das, B. (2022, December). Application of artificial intelligence in mental health. In International Conference on Intelligent Systems Design and Applications (pp. 128-141). Cham: Springer Nature Switzerland.

[28] Kattenborn, T., Leitloff, J., Schiefer, F., & Hinz, S. (2021). Review on Convolutional Neural Networks (CNN) in vegetation remote sensing. ISPRS journal of photogrammetry and remote sensing, 173, 24-49.

[29] Das, A., Paul, R., Nag, A., & Das, B. (2024). A Study of Cloud of Things Enabled Machine Learning-Based Smart Health Monitoring System. In Sustainability in Industry 5.0 (pp. 156-176). CRC Press.

[30] Nag, A., Das, B., Sil, R., Hameed, A. A., Bhushan, B., & Jamil, A. (2023, July). A Survey on Image-Based Cardiac Diagnosis Prediction Using Machine Learning and Deep Learning Techniques. In International Conference on Computing, Intelligence and Data Analytics (pp. 478-491). Cham: Springer Nature Switzerland.

[31] Das, M., Nag, A., Hassan, M. M., Santra, A., Chand, N., Yasmin, F., ... & Alkhayyat, A. Synergy of 6G technology and IoT networks for transformative applications. International Journal of Communication Systems, e5869.

[32] Makin, J. G., Moses, D. A., & Chang, E. F. (2020). Machine translation of cortical activity to text with an encoder–decoder framework. Nature neuroscience, 23(4), 575-582.

Index

About the Editors

Md. Mehedi Hassan (Member, IEEE) is a dedicated and accomplished researcher and completed a Master of Science (M.Sc.) degree in computer science and engineering at Khulna University, Khulna, Bangladesh. He is a member of the prestigious Institute of Electrical and Electronics Engineers (IEEE). Mehedi completed his B.Sc. degree in computer science and engineering from North Western University, Khulna in 2022, where he excelled in his studies and demonstrated a strong aptitude for research. As the founder and CEO of The Virtual BD IT Firm and VRD Research Laboratory, Bangladesh, Mehedi has established himself as a highly respected leader in the fields of biomedical engineering, data science, and expert systems. Mehedi's research interests are broad and include important human diseases, such as oncology, cancer, and hepatitis, as well as human behavior analysis and mental health. He is highly skilled in association rule mining, predictive analysis, machine learning, and data analysis, with a particular focus on the biomedical sciences. As a young researcher, Mehedi has published more than 70 articles and 5 books in various international top journals and conferences, which is a remarkable achievement. His work has been well-received by the research community and has significantly contributed to the advancement of knowledge in his field. Overall, Mehedi is a highly motivated and skilled researcher with a strong commitment to improving human health and well-being through cutting-edge scientific research. His accomplishments to date are impressive, and his potential for future contributions to his field is very promising. Additionally, he serves as a reviewer for 60 prestigious journals. Md. Mehedi Hassan has filed more than 3 patents out of which 2 have been granted to his name.

Farhana Yasmin, currently pursuing her Ph.D. at the School of Computer and Software, Nanjing University of Information Science and Technology, Nanjing, China (210044), earned her B.Sc. degree from the Department of Computer Science and Engineering, Eastern University, Bangladesh. She furthered her academic journey by completing her M.Sc. degree in computer application and technology at Changzhou University, China. She has expertise spanning computer vision, biomedical engineering, and data science all under the expansive domain of artificial intelligence (AI). Farhana's research focuses primarily on predictive analysis and the advancement of expert systems. Her scholarly contributions include approximately 9 publications in esteemed international journals such as IEEE, Elsevier, and Springer. Collaborating with researchers globally, she actively seeks to broaden her knowledge and enhance her proficiency in her field.

Shariful Islam is a physician scientist at the Institute for Physical Activity and Nutrition, Deakin University. He is Honorary Senior Lecturer at Sydney Medical School, University of Sydney and Honorary Senior Fellow at the George Institute for Global Health, UNSW. Previously he led the diabetes research program at the Center for Control of Chronic Diseases, ICDDR,B and has experience working with United Nations Development Programme, World Health Organization, German Technical Cooperation and as a resident cardiologist in Bangladesh. He has more than 99,000 citations to his scholar profile.

Anupam Kumar Bairagi (Senior Member, IEEE) received B.Sc. and M.Sc. degrees in computer science and engineering from Khulna University (KU), Bangladesh, and a Ph.D. degree in computer engineering from Kyung Hee University, South Korea. He is a Professor of Computer Science and Engineering Discipline, KU. He has authored or coauthored around 60 publications, including refereed IEEE/ACM journals and conference papers. His research interests include wireless resource management in 5G and beyond, healthcare, the IIoT, cooperative communication, and game theory. He has served as a technical program committee member for different international conferences.

Si Thu Aung received a B.Eng. degree from Technological University, Myanmar in 2014, a Master of Engineering in Electronics from Mandalay Technological University, Myanmar in 2017, and a Ph.D. in Biomedical Engineering from Faculty of Engineering, Mahidol University, Thailand, in 2021. He is currently a post-doctoral researcher at Rail and Modern Transports Research Center under National Science and Technology Development Agency, Thailand Science Park, Pathum Thani, Thailand. His current research interests include biomedical signal processing, digital image processing, machine learning, and deep learning.